Energy from Nuclear Fusion

Energy from Nuclear Fusion

Richard A Dunlap
Department of Physics and Atmospheric Science, Dalhousie University, Halifax, Nova Scotia, Canada

IOP Publishing, Bristol, UK

ISBN 978-0-7503-3307-8 (ebook)
ISBN 978-0-7503-3305-4 (print)
ISBN 978-0-7503-3308-5 (myPrint)
ISBN 978-0-7503-3306-1 (mobi)

DOI 10.1088/978-0-7503-3307-8

Version: 20210901

IOP ebooks

British Library Cataloguing-in-Publication Data: A catalogue record for this book is available from the British Library.

Published by IOP Publishing, wholly owned by The Institute of Physics, London

IOP Publishing, Temple Circus, Temple Way, Bristol, BS1 6HG, UK

US Office: IOP Publishing, Inc., 190 North Independence Mall West, Suite 601, Philadelphia, PA 19106, USA

Contents

Preface

Humans rely on energy to fulfill their basic needs, including heating, cooking, transportation and powering electronic devices. Since the end of the 19th century the majority of our energy has come from the combustion of fossil fuels. In the past half a century or so it has become increasingly apparent that our dependence on fossil fuels cannot continue much longer. Our extensive use of fossil fuels has not only substantially depleted the available resources but has been responsible for detrimental global climate changes. The development of alternative, carbon-free energy sources is necessary to allow us to fulfill our future energy needs and to mitigate the adverse effects of our fossil fuel use.

Most renewable energy sources, such as wind and solar, are intermittent. Even those, such as hydroelectric energy, which do not show short term variability are often subject to seasonal changes. Thus, the full implementation of renewable carbon-free energy sources requires the parallel development of efficient energy storage methods. Nuclear energy, on the other hand, provides a convenient replacement for fossil fuel energy facilities such as coal fired electric generating stations, which currently provide the largest fraction of base load electric power worldwide.

Energy may be obtained from nuclei in two ways: fission, that is the splitting of heavy nuclei or fusion, that is the fusing of two light nuclei. Current nuclear energy comes exclusively from nuclear fission. Although there a number of attractive features of nuclear fission energy, there are some serious drawbacks, as well. These include:

- Safety concerns due to the possibility of nuclear accidents.
- Security concerns due to the potential proliferation of weapon grade nuclear material.
- Environmental concerns due to the disposal of spent fuel.

It is also important to consider that nuclear fission energy is not renewable. Most current facilities utilize thermal neutron reactors, which extract energy from less than 1% of uranium resources. Known uranium reserves can satisfy all of our energy needs for a few decades using this technology. A small number of facilities utilize fast breeder reactors which extract nearly all the energy from uranium. As well, in some parts of the world thorium reactors are under development. The implementation of fast breeder reactors and thorium reactors would extend the longevity of fission reactor resources by about two orders of magnitude.

The production of energy through nuclear fusion reactions is, to some extent, free from these drawbacks. Unlike the reaction in a fission reactor, which becomes an uncontrolled chain reaction unless proper feedback mechanisms are maintained, the reaction in a fusion reactor will naturally die out without active control. As well, the radioactive nuclides that are present in a fusion reactor are non-fissile and are, therefore not appropriate for weapons use. Tritium, which is a fuel in current fusion reactor designs, can be used to enhance the yield of fusion weapons, but this is generally considered to be a situation that is easier to regulate than that for fissile materials from fission reactors.

In principle, fusion reactions such as those that produce the majority of energy in the sun, fuse four hydrogen nuclei together to produce one helium nucleus and do not produce any radioactive by-products. In practice, however, fusion reactions that are likely candidates for commercial fusion reactors are not necessarily so benign. Most serious fusion research is centered around the deuterium–tritium fusion reaction. This reaction utilizes radioactive tritium as one of the reactants and produces neutrons as a by-product (along with helium). The neutrons interact with materials in the reactor, and in many cases produce radioactive by-products. Thus, it is not the reaction by-products themselves that pose a radiation concern, but the un-reacted fuel and the irradiated reactor materials. A similar situation exists for deuterium–deuterium fusion (if it is ever used as a commercial reactor process) as it also produces neutrons. The hazards associated with fusion reactor materials are much less worrisome than those associated with fission reactors. Tritium produces low energy electrons during its decay, compared to high energy gamma rays produced by most fission reactor waste products. As well, reactor materials can be carefully designed so as to minimize the hazards associated with neutron activated materials.

Finally, the availability of fusion fuels is an important consideration in the longevity of fusion energy technology. Deuterium–tritium fusion fuel is limited by the world's availability of lithium, which is used to produce tritium. Even this technology could likely satisfy our energy needs for millions of years. Deuterium–deuterium fusion, on the other hand, utilizes only components of naturally occurring hydrogen and is essentially infinite on any conceivable human timescale.

As an energy source, nuclear fusion would appear to be attractive as a base-load contribution to renewable sources and as a safe long-term alternative to nuclear fission energy. Efforts to harness energy from nuclear fusion have existed for well over half a century, but this technology has not yet reached the point of commercialization.

The present book provides an overview of the ways in which research has attempted to use nuclear fusion as a viable energy source and why this task has thus far been elusive. The book begins with an introduction to our energy needs, our current energy production methods and possible future energy technologies. This introduction provides a rationale for the development of energy facilities based on nuclear fusion. Chapter 2 gives an overview of nuclear physics and focusses on those aspects of the field that relate directly to the use of nuclear reactions for the commercial production of energy. The remainder of the book covers the development of our knowledge of nuclear fusion with an emphasis on its use as a viable source of energy for society. The material in this book is presented at a level appropriate for readers with a knowledge of introductory university physics. A detailed understanding of nuclear physics or plasma physics is not necessary although a general knowledge of introductory quantum mechanics and electrodynamics will be useful. The book, therefore, provides an overview of nuclear fusion as an energy source that is accessible to typical university science or engineering students but is also informative for science and engineering professionals who are not fusion researchers.

Author biography

Richard A Dunlap

Richard A Dunlap received a B.S. from Worcester Polytechnic Institute (Physics 1974), an A.M from Dartmouth College (Physics 1976) and a PhD from Clark University (Physics 1981). Since 1981 he has been a professor in the Department of Physics and Atmospheric Science at Dalhousie University and currently holds a position as Research Professor. He was Faculty of Science Killam Research Professor from 2001 to 2006 and Director of the Dalhousie University Institute for Research in Materials from 2009 to 2015. Professor Dunlap's research interests include nuclear spectroscopies, magnetic materials, quasicrystals, critical phenomena and advanced batteries materials. He has published more than 300 refereed research papers and his previously published books include; *Experimental Physics: Modern Methods* (Oxford 1988), *The Golden Ratio and Fibonacci Numbers* (World Scientific 1997), *An Introduction to the Physics of Nuclei and Particles* (Brooks/Cole 2004), *Sustainable Energy* (Cengage, 1st edn 2015, 2nd edn 2019), *Novel Microstructures for Solids* (IOP/Morgan & Claypool 2018), *Particle Physics* (IOP/Morgan & Claypool 2018), *The Mössbauer Effect* (IOP/Morgan & Claypool 2019), *Lasers and Their Application to the Observation of Bose-Einstein Condensates* (IOP/Morgan & Claypool 2019), *Electrons in Solids—Contemporary Topics* (IOP/Morgan & Claypool 2019) and *Renewable Energy* (Morgan & Claypool 2020).

IOP Publishing

Energy from Nuclear Fusion

Richard A Dunlap

Chapter 1

Energy for the future

1.1 Introduction

Energy is an essential resource of our society. Since the late 19th century, the majority of our energy has come from fossil fuels, i.e. coal, oil and natural gas. It is now apparent that this situation cannot continue very far into the future for two reasons. Firstly, fossil fuel resources are limited and at our current rate of use could not provide the majority of our energy much past the end of the current century. Secondly, our continued use of fossil fuels will have serious and perhaps even irreversible adverse environmental consequences, the most serious of these being climate change resulting from greenhouse gas emissions. It is, therefore, crucial that we transition from our fossil fuel-based energy system to a carbon-free energy system. While some (more-or-less) carbon-free energy sources have been in use for a number of years, notably hydroelectricity, geothermal energy and nuclear (fission) energy, it has only been in fairly recent years that there has been any significant development of other renewable sources, principally solar and wind.

Our current energy comes from a mix of sources, as different sources are more suitable to fulfilling different needs. Our future energy sources, no doubt, will follow this same approach. Decisions concerning the path to carbon-free energy require the consideration of a number of factors. Among the factors for possible carbon-free energy resources are the following:

- Amount of primary energy that will be required by society in the future.
- The availability of specific carbon-free resources. This includes not only the amount of primary energy that is available and its future availability, but the availability of materials that are necessary to utilize the resource.
- The cost of developing and utilizing particular resources.
- Potentially adverse environmental consequences.

While a number of carbon-free energy sources, for example, solar, wind, geothermal, hydroelectric and nuclear fission, have already demonstrated their usefulness, the

doi:10.1088/978-0-7503-3307-8ch1

development of other potential future energy sources has been an active area of research in recent years. The possibility that nuclear fusion could be a major contributor to our future energy needs has been promoted. It has been suggested that fusion may provide limitless inexpensive energy from seawater, without the adverse environmental, safety and security concerns of energy from nuclear fission.

In order to consider the suitability of future energy sources including nuclear fusion, it is important to know our future energy needs and to consider the viability of the various options. In the present chapter, we begin with an analysis of our current energy use along with an assessment of the future availability of our current energy sources and an evaluation of the environmental concerns associated with our energy use. This is followed by consideration of future energy needs. The various potential energy sources for the future are considered, including the role that fusion energy could play in a future energy mix.

1.2 Current world energy use

Currently primary energy use worldwide is about 6×10^{20} J per year. This amounts to an average power consumption of 1.9×10^{13} W. Primary energy is the energy that we extract from nature. It is not always the energy that is used by the consumer. For example, oil is obtained from nature and the energy content of the oil is the primary energy that we harvest. If a consumer uses the oil to provide heat by burning it in a furnace, then the heat that is obtained is typically 85%–90% of the primary energy of the oil. If, on the other hand, oil is used in a thermal generating station to produce electricity which is supplied to the consumer then the electricity amounts to only about 40% of the primary energy content of the oil because of the efficiency of the generating station. Finally, if the oil is refined into gasoline and burned in an automobile engine, then energy provided to the wheels is limited to the efficiency of the internal combustion engine, about 20%. Therefore, in a detailed analysis, it is not only important how much energy we use but also how we use that energy.

Table 1.1 gives a breakdown of our current primary energy sources worldwide. It is clear that fossil fuels account for more than 85% of our primary energy. The ways

Table 1.1. Breakdown of annual world primary energy consumption by source for 2017. Data adapted from https://www.bp.com/content/dam/bp/business-sites/en/global/corporate/pdfs/energy-economics/statistical-review/bp-stats-review-2018-full-report.pdf.

Energy source	Total consumption (GJ)	% total
Oil	1.935×10^{11}	34.2
Coal	1.562×10^{11}	27.6
Natural gas	1.321×10^{11}	23.4
Hydroelectricity	3.846×10^{10}	6.8
Nuclear (fission)	2.497×10^{10}	4.4
Other renewables	2.039×10^{10}	3.6
Total	5.656×10^{11}	100.0

in which this energy is used (in industrialized countries) is summarized in table 1.2. Certainly, the efficiency of converting primary energy to end user energy is an important factor in assessing the energy needs of society. Because of the efficiency of energy conversions, approximately two thirds of the primary energy that is obtained from nature is rejected energy and only about one third is energy that is supplied in the form that is used by the consumer.

Based on the current world population of 7.5×10^9, it can be found that the average per capita annual energy consumption is 8×10^{10} J, corresponding to an average power use of 2.5 kW per capita. There are, however, considerable variations in energy use in different regions of the world and in order to fully understand our future energy needs, it is important to consider these differences.

Total energy use is the product of per capita energy use and population. It is, therefore, appropriate to begin with a consideration of per capita energy use. Figure 1.1 shows the annual per capita energy consumption for different countries as

Table 1.2. Approximate breakdown of the uses of primary energy sources in industrialized countries.

Use	%
Electricity generation	38
Transportation	29
Industrial processes	22
Space heating/cooking	11

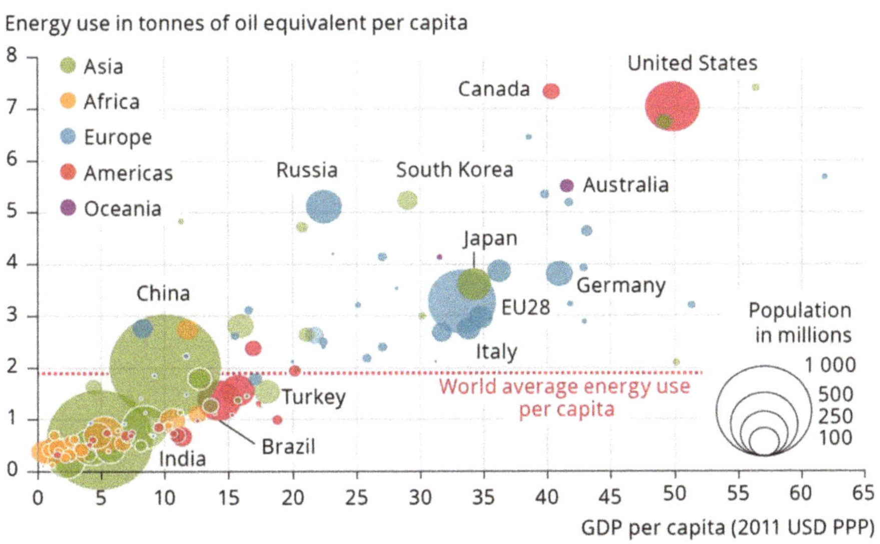

Figure 1.1. Annual per capita energy use as a function of per capita GDP for various countries for the year 2011. Note that energy on the vertical axis is measured in tonnes of oil equivalent (abbreviated toe) where 1 toe = 4.187×10^{10} J. Reproduced from https://www.eea.europa.eu/data-and-maps/figures/correlation-of-per-capita-energy. CC BY 2.5.

a function of per capita gross domestic product (GDP). While there is a general relationship between energy consumption and GDP, there is some scatter in the data. Some countries, such as Canada and Russia, fall above the general trend while others, such as Germany and Italy, fall below the trend. Variations may exist for a variety of reasons including, climate, population density, prevalent industries and national energy policy. In general, countries with colder climates use more energy than countries with warmer climates as a result of increased heating requirements. As well, countries with low population densities require more energy because of greater transportation needs.

It is seen in figure 1.1 that the per capita energy consumption in developing countries in 2011 was much less than in more industrially developed nations. The change in energy consumption as a function of time in some countries is shown in figure 1.2. It is clear that the total energy consumption in the United States and Europe has increased only slightly over the past 35 years. This is in contrast to the

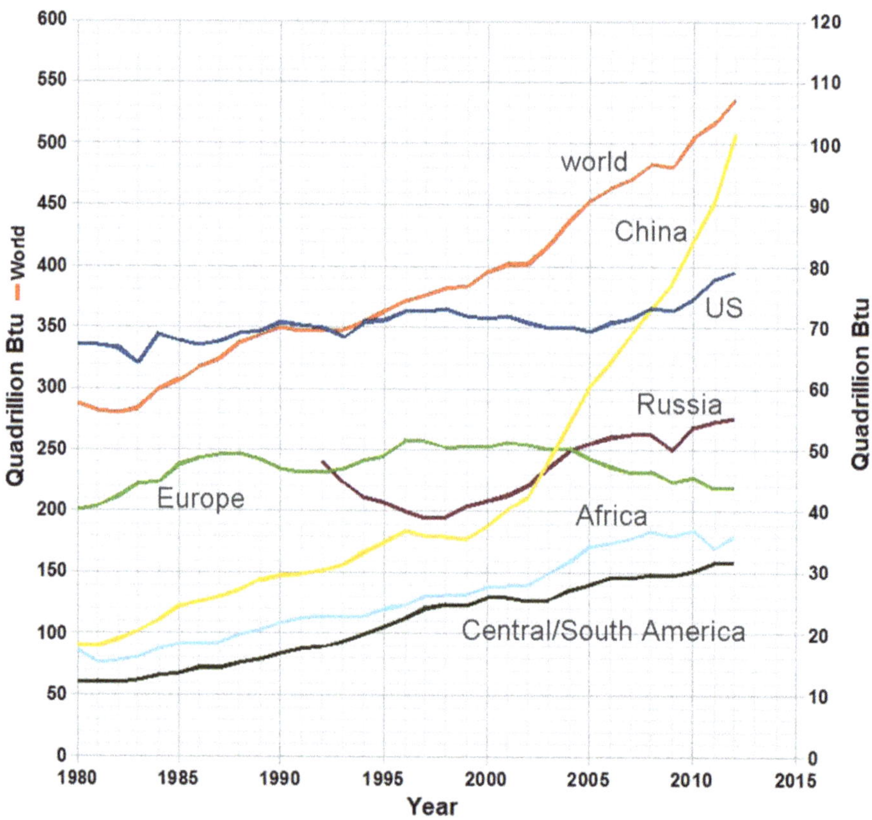

Figure 1.2. Primary energy consumption for different geographical regions from 1980 to 2012. This World Primary Energy image has been obtained by the author from the Wikimedia website https://commons.wikimedia.org/wiki/File:World_total_primary_energy_production_chart_only.png, where it is stated to have been released into the public domain. It is included within this book on that basis.

Table 1.3. Annual per capita energy use for some selected countries for the years 2007 and 2017 and the total change over that period. Data are adapted from. https://www.bp.com/content/dam/bp/business-sites/en/global/corporate/pdfs/energy-economics/statistical-review/bp-stats-review-2018-full-report.pdf and http://www.worldometers.info/world-population/population-by-country/.

Country	Annual per capita energy use 2007(GJ)	Annual per capita energy use 2017(GJ)	Change (%)
Australia	253.6	238.7	−5.9
Brazil	50.3	58.9	+17.0
Canada	407.5	398.7	−2.2
China	67.3	93.0	+38.2
France	175.8	153.3	−12.8
Germany	170.8	170.9	+0.1
India	16.0	23.6	+47.4
Japan	170.9	149.9	−12.3
Norway	407.2	374.9	−7.9
Saudi Arabia	277.5	341.4	+23.0
United States	323.2	288.4	−10.8

situation in China where the total energy consumption has increased consistently and significantly over the same period. As the total energy is a function of both the population and the per capita energy consumption, it is important to separate these two factors. Table 1.3 shows the change in the per capita energy for several countries between 2007 and 2017. In industrially developed countries the per capita energy consumption over this period either remained the same or decreased, while in developing countries the per capita energy consumption increased, in some cases quite significantly. The former feature is the result of increased conservation efforts while the latter is the result of increase energy demands that result from the greater availability of technology.

Returning to figure 1.2, it is obvious that the slight increase in total energy consumption for, e.g. the United States, is the result of the gradual population increase while the significant energy consumption in, e.g. China, is largely the result of the significant per capita energy consumption. It is, therefore, mainly the increase level of industrialization in primarily China and India, as well as other developing countries that is responsible for the increased energy consumption worldwide.

The features observed above for recent trends in energy consumption provide guidance for estimating future world energy needs. From figure 1.1 it is clear that increasing the technological level, and hence the per capita energy consumption, in developing countries, to that which is commensurate with typical developed countries would increase world energy consumption by a factor of two to three, independent of population growth. This is discussed further in the next section with specific reference to future energy sources.

1.3 Future energy needs

Since the total world energy use is merely the total world population multiplied by the average world per capita energy use, projected future energy needs can be based on an estimate of these two quantities. The world population, which currently stands at about 7.7 billion, will, inevitably, increase further, at least in the near future. Populations of animal species in nature are self-limiting, based on the availability of resources, primarily, food, water and shelter. Other factors that can limit population growth include competition for resources, predation, disease and loss of suitable habitat. Human population is determined by many of the same factors and cannot continue to grow without limit. Figure 1.3 shows some models of human population up to year 2100. While some models show the human population continuing to grow for the next 80 years or so, others show the population peaking at 9 billion around 2055 and then decreasing. On the average, models show the population beginning to level off at around 11 billion by the end of this century.

Since we would expect (based on recent trends) that the per capita energy consumption in industrialized countries would remain relatively constant over the next few decades, the increase in global energy use would be dominated by the increase in population and the increase in per capita energy use in developing countries. Based on figure 1.1, we might expect a global average per capita energy use that falls around the middle of the range of the graph. Using 4 toe y^{-1} per capita

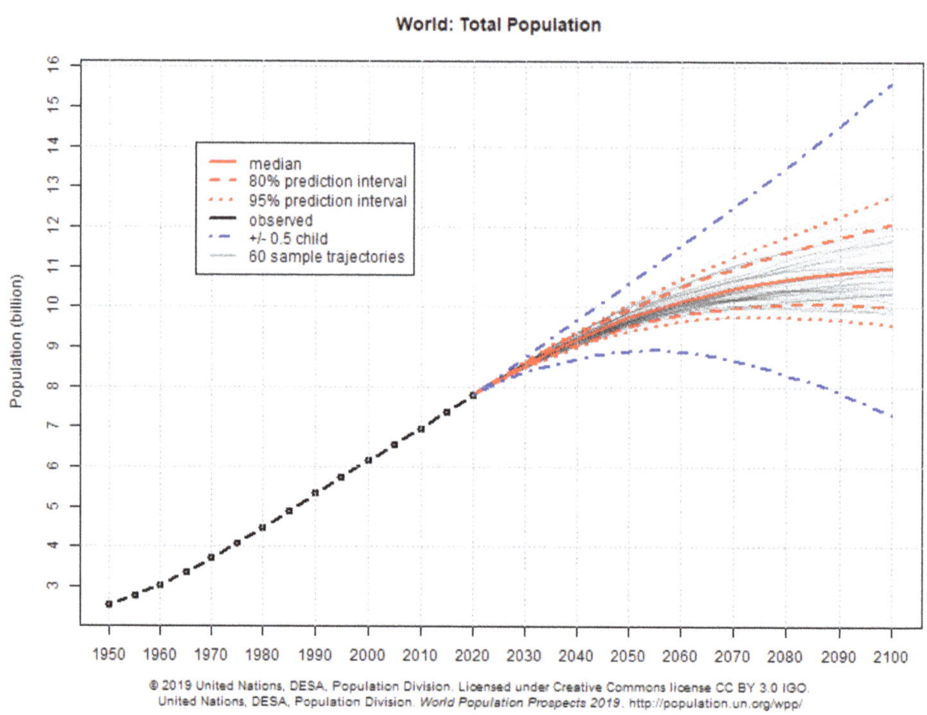

Figure 1.3. Projected world population from the United Nations, showing projections based on several models. Reproduced from https://population.un.org/wpp/Graphs/Probabilistic/POP/TOT/900. CC BY 3.0.

(or about twice the current world average) as a rough estimate gives a total world energy consumption in terms of the current use of 6×10^{20} J per year,

$$2 \times (11 \times 10^9/7.7 \times 10^9) \times 6.1 \times 10^{20}\ \mathrm{J\ y^{-1}} = 1.7 \times 10^{21}\ \mathrm{J\ y^{-1}} \qquad (1.1)$$

There is considerable uncertainty in the estimate presented above. Other projections predict an even greater energy need, while still others predict a much-decreased energy need (more on the reasons for this below). While the prediction above can be viewed as optimistic as it indicates continued human population growth and some progress towards global energy equity, it also establishes a formidable goal for the future use of our energy resources.

It is also important to note that, in addition to the factors discussed above, i.e. population growth and per capita energy use, the details of our energy utilization will play a role in overall primary energy requirements. For example, if biofuels (e.g. ethanol and biodiesel) were to replace fossil fuels such as gasoline and diesel fuel for transportation, then the total primary energy requirement would remain about the same because the efficiency of converting primary energy (i.e. the energy content of the fuel) to end user energy (i.e. energy to the vehicle's wheels), would be similar. On the other hand, if we were to replace fossil fuel powered vehicles with electric vehicles, the overall primary energy requirement could decrease. This is because the efficiency of an electric vehicle is greater than the efficiency of a fossil fuel powered vehicle. That is, the ratio of the end user energy utilized to the primary energy consumed is greater for the electric vehicle than for the fossil fuel vehicle. In the case of a gasoline vehicle, the overall efficiency might be 15%–20%, while for the electric vehicle the efficiency, i.e. energy content of coal burned to produce electricity to recharge the batteries compared to the energy to the vehicle wheels might be 35%. However, it is important to note that in a future energy scenario where electric vehicles replace fossil fuel vehicles, the total primary energy might decrease although the electricity generating capacity would have to increase. Nevertheless, the result of equation (1.1) is a reasonable first estimate of our future energy requirements.

1.4 Fossil fuel resources

Reserves-to-production ratios

As noted above in section 1.2, about 80% of our current energy comes from fossil fuels; either oil, coal or natural gas. Since fossil fuels are not renewable, we need to consider their longevity in order to establish their possible significance as a future energy source. This type of analysis can be done individually for oil, coal and natural gas. The simplest approach to understanding the future availability of a non-renewable energy resource is the reserves-to-production ratio. That is the ratio of the annual production of a resource to the known reserves of that resource. This ratio gives the number of years before the resource's depletion, assuming a constant rate of use. Table 1.4 provides data for a reserves-to-production analysis for oil, coal and natural gas. This analysis shows that oil and natural gas reserves will be depleted in about 50 years while coal can continue to provide energy until the mid-22nd century.

Table 1.4. Reserves-to-production ratios (R/P) and multi-Hubbert analysis results for oil, coal and natural gas. Data for R/P analysis are for 2018; adapted from. https://www.bp.com/content/dam/bp/business-sites/en/global/corporate/pdfs/energy-economics/statistical-review/bp-stats-review-2018-full-report.pdf and data for the multi-Hubbert analysis follows from the data in figure 1.5 as described in the text (t = tonne).

Fuel	Reserves	Current annual production	R/P (years)	R/P end year	Multi-Hubbert 85% year
Oil	2.44×10^{11} t	4.66×10^{9} t	52	2070	2053
Coal	1.054×10^{12} t	7.99×10^{9} t	132	2150	2111
Natural gas	1.97×10^{14} m^3	3.87×10^{12} m^3	51	2069	2077

The Hubbert model

A more detailed analysis of resource utilization is based on the Hubbert model which was developed by M King Hubbert in the 1950s to predict the longevity of United States oil resources. The model assumes that when a particular resource is deemed to be useful, production will begin slowly and will gradually increase. This is because it will take time to identify the deposits of the resource and to develop the infrastructure for its use. When the resource begins to become depleted, production will decrease and eventually end when the resource is exhausted. This gives rise to a Gaussian-like production curve, where the area under the production curve gives the total quantity of resource available.

The validity of the Hubbert model may be demonstrated by an analysis of a specific, well-defined subset of a resource. An example is shown in figure 1.4 for Norwegian oil. Norwegian oil production from offshore drilling began in the early 1970s and at present 80%–85% of the known resource has been exploited. The Hubbert model is seen to provide a good description of these data as shown in the figure. This type of analysis is useful in predicting the future availability of the resource.

A Hubbert analysis of world fossil fuel production is slightly more complex, as shown in figure 1.5. A single Hubbert curve is not necessarily sufficient to describe all features of the data. This is seen, for example, for oil production in the 1970s where there is a small peak in production. In this case, the peak corresponds to an increase in oil production in the late 1960s from a number of newly discovered oil deposits followed by a drop in production in 1973 that resulted from an oil embargo imposed by Middle East nations for political reasons. Additional structure in the production curve can also result from the utilization of newly discovered deposits or by the implementation of new extraction and/or refining technologies. In such cases a multi-Hubbert analysis involving more than a single Hubbert curve may be appropriate.

A multi-Hubbert analysis of world fossil fuel production, as shown in figure 1.5, allows for an analysis of future resource availability. Table 1.4 gives the year in which 85% of the known resources will have been utilized. These results show the same trends as the results from the simpler R/P analysis.

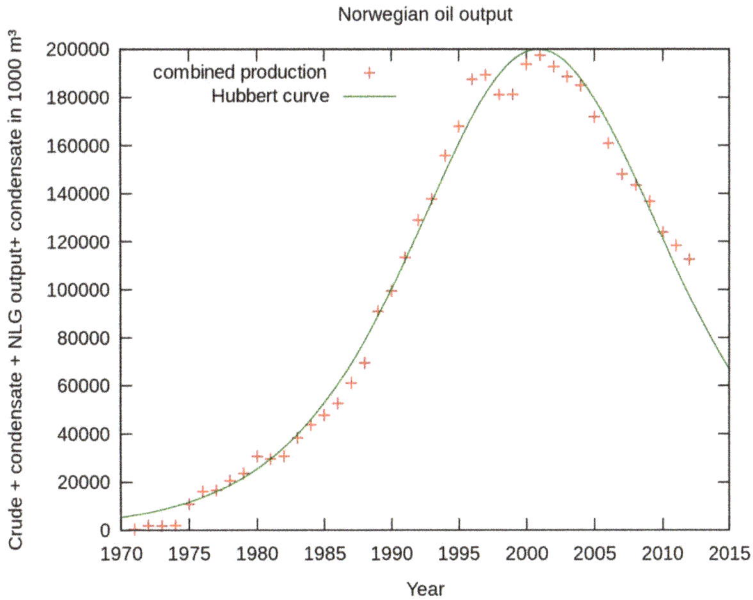

Figure 1.4. Hubbert model applied to oil production in Norway from 1970 to 2013. This Norway Oil Production image has been obtained by the author from the Wikimedia website https://commons.wikimedia.org/wiki/File:NorwayoilproductionandHubbert2013.svg, where it was made available by Krockmeyer under a CC BY 3.0 license. It is included within this book on that basis. It is attributed to Krockmeyer.

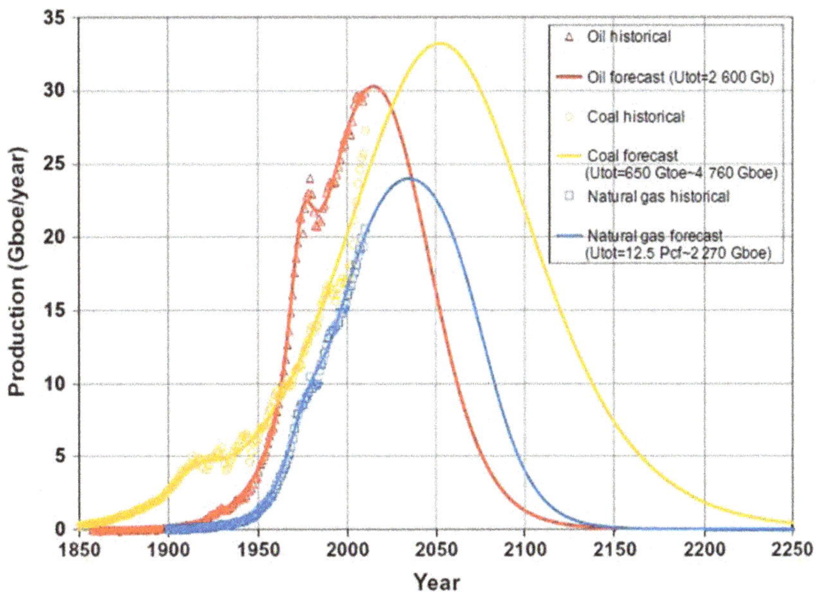

Figure 1.5. Multiple-Hubbert analysis of world oil, natural gas and coal. Reprinted from Maggio and Cacciola (2012). Copyright (2012), with permission from Elsevier.

The analyses presented above indicate that, strictly from an available resource standpoint, we can expect that within 40–50 years oil and natural gas availability will be severely limited and that this will be the case for coal within about 100 years.

1.5 Environmental aspects of fossil fuel use

Greenhouse gas emissions

In addition to the need to develop alternative energy sources to deal with the diminishing fossil fuel supplies, there is compelling evidence to warrant a change in our approach to energy for environmental reasons. The extensive use of fossil fuels as sources of energy is clearly seen in table 1.1. The use of fossil fuels produces chemical and thermal pollution. Most importantly, however, the combustion of fossil fuels produces carbon dioxide, which is a greenhouse gas. A simple example of carbon dioxide production from fossil fuel combustion is the burning of carbon, the principal component of coal, by the reaction

$$C + O_2 \rightarrow CO_2 \tag{1.2}$$

Other examples of carbon dioxide production include the burning of methane, the principal component of natural gas, by the reaction

$$CH_4 + 2O_2 \rightarrow CO_2 + 2H_2O \tag{1.3}$$

and the burning of octane, an important component of gasoline, by the reaction

$$2C_8H_{18} + 25O_2 \rightarrow 16CO_2 + 18H_2O \tag{1.4}$$

Historically, the concentration of CO_2 in the Earth's atmosphere is as illustrated in figure 1.6. This graph shows that over the past 400 000 years the CO_2 concentration has oscillated between about 200 ppm and 280 ppm with a period of about 100 000 years. The figure shows that variations in the surface temperature of the Earth correlate very well with atmospheric CO_2 concentrations, confirming the importance of the greenhouse effect on the Earth's surface temperature. Although the reasons for these natural periodic variations in CO_2 concentrations are not known with certainty, it has been hypothesized that they may be the result of variations in cosmic

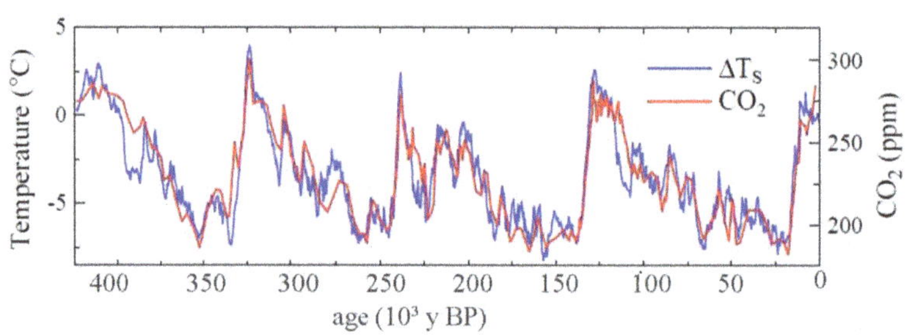

Figure 1.6. Atmospheric CO_2 concentration for the past 400 000 years compared with surface temperature change as obtained from the analysis of Vostok, Antarctic ice cores (https://www.giss.nasa.gov/research/briefs/hansen_11/fig1.pdf). Courtesy of NASA.

ray flux. Cosmic rays, which largely originate from sources outside our solar system, are high energy charged particles, primarily protons. Ionization in the atmosphere, that results from the interaction of cosmic rays with atmospheric atoms, leads to cloud formation and a corresponding decrease in the Earth's surface temperature. The period of the oscillations in figure 1.6 is approximately half the period of the Sun's orbit around the center of the Galaxy and is consistent with periodic variations in cosmic ray flux that correlates to the solar system's position relative to the galactic core.

Since the beginning of industrialization in the 19th century, atmospheric CO_2 levels have risen consistently to a level well above the maximum shown in figure 1.6. Figure 1.7 shows measured CO_2 concentrations for the years 1750–2019 and indicates that the current concentration of this greenhouse gas is in excess of 410 ppm. The graph also shows anthropogenic CO_2 emissions for the same period of time and illustrates the correlation between atmospheric concentrations of CO_2 and human emissions.

The effects of increased greenhouse gases in the Earth's atmosphere are manifested in several ways and a few of the more important consequences of our fossil fuel use are described below.

Global warming

As illustrated in figure 1.6 for historical data, the increase in atmospheric CO_2 concentration that results from our fossil fuel use, is accompanied by an increase in

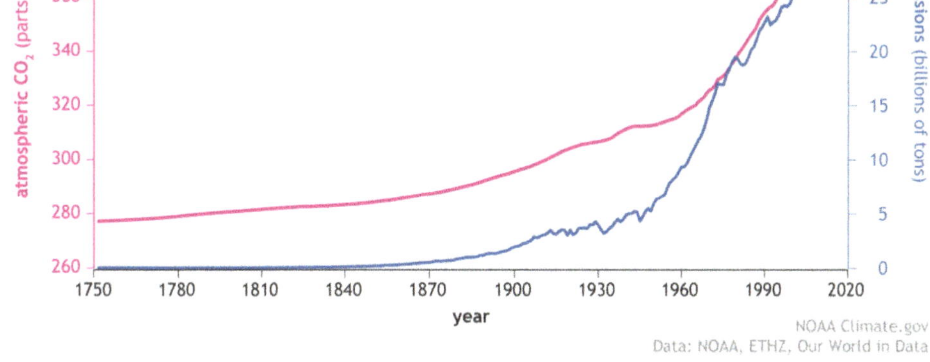

Figure 1.7. Atmospheric CO_2 concentrations since 1750 (red curve). The blue curve shows anthropogenic CO_2 emissions (https://www.climate.gov/sites/default/files/CO2_emissions_vs_concentrations_1751–2019_lrg.gif). Source: NOAA.

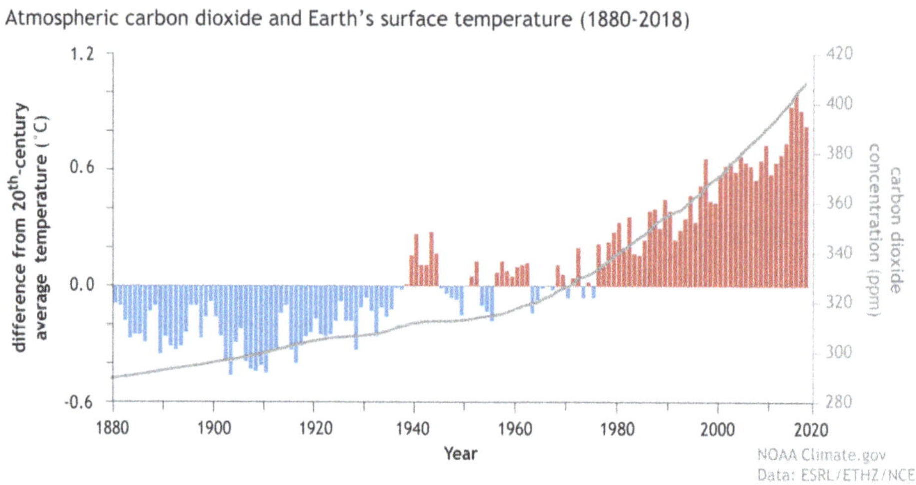

Figure 1.8. Difference in average temperature of the Earth's surface from the average for the years 1900–99 compared with the concentration of atmospheric carbon dioxide (https://www.climate.gov/sites/default/files/global_temp_vs_carbon_dioxide_graph_lrg.gif). Source: NOAA.

the average temperature at the Earth's surface. This correlation is clearly shown in the recent data in figure 1.8. During the past century average global temperatures have increased by about 1.2 °C.

Glacial extent

A 1901 geological map showed that Okjökull (Icelandic for Ok glacier) in Iceland covered an area of about 38 km^2. Today, Okjökull covers an area of less than 1 km^2. This drastic reduction in glacial area is very obvious just in the past 30 years or so. This is illustrated in figure 1.9. In 2019 this ice mass was no longer of sufficient extent to warrant its classification as a glacier.

Numerous glaciers in other parts of the world have followed this same trend over the past century as a result of rising global temperatures.

Polar ice masses

The masses of large ice sheets near the north and south poles have decreased consistently in recent years as a result of increasing average temperatures. The change in the mass of the Greenland ice sheet is shown in figure 1.10 and illustrates one of the effects of global warming in the polar regions.

Sea ice extent anomaly

Sea ice in the polar regions undergoes seasonal variations in its extent. In the Arctic regions the maximum extent of the sea ice occurs in March of each year and the minimum extent occurs in September of each year. Observations of Arctic sea ice extent over the past 40 years, as illustrated in figure 1.11, have shown a decrease in

Figure 1.9. Satellite images of Okjökull in Iceland show the decline of the glacier in recent years; September 14, 1986, (top) and August 1, 2019 (bottom) (https://earthobservatory.nasa.gov/images/145439/okjokull-remembered). Source: NOAA.

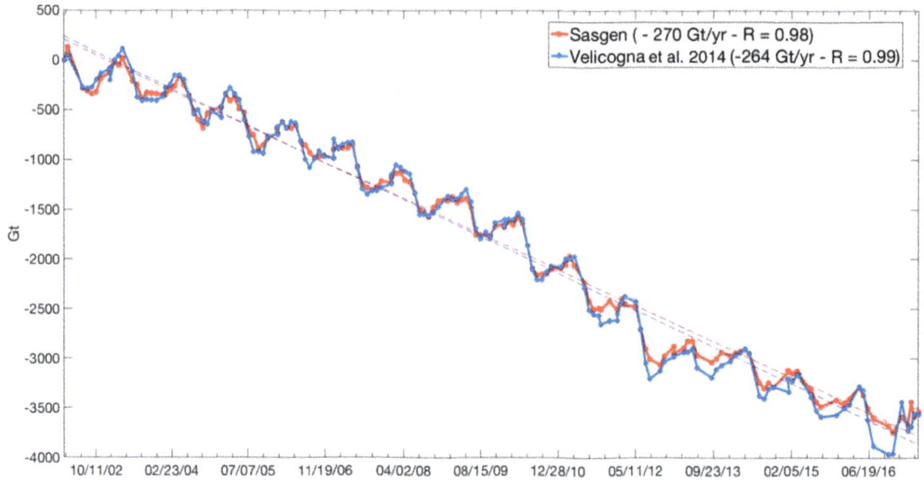

Figure 1.10. Changes in the Greenland ice mass referenced to the 2002 value in Gt = 10^{12} kg (https://arctic.noaa.gov/Report-Card/Report-Card-2017/ArtMID/7798/ArticleID/697/Greenland-Ice-Sheet). Source: NOAA.

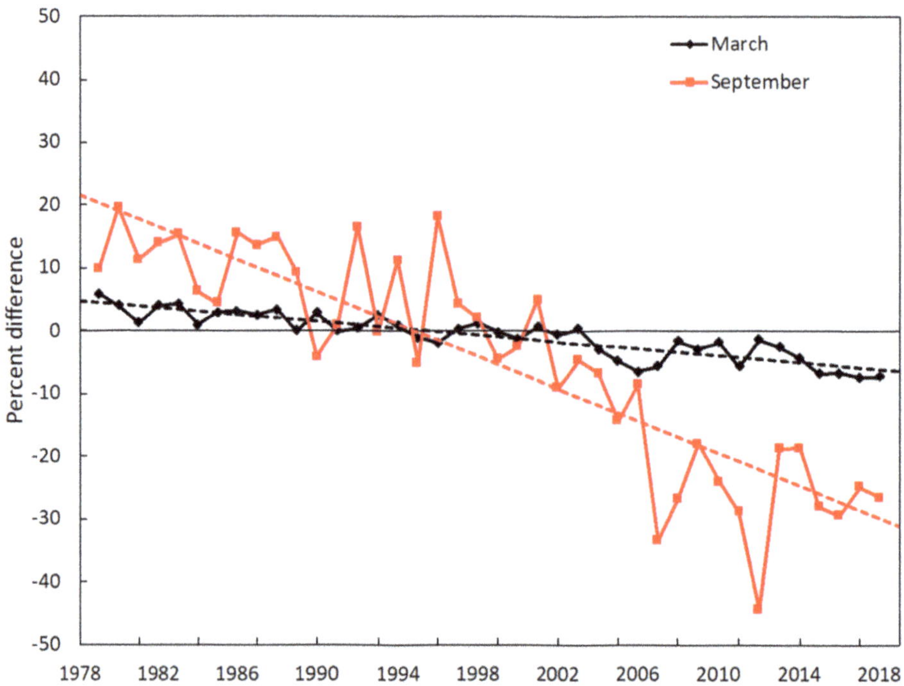

Figure 1.11. Artic sea ice extent anomaly in March and September for the past 40 years. The graph shows the quantity of Arctic sea ice (in %) above or below the average for the period of 1981–2010 in units of the standard deviation of the data from this period. Data are shown for the months of March (the month of maximum sea ice) and September (the month of minimum sea ice). The red and black dashed lines are linear least squares fits to the data (https://arctic.noaa.gov/Report-Card/Report-Card-2018/ArtMID/7878/ArticleID/780/SeanbspIce). Source: NOAA.

maximum coverage. This change is most apparent for the September measurements where a decrease of more than 40% is observed. Data for Antarctic sea ice extent have shown similar trends in recent decades.

Sea level rise

A direct consequence of global warming and the melting of ice masses worldwide is an increase in sea level. In fact, two different aspects of global warming contribute to sea level. Firstly, the melting of polar ice increases sea water amount and, secondly, increasing global temperatures increases water volume as a result of thermal expansion. Figure 1.12 shows the effects of global warming on ocean temperature. Figure 1.13 shows the total observed sea level rise from 1880 to 2020. In recent years (i.e. 2006–15) an analysis of sea level data shows an average increase of 3.6 mm per year. This may be compared to an average rise of 1.4 mm per year observed during much of the 20th century. These results are an illustration of the acceleration of effects due to global warming that has occurred in the past 20 years or so.

The discussion above shows that dwindling fossil fuel resources along with the adverse environmental consequences of continued greenhouse gas production provides a compelling motivation to pursue the development of alternative carbon-free energy technologies. The most important of these new technologies are overviewed in the next three sections.

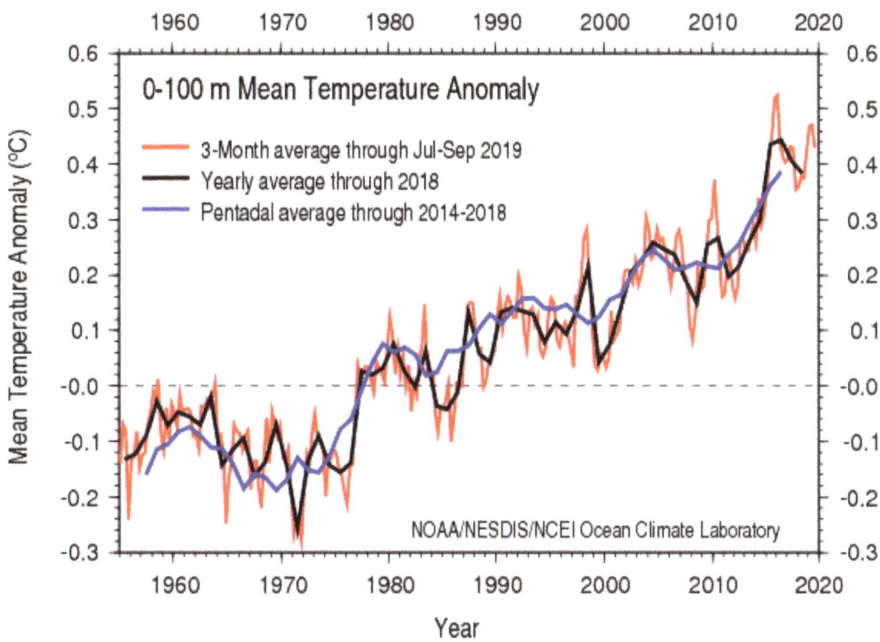

Figure 1.12. Mean temperature difference of the top 100 m of the oceans referenced to the average of the period 1955–2006 (https://www.nodc.noaa.gov/OC5/3M_HEAT_CONTENT/index3.html). Source: NOAA.

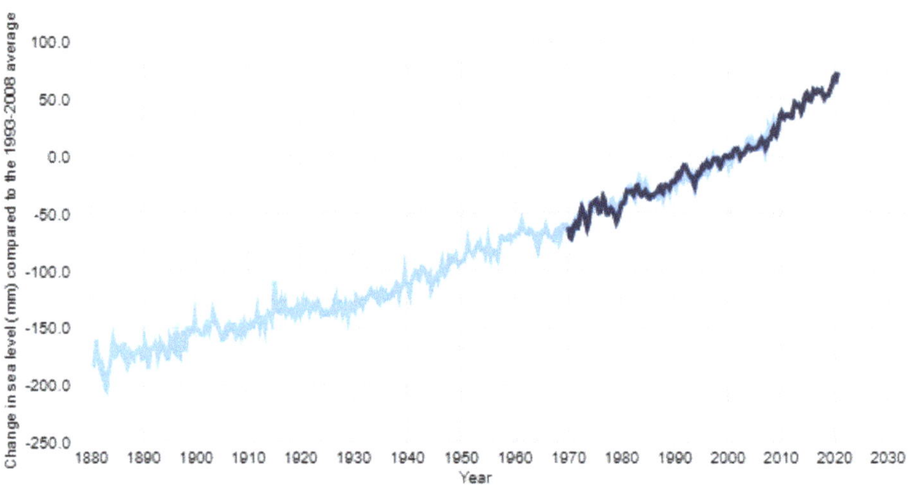

Figure 1.13. Changes in sea level from 1880 to 2020 and the breakdown into components from thermal expansion and increased water content (https://www.climate.gov/news-features/understanding-climate/climate-change-global-sea-level). Source: NOAA.

1.6 Future energy sources I—renewable energy

In general, non-fossil fuel-based energy sources do not contribute directly to the production of greenhouse gases. They can be divided into two categories, renewable and non-renewable. The former includes, for example, wind, solar and hydroelectric energy, which are, in principle, renewable indefinitely, at least on human time scale. The latter category includes nuclear energy. Although, in principle, nuclear energy is not renewable, as it relies on the consumption of naturally occurring resources, it has sometimes been argued that the availability of these resources is sufficiently large to guarantee reliable energy well into the future.

In the present section the most important sources of renewable energy are described. These energy sources are those that have already been commercialized successfully and include solar energy, wind energy, hydroelectricity, geothermal energy and biofuels. Other potential sources of renewable energy such as tidal energy, wave energy, ocean thermal energy conversion and osmotic energy have not been shown to be commercially viable on a large scale and will not be discussed in the present volume. As well, only grid-scale applications of renewable energy are discussed in order to facilitate a comparison of different energy sources.

Solar energy

Basics of solar energy

The Sun produces energy by nuclear fusion as described in detail in section 3.3. The total power produced by the Sun is 3.8×10^{26} W. This is emitted isotopically so that the power density (power per unit area, P/A) at the Earth can be determined by dividing the total power by the surface area of a sphere with a radius equal to the mean distance between the Sun and Earth. For the average Sun to Earth distance of 1.49×10^{11} m this calculation gives

$$\frac{P}{A} = \frac{3.8 \times 10^{26}\ \text{W}}{4\pi \times (1.49 \times 10^{11}\ \text{m})^2} = 1367\ \text{W m}^{-2} \tag{1.5}$$

This is the power density of sunlight outside of the Earth's atmosphere and is referred to as the 'solar constant'. The total sunlight incident on the Earth is given by the solar constant times the area of a disk with the radius of the Earth (6.371×10^6 m), that is

$$P = 3.1415 \times (6.371 \times 10^6\ \text{m})^2 \times (1367\ \text{W m}^{-2}) = 1.75 \times 10^{17}\ \text{W} \tag{1.6}$$

sunlight that is incident on the Earth may be reflected back into space, absorbed by the Earth's atmosphere or arrive at the Earth's surface. This last component, referred to as the insolation at the Earth's surface, amounts to about 49% of the solar power incident outside of the atmosphere (or about 8.56×10^{17} W). Obviously, this radiation is not distributed uniformly over the surface of the Earth, nor is it constant in time. We can easily calculate the insolation averaged over time and surface location on the Earth by dividing the total insolation incident on the Earth's surface by the total surface area of the Earth

$$\frac{P}{A} = \frac{8.56 \times 10^{17}\ \text{W}}{4 \times 3.14 \times (6.371 \times 10^6\ \text{m})^2} = 168\ \text{W m}^{-2} \tag{1.7}$$

Of course, the insolation varies as a function of latitude, day of the year, time of day and local weather conditions. In the United States, for example, typical average insolation ranges from about 150 W m^{-2} in the Northeast and Northwest, to about 250 W m^{-2} in the Southwest. The utilization of solar energy is most appropriate for locations that are at a low latitude and for which average cloud cover is minimal.

Solar energy has found a wide variety of applications as a carbon-free energy source. These include small-scale applications such as residential thermal energy (e.g. domestic heat and hot water) and residential electricity. While some larger scale installations, such as the Drake Landing Solar Community in Okotoks, Alberta, Canada and Brædstrup Fjernvarme in Brædstrup, Denmark, utilize solar thermal energy for community heating, most large-scale solar energy facilities are designed for grid-scale electricity production. These latter systems utilize either thermal generation along the lines of a coal-fired generating station or they utilize photovoltaics. These two approaches are described briefly below.

Solar thermal electricity

Electricity may be generated using solar energy by concentrating the solar radiation onto a working fluid which, in turn, transfers heat to water to generate steam to drive a turbine/generator. Several different geometries have been used to concentrate the solar radiation. The most successful of these have been parabolic trough reflectors and central receivers.

Figure 1.14. Parabolic trough solar collector near Harper Lake in California's Mojave Desert. This Parabolic Trough at Harper Lake image has been obtained by the author from the Wikimedia website https://commons.wikimedia.org/wiki/File:Parabolic_trough_at_Harper_Lake_in_California.jpg where it was made available Z22 under a CC BY-SA 3.0 license. It is included within this book on that basis. It is attributed to Z22.

Parabolic trough systems use a parabolic reflector, as shown in figure 1.14, which tracks the Sun on one axis and focusses the solar radiation on a tube carrying the working fluid (typically synthetic oil) which is heated to ~400 °C. Steam, produced from water heated by the oil through a heat exchanger, is typically used to drive a Rankine cycle turbine. The efficiency of such systems is limited to the thermodynamic efficiency of the heat engine (i.e. turbine) used to convert thermal energy to mechanical energy and is typically in the range of about 25%.

United States and Spain have been the most active in developing parabolic trough solar generating stations. The largest operational systems worldwide are summarized in table 1.5. The capacity given in the table represents the maximum rated electrical output.

Central receiver systems use a large array of individually positioned mirrors to focus sunlight on a receiver located at the top of a central tower, referred to as a 'power tower'. The individual mirrors, referred to as 'heliostats', are oriented by computer to track the Sun. An example of a commercial central receiver solar generating station is shown in figure 1.15. At present the largest station of this type is the Ivanpah Solar Power Facility in San Bernardino County, California, which has

Table 1.5. Largest parabolic trough solar generating stations.

Name	Location	Capacity (MW)
Solar Energy Generating Systems	Mojave Desert, California	361
Ouarzazate Solar Power Station	Ghassate, Spain	360
Mojave Solar Project	Barstow, California	280
Solana Generating Station	Gila Bend, Arizona	280
Genesis Solar Energy Project	Blythe, California	280
Solaben Solar Power Station	Logrosán, Spain	200

Figure 1.15. Solucar PS10 in Sanlúcar la Mayor, Seville Province, Andalusia, Spain was commissioned in 2007 and was the first central tower solar power plant in the world to generate electricity commercially. This Solucar PS10 image has been obtained by the author from the Wikimedia website https://commons.wikimedia.org/wiki/File:PS10_solar_power_tower.jpg where it was made available by afloresm under a CC 2.0 license. It is included within this book on that basis. It is attributed to SOLUCAR PS10.

an electrical capacity of 392 MW. It is important to note that the nameplate capacity is the maximum electrical output that the facility can produce.

In 2019 the total world electrical capacity of commercial (grid-connected) solar thermal generating stations was 4815 MW.

Photovoltaics

The second major approach to converting solar radiation into electricity is through the use of photovoltaics. A photovoltaic cell is a semiconducting junction that produces an electric current when photons of sufficient energy are incident upon it.

A simple semiconducting junction is formed when two semiconducting materials with different types of impurities are in electrical contact. Here we give a brief example of a silicon based semiconducting junction.

Pure silicon (often referred to as intrinsic silicon) has 14 electrons with the electronic structure

$$1s^2 2s^2 2p^6 3s^2 3p^2 \tag{1.8}$$

In an isolated silicon atom these energy levels are discrete. However, in a solid with interacting silicon atoms, the energy levels are smeared out into bands as a result of the Pauli exclusion principle. In addition, levels with angular momentum quantum numbers $l > 0$ (i.e. p, d, f… states) are split into $2l + 1$ sublevels. These properties give rise to the so-called energy gap, E_g, between the uppermost occupied energy levels, the valence band, and the lowermost unoccupied energy levels, the conduction band.

The behavior of a free electron in an electric field, $\boldsymbol{\varepsilon}$, is determined by the Lorentz force (note the bold font is used to indicate vectors);

$$F = -e\boldsymbol{\varepsilon} \tag{1.9}$$

where e is the magnitude of the electronic charge. The Lorentz force causes the electron to accelerate in a direction opposite to the electric field vector. The movement of electrons through a material in response to the application of an electric field is a simple description of electrical conduction in a conductor.

The electrical properties of silicon are determined by the distribution of electrons in the valence and conduction bands. At zero temperature, the states in the valence band are filled while the states in the conduction band are empty, leading to the insulating behavior exhibited by silicon at low temperature. We can increase the electrical conductivity of silicon in one of two ways. Firstly, we can increase the temperature above absolute zero, thereby exciting some of the electrons from the valence band into unoccupied states in the conduction band. This increases electrical conductivity in two ways; by creating electron states in the conduction band and by leaving an equal number of vacant states in the valence band. These vacant states in the valence band are referred to as 'holes' and behave, in many ways as charge carriers with a positive charge. The electrons in the conduction band can conduct electricity by responding to interactions with an applied electric field because there are vacant states above them in the conduction band that they can move into. As well, electrons near the top of the occupied states in the valence band can increase their energy to fill vacant hole states. This can also be viewed as positively charged hole states moving to lower energy. Thus, at non-zero temperature silicon can conduct electricity (although only weakly) and is referred to as a semiconductor. The number of electron states in the conduction band (and the equal number of hole states in the valence band) is quite small. This is because in silicon (and other similar materials) the energy gap, E_g (1.11 eV) is very large compared to the thermal energy of the electrons at (say) room temperature, $k_B T$ (0.025 eV). We can calculate the number of electrons in the conduction band at a temperature T as

$$n_e(T) = 2UT^{3/2}\exp\left[-\frac{E_g}{2k_BT}\right] \tag{1.10}$$

where U is a parameter that depends on the details of the interactions in the material. The sensitivity of the number of electrons to temperature and energy gap is seen in equation (1.10) by the presence of these quantities in the exponential.

Secondly, we can increase the conductivity of silicon by adding impurities. We consider two possibilities; adding aluminum to silicon and adding phosphorus to silicon. Aluminum, referred to as an acceptor impurity, has 13 electrons and the electronic structure

$$1s^2 2s^2 2p^6 3s^2 3p^1 \tag{1.11}$$

Phosphorus, referred to as a donor impurity, has 15 electrons and the electronic structure

$$1s^2 2s^2 2p^6 3s^2 3p^3 \tag{1.12}$$

The effects of aluminum and phosphorus on the electron states in silicon are shown in figure 1.16. The missing electron in aluminum creates a hole state in the valence band. We see that the extra electron from phosphorus has no place to go in the valence band and occupies a state in the conduction band. Both the hole state in aluminum doped silicon and the electron in the conduction band in phosphorus doped silicon can contribute to the electrical conductivity. Aluminum is referred to as a p-type semiconductor, since the charge carriers are positively charged holes, while phosphorus doped silicon is referred to as an n-type semiconductor, since the charge carriers are negatively charged electrons. Note in figure 1.16 that, because of Coulomb interactions, the energy levels in the p-type semiconductor are shifted upward while the energy levels in the n-type semiconductor are shifted downward.

There are substantial differences between charge carriers in silicon that are created by doping and the charge carriers in silicon that are created by increasing the temperature. Firstly, the hole charge carriers in a p-type semiconductor exist without

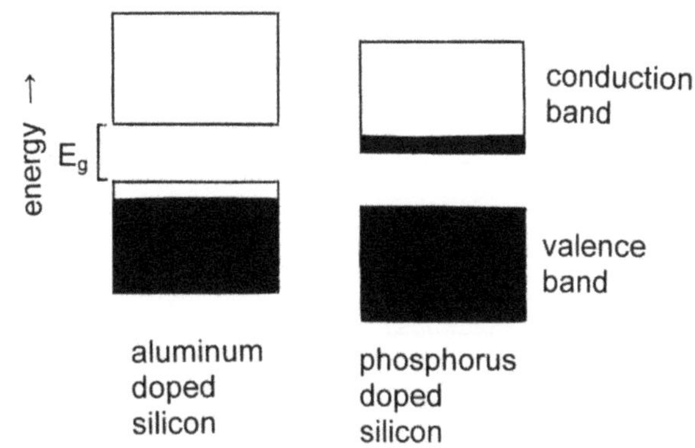

Figure 1.16. Band structure of silicon with (left) phosphorus impurities and (right) aluminum impurities showing occupied electron states (dark area) and unoccupied electron states (light area).

the creation of corresponding electron carriers. Similarly, the electron charge carriers in an n-type semiconductor exist without the creation of corresponding hole carriers. Secondly, even at very low doping levels, charge carriers created by impurities generally greatly outnumber the small number of carriers that are thermally created. Thus, in a p-type semiconductor at (say) room temperature, the holes, referred to as majority carriers, greatly outnumber the thermally created electrons, referred to as minority carriers, and vice versa for an n-type semiconductor.

A p–n junction is formed when a p-type semiconductor and an n-type semiconductor are in electrical contact, as shown in figure 1.17. The properties of this junction can be described by an analysis of the charge carriers in the region around the junction. For simplicity, we consider only the behavior of the majority carriers.

The majority hole carriers in the valence band on the p-side of the junction see a much lower concentration of holes in the valence band on the n-side of the junction. Similarly, the majority electrons in the conduction band on the n-side of the junction see a much lower concentration of electrons in the conduction band on the p-side of the junction. These concentration gradients are the driving force for diffusion of majority charge carriers across the junction, as shown in figure 1.17. When charges diffuse across the junction, they create an electric field that opposes the further diffusion of charges. An equilibrium condition is established where the electric field exactly compensates for the driving force due to the concentration gradient where the energy levels align, as in figure 1.16. The spatial extent of the electric field, as seen in the figure, is referred to as the depletion region. That is because any charges that happen to wander into this region (due to their thermal motion) are carried by the electric field to the appropriate side of the junction to maintain the equilibrium condition.

The behavior of the p–n junction, as shown in figure 1.17 and as described above, allows us to understand the basics of the operation of a photovoltaic device. A photovoltaic cell is constructed so that optical photons can be incident on the depletion region. This can be accomplished by using a wire mesh for making electrical contacts to one of the semiconducting regions and making this semiconductor very thin. When a photon of sufficient energy is incident on the depletion

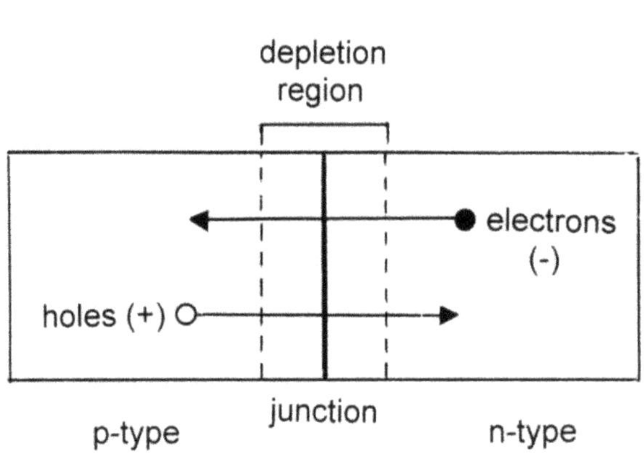

Figure 1.17. Diffusion of majority carriers across a p–n junction.

region an electron from the valence band can be excited up into the conduction band leaving a hole behind. The electron–hole pair that is thus formed interacts with the electric field across the depletion region. The electron is attracted to the n-side where it joins other majority electron carriers, while the hole is attracted to the p-side where it joins the other majority hole carriers. The movement of charges across the depletion region constitutes a current which can do work on an external circuit.

In order to create an electron–hole pair in the depletion region, a photon's energy must be at least equal to the energy gap of the semiconductor. This requires that the wavelength of the light, λ, satisfies the relation

$$\lambda < \frac{hc}{E_g} \tag{1.13}$$

where h is the Planck constant and c is the speed of light. For the energy gap of silicon (1.11 eV) a photovoltaic cell is sensitive to wavelengths of less than 1130 nm. This condition includes all of the visible region of the electromagnetic spectrum (~400 nm to ~700 nm) and a large fraction of the infrared portion of the solar spectrum that reaches the Earth's surface. Overall silicon based photovoltaic cells are sensitive to about 77% of incident sunlight. The maximum theoretical efficiency for converting photon energy into electrical energy is about 25% for a silicon photovoltaic cell. Actual commercial silicon cells have typical efficiencies of around 17%. While other materials may yield somewhat higher efficiencies, silicon is the most cost-effective semiconducting material for photovoltaic cell construction.

Photovoltaic cells can be implemented on a wide range of scales. These applications include, on the small end of the scale, personal devices such as wristwatches and calculators and, on the large end of the scale, grid-connected solar farms. A typical grid-connected photovoltaic installation is shown in figure 1.18. Installed grid-connected photovoltaic capacity worldwide was approximately 510 GW as of 2018.

Solar energy resources

As given above, the total insolation worldwide is 8.56×10^{17} W. This value may be compared with our expected future energy needs as given by equations (1.1), 1.7×10^{21} J y^{-1}, which corresponds to a power of

$$P = \frac{1.7 \times 10^{21}\ \mathrm{J\ y^{-1}}}{3.16 \times 10^{7}\ \mathrm{s\ y^{-1}}} = 5.4 \times 10^{13}\ \mathrm{W} \tag{1.14}$$

It is clear that less than 0.1% of the solar radiation that is incident on the surface of the Earth would be sufficient to satisfy our energy requirement now and for the foreseeable future. If fact, solar energy is the only renewable source that can provide sufficient power to fulfill all of our needs. Assuming a conversion efficiency from solar radiation to electricity of about 20%, as discussed above for thermal conversion or photovoltaics, solar collectors covering about 0.2% of the Earth's surface and positioned major installations in locations with an average insolation of about 250 W m^{-2} would provide all of society's energy needs. This corresponds to about 1×10^{12} m^2, or ten facilities about 300 km × 300 km, located in reasonably

Figure 1.18. Solarkraftwerk Waldpolenz a 40 MW photovoltaic farm in Brandis, Germany. This Juwi PV Field image has been obtained by the author from the Wikimedia website https://commons.wikimedia.org/wiki/File:Juwi_PV_Field.jpg where it was made available JUWI Group under a CC BY-SA 3.0 license. It is included within this book on that basis. It is attributed to JUWI Group.

high insolation areas. In 2017 solar generated electricity accounted for 1.73% of total worldwide electricity generation. This corresponded to about 0.7% of primary energy use. As noted above, the vast majority of this comes from photovoltaic installations, with a minor amount coming from concentrated solar thermal facilities.

Wind energy

Basics of wind energy

The use of wind energy by humans dates back to antiquity. Perhaps the earliest use of wind energy was for sailing ships. During the renaissance, the use of windmills for grinding grain became common. In the 20th century there was developing interest in the use of wind turbines to generate electricity.

A simple analysis of the dynamics of moving air provides a description of the operation of a wind turbine. The kinetic energy of a parcel of air with a mass m moving with a velocity v is given as

$$E = \frac{1}{2}mv^2 \tag{1.15}$$

The mass of the air parcel may be expressed as the product of its length, l, parallel to its direction of motion, its cross-sectional area, A, perpendicular to its direction of motion, and its density, ρ. Substituting in equation (1.15) gives the kinetic energy as

$$E = \frac{1}{2} l A \rho v^2 \tag{1.16}$$

The power available is given as the time derivative of the energy as

$$P = \frac{1}{2}\frac{\mathrm{d}l}{\mathrm{d}t} A \rho v^2 = \frac{1}{2} A \rho v^3 \tag{1.17}$$

This expression gives the total power available over an area A as a function of wind velocity, v. It is important to note that the power is a function of the cube of the wind velocity, and, therefore, small increases in wind velocity represent substantial increases in power.

Wind turbines

Figure 1.19 shows some typical modern wind turbines. The power available from the wind is given by equation (1.17). The area, A, is the swept area of the rotor and is given in terms of the diameter, d, as $\pi d^2/4$. The power produced by the turbine is given in terms of equation (1.16) and the turbine efficiency, η,

$$P = \frac{\pi \rho}{8} \eta d^2 v^3 = 0.473 \eta d^2 v^3 \tag{1.18}$$

where the power is in W, the diameter is in m and the air velocity is in m s^{-1}. The standard air density has been taken to be 1.204 kg m^{-3}.

Wind turbines are not 100% efficient, as in order to extract all of the kinetic energy from a parcel of air, the air would have to be stopped. The theoretical maximum efficiency of a wind turbine is given by the 'Betz limit' as 59% ($\eta = 0.59$). In practice, the actual efficiency of a wind turbine depends on its design, the angular

Figure 1.19. Four Suzlon S97 2.1 MW wind turbines. These turbines are part of a wind farm in Amherst, Nova Scotia, Canada incorporating 15 turbines. Each turbine has a hub height of 90 m and a rotor diameter of 97 m. (Image: Richard A Dunlap.)

rotational frequency of the rotor, ω, and the wind velocity. The tip speed ratio, T, is defined as the ratio of the velocity of the tip of the rotor to the velocity of the wind. This is expressed as

$$T = \frac{d\omega}{2v} \tag{1.19}$$

Figure 1.20 shows the theoretical efficiency of a modern three-blade wind turbine as a function of T. The maximum efficiency of about 47% ($\eta = 0.47$) occurs for a value of T of around 7.5.

The utilization of wind energy requires the placement of wind turbines in locations that have the most advantageous wind conditions. As equation (1.18) shows that the power is related to the cube of the wind velocity, the distribution of wind speeds, $f(v)$, is an important factor. The Weibull distribution is a suitable expression for modeling the wind speed distribution at many locations. This distribution has the functional form

$$f(v) = \left[\frac{k}{c}\right]\left[\frac{v}{c}\right]^{k-1} \exp\left[-\left(\frac{v}{c}\right)^{k}\right] \tag{1.20}$$

where the parameters k and c are defined as

$$k = \left[\frac{\sigma}{\langle v \rangle}\right]^{-1.086} \tag{1.21}$$

and

$$c = \frac{\langle v \rangle}{\Gamma\left(1 + \frac{1}{k}\right)} \tag{1.22}$$

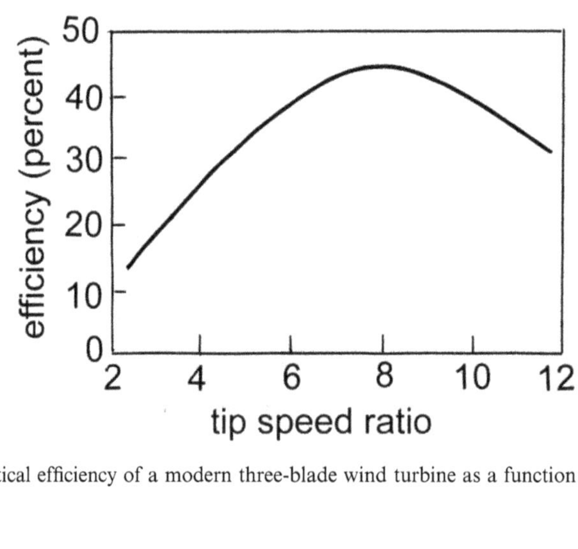

Figure 1.20. Theoretical efficiency of a modern three-blade wind turbine as a function of the tip speed ratio.

respectively. Here $\langle v \rangle$ is the mean wind speed and σ is the standard deviation of the speed distribution. A simplified form of equation (1.20) is the Rayleigh distribution given as

$$f(v) = \frac{2v}{c^2} \exp\left[-\left(\frac{v}{c} \right)^k \right] \tag{1.23}$$

Figure 1.21 shows a comparison of the Weibull and Rayleigh distribution analyses of some experimental wind speed distribution data. The figure shows that, in general, wind speed distributions are well described by these distribution functions.

Another important factor in determining the effective placement of wind turbines is related to the height dependence of the wind speed. Typically, at least up to some point, wind speed increases with height, z, according to the expression

$$v(z) = v(z_0) \frac{\ln\left[\frac{z}{\delta}\right]}{\ln\left[\frac{z_0}{\delta}\right]} \tag{1.24}$$

Here the wind speed as a function of height, $v(z)$, is related to the wind speed, $v(z_0)$, at a reference height, z_0. The parameter δ is referred to as the roughness length and is a measure of the surface features of the land around the turbine. Typical values of δ range from 0.0002 for very smooth surfaces such as water to 1.6 for very irregular surfaces such as cities with tall buildings. Figure 1.22 shows the height dependence of wind speeds for a typical case of a turbine located on flat agricultural land, as would be the situation in the Midwest of the United States. It is clear that increasing the height of a wind turbine will provide increased power output as a result of increased

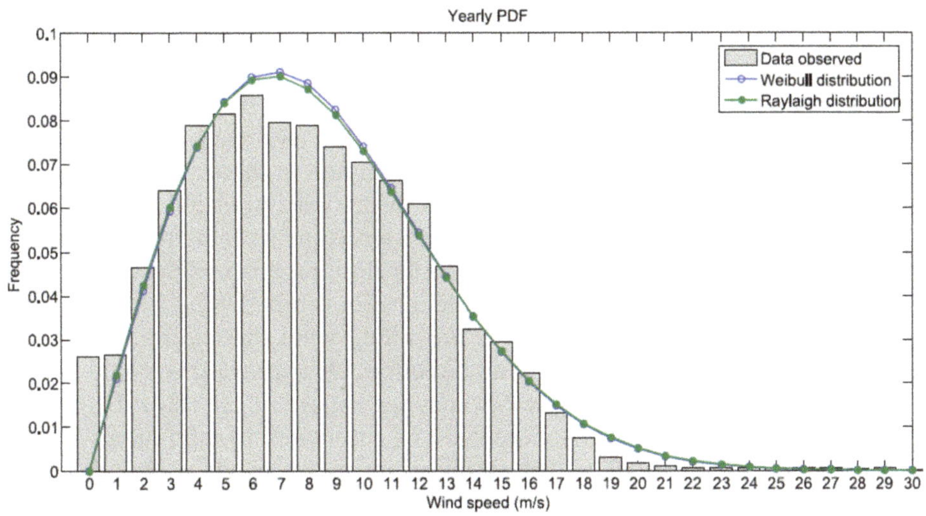

Figure 1.21. Weibull and Rayleigh distributions for annual wind speeds at Tetouan, Morocco. Reproduced with permission from Ouahabi et al (2017).

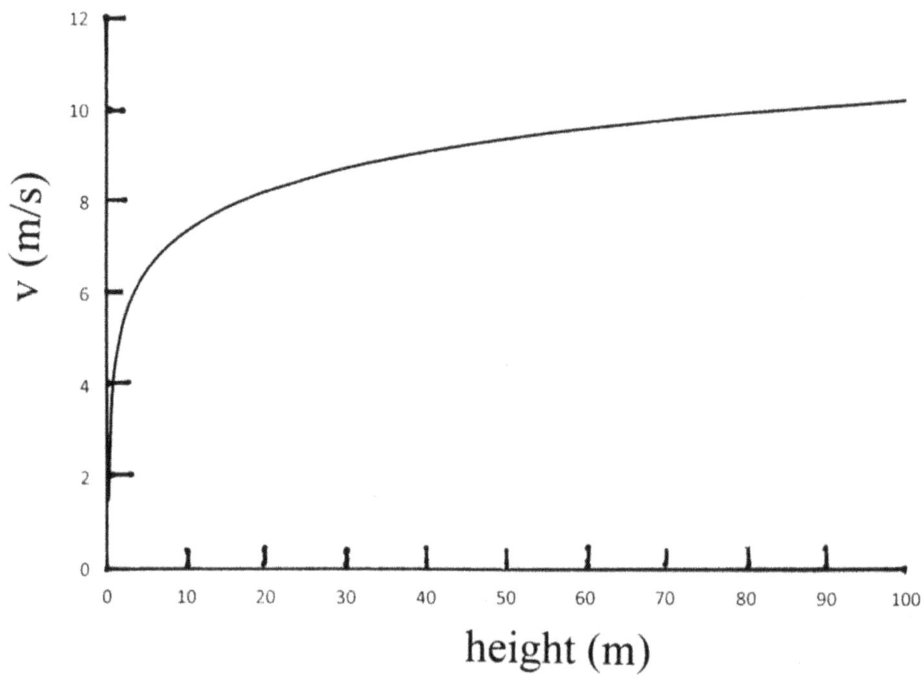

Figure 1.22. Height dependence of wind speed for roughness parameters $\delta = 0.03$ (appropriate for flat agricultural land) and a velocity of 10 m s^{-1} at 80 m.

wind speed. Obviously, there is a trade-off between power production and the practicality and economy of construction. This is considered further below.

Wind energy resources

Because the power available from wind is a sensitive function of the wind speed, it is important to locate wind turbines where wind speeds are the most advantageous. In North America, average wind speeds are the greatest along the Atlantic and Pacific coasts, on the great plains and around the Great Lakes. Average annual wind speeds in the range of 8–9 m s^{-1} (at 80 m height) are common in these areas. Major wind energy initiatives generally consist of wind farms with multiple wind turbines. In order to best make use of the land area of locations with advantageous wind conditions, it is important to space turbines as close together as possible without degrading their performance. Typical guidelines for optimal wind turbine spacing suggest 8–12 rotor diameters along the prevailing wind direction and 2–4 rotor diameters normal to the prevailing wind direction. On average this would indicate a land area per turbine of $3 \times 10 = 30$ times the square of the rotor diameter, d. Since the rotor area is $\pi d^2/4$, then the ratio of rotor area to land area would be $(\pi d^2/4)/(30d^2) = \pi/120 = 0.026$. The implication of this result is that the power available from a wind farm with optimal spacing depends only on the land area utilized and that a wind farm consisting of a large number of small turbines would be equivalent to a wind farm consisting of a smaller number of large turbines.

This is not exactly true as larger turbines are mounted on taller towers and therefore, in accordance with figure 1.22, access greater wind speeds.

Relating the above discussion to the power generated by a wind turbine in accordance with equation (1.18) gives the power per unit land area, A_L, as

$$P = \frac{0.026}{2}\eta\rho A_L v^3 \approx 3A_L \tag{1.25}$$

for power in W and land area in m^2. This is slightly less than 10% of the power per land area from solar photovoltaics (at 17% efficiency), suggesting that substantially larger land area would need to be utilized for wind turbines compared to solar photovoltaics in order to fulfill our energy needs. From a practical standpoint wind is probably not likely to satisfy more than about 10% of our future energy requirements. This is a rather simplistic analysis, as it considers only the land area need for photovoltaic cells or wind turbines and does not address the requirements for producing required materials or manufacturing infrastructure, etc.

Wind power has grown substantially in the past two decades or so. Total worldwide installed wind capacity has increased from 6.1 GW in 1996 to 592 GW in 2018. Currently, the largest utilization of wind power is in China (installed capacity of 211 GW in 2018) and the United States (installed capacity of 97 GW in 2018). The greatest penetration of wind power, however, has been in some European Union countries, as illustrated in table 1.6.

Hydroelectric energy

Basics of hydroelectric energy

Like wind power, hydropower has been used since well before the industrial revolution as a source of mechanical power. The energy content of water may be related to gravitational potential energy for water that undergoes a change in elevation or kinetic energy related to its velocity. Current commercial hydroelectric facilities fall basically into two categories; high head facilities that utilize the gravitational potential energy of water that is impounded behind a dam and low head or run-of-the-river generating stations, which do not impound water behind a dam and which, to a large extent, utilize the kinetic energy of flowing water.

Table 1.6. Wind generated electricity as a fraction of total domestic electricity production in 2018.

Country	Fraction of domestic electricity production
Denmark	0.41
Ireland	0.28
Portugal	0.24
Germany	0.21
Spain	0.19

An analysis of the basic physics of a high head hydroelectric facility considers the gravitational potential energy of water that is impounded in a reservoir behind a dam. The 'head', h, is the height through which the water falls from the reservoir to turbines connected to generators located at the base of the dam. The gravitational potential energy of a mass of water, m, is given as

$$E = mgh \tag{1.26}$$

where g is the gravitational acceleration ($g = 9.8$ m s^{-2}). The power that is generated by water flowing through the turbine is given by the time derivative of equation (1.26) as

$$P = \frac{\mathrm{d}E}{\mathrm{d}t} = \frac{\mathrm{d}m}{\mathrm{d}t}gh \tag{1.27}$$

The time derivative of mass may be related to the volume, V, of water passing through the turbines to give

$$P = \frac{\mathrm{d}V}{\mathrm{d}t}\rho gh \tag{1.28}$$

where the density of water is typically $\rho = 1000$ kg m^{-3}. The electrical power produced is given by equation (1.28) and the efficiency of the turbine/generator, η, as

$$P = \frac{\mathrm{d}V}{\mathrm{d}t}\eta\rho gh \tag{1.29}$$

Hydroelectric turbine/generator systems are typically very efficient; $\eta = 0.85$–0.90.

The term low head or run-of-the-river hydroelectric generally refers to any system where a reservoir is not created by impounding water behind a dam. Although this terminology is often not well defined and various hydroelectric generating station designs may fall in this category, these facilities are typically those in which the kinetic energy associated with moving water represents a significant component of the total energy production. The analysis of the kinetic energy of moving water follows directly along the lines of the analysis of a wind turbine as given in equation (1.17). The power available from moving water is, therefore, given by

$$P = \frac{1}{2}\eta A\rho v^3 \tag{1.30}$$

where η is the generating efficiency and A is the cross-sectional area of the channel. Although typical water velocities may be less than typical wind speeds, hydroelectric power benefits in two ways. Most importantly, the density of water, 1000 kg m^{-3}, is more than 800 times the density of air and efficiencies of hydroelectric generating facility can be greater than those of wind turbines.

High head hydroelectric energy

The largest capacity hydroelectric generating stations worldwide fall under the category of high head hydroelectric. At present, the world's largest hydroelectric generating facility in terms of rated capacity is the Three Gorges Hydroelectric Dam in Sandouping, Hubei Province, China shown in figure 1.23. The largest

Figure 1.23. The Three Gorges Hydroelectric Dam in Sandouping, Hubei Province, China. This Three Gorges Hydroelectric Dam image has been obtained by the author from the Wikimedia website https://commons.wikimedia.org/wiki/File:200407-sandouping-sanxiadaba-4.med.jpg, where it was made available Nowozin under a CC BY-SA 3.0 license. It is included within this book on that basis. It is attributed to Nowozin.

Figure 1.24. The Itaipu Dam on the border between Brazil and Paraguay. This Itaipu Dam image has been obtained by the author from the Wikimedia website https://commons.wikimedia.org/wiki/File:Itaipu_geral.jpg, where it was made available by Jonas de Carvalho under a CC BY-SA 2.0 license. It is included within this book on that basis. It is attributed to Jonas de Carvalho.

hydroelectric facility in terms of electricity produced is the Itaipu Dam on the border between Brazil and Paraguay shown in figure 1.24. Properties of these two dams are summarized in table 1.7.

Low head and run-of-the-river hydroelectric energy

The capacity of low head and run-of-the-river hydroelectric facilities is generally less than that of high head facilities, although a few fairly large facilities exist, such as the

Table 1.7. Properties of the world's two largest hydroelectric dams.

Dam	Country	River	Height (m)	Length (m)	Installed capacity (GW)	Maximum annual generation (TWh)	Capacity factor
Three Gorges	China	Yangtze	181	2335	22.5	98.8	0.50
Itaipu	Brazil/Paraguay	Paraná	196	7919	14.0	103.1	0.84

Jirau Dam in Brazil, with a rated capacity of 3.75 GW. All have the common feature that they do not impound any significant quantity of water and therefore do not create a substantial reservoir upstream from the station. Several different designs generally fall under this classification. Some facilities utilize a low head dam constructed across the width of a river. Such a design makes use of the entire flow of water in the river. Another common design is referred to as a 'diversion system'. A portion of the river flow is diverted through pipes referred to as 'penstocks'. The penstocks carry water downhill to a generating facility at a lower elevation. After passing through the generating station the water flow is returned to the river or, in some cases, output to the ocean or to a lake. Such a system may avoid the environmental impact of a dam and may be appropriate in cases where a suitable dam location does not exist.

Figure 1.25 shows the Mill Cove Hydroelectric Plant in Nova Scotia, Canada. This is a small hydroelectric facility that utilizes water carried by penstocks (visible just above ground level to the left of the generator building) from two rivers to operate two turbine/generators with a total capacity of 12 MW. The outflow from the turbines empties into Mill Lake, shown in the foreground in the photograph. Water is carried about a kilometer through a penstock from this lake to the 5 MW Tidewater Hydroelectric Station on the shore of St. Margaret's Bay. The two elevated tanks on the left in the photograph are surge tanks (sometimes also referred to as surge towers or surge drums), one for each penstock. The purpose of these tanks is to even out the flow of water to the turbines. Specifically, during start up the surge tank provides water in addition to the flow from the penstock to get the turbine started. Also, during shutdown the surge tank accommodates additional water due to a build up of pressure in the penstock.

Hydroelectric energy resources

Hydroelectric power has long been an important component of electricity generation since the first facility was constructed at Cragside in Northumberland, UK in 1878. At present hydroelectric power represents 16.6% of global electricity production (or about 6% of total world energy use) and about 70% of all renewable electricity. Table 1.8 summarizes the hydroelectric capacity and production of the seven leading countries. The table shows that for some countries (e.g. Canada, Brazil and Norway) hydroelectricity provides the largest portion of total electricity, while for other countries (e.g. China, United States, Russia and India) it represents a much smaller component.

Figure 1.25. Mill Cove Hydroelectric Plant near St. Margaret's Bay, Halifax Regional Municipality, Nova Scotia, Canada. The facility consists of two 6 MW generators. (Image: Richard A Dunlap.)

Table 1.8. Hydroelectric production of the leading countries in 2018.

Country	Installed capacity (GW)	Annual production (TWh)	Capacity factor	Fraction of domestic electricity
China	352	1232	0.37	0.17
Brazil	104	418	0.56	0.66
Canada	81	381	0.59	0.59
United States	103	292	0.42	0.06
Russia	49	184	0.42	0.17
Norway	32	140	0.49	0.96
India	50	130	0.43	0.09

While a substantial amount of hydroelectric energy has been developed in its 140 years of history, there is much more available worldwide that has not yet been developed. An analysis of the role of hydroelectricity in fulfilling our future energy needs requires an evaluation of the remaining commercially viable but unutilized resources. An interesting feature of our hydroelectric resource utilization is illustrated in figure 1.26. This shows the total hydroelectric generation and the fraction of domestic United States electricity that it represents. The graph shows a three-fold increase in hydroelectric generation from the late 1940s to the mid-1970s. Although there have been year-to-year fluctuations in generation in the past 40 years or so,

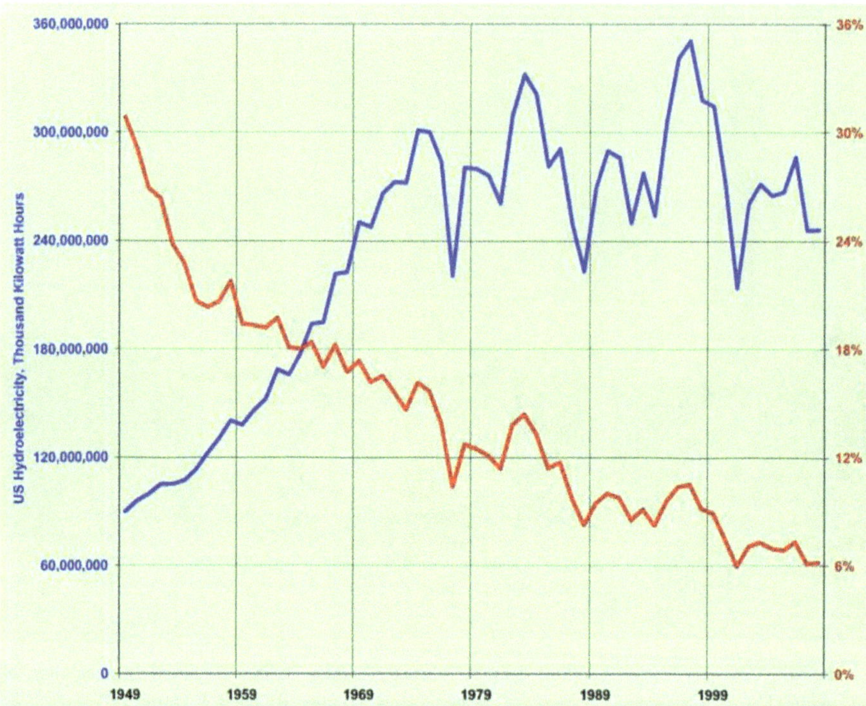

Figure 1.26. Total United States hydroelectric generation (blue line) and hydroelectricity as a fraction of total domestic electricity (red line). This United States Hydroelectric Power image has been obtained by the author from the Wikimedia website https://commons.wikimedia.org/wiki/File:USHydroPower.jpg where it was made available by Plazak under a CC BY-SA 3.0 license. It is included within this book on that basis. It is attributed to Plazak.

there has not been any measurable overall increase. As the total electricity use in the United States has increased over that period, hydroelectricity has played a decreasingly important role. The reasons for this behavior are quite straightforward. Hydroelectricity is a mature technology and in the United States a substantial fraction of the available hydroelectric resources have already been developed. The availability of hydroelectric resources that are technically and economically feasible as well as environmentally viable is limited. This is particularly true of high head resources, where geographical requirements are more demanding and environmental concerns are more serious than for low head or run-of-the-river resources. The situation in most industrialized countries (e.g. Canada and European Union countries) is similar. In developing countries, the situation is somewhat different. China is a relevant case. Figure 1.27 shows the various contributions to China's electricity production from 1980 to 2014. The increase in hydroelectric power parallels the general increase in energy requirements in China. In other developing countries, particularly in Africa, hydroelectric resources that can be developed economically and in an environmentally conscious manner are still readily available.

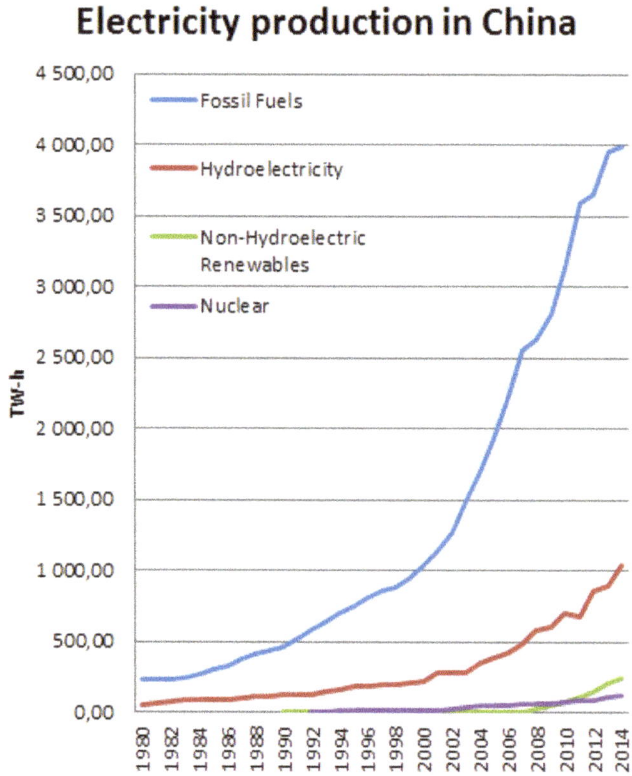

Figure 1.27. Sources of electricity in China from 1980 to 2014. This Electricity Production in China image has been obtained by the author from the Wikimedia website https://commons.wikimedia.org/wiki/File:Electricity_Production_in_China.png where it was made available by Tallungs under a CC BY-SA 4.0 license. It is included within this book on that basis. It is attributed to Tallungs.

At present hydroelectric power contributes about 6% of our total primary power. Available hydroelectric resources for future development are limited, worldwide. These will, no doubt, be developed elsewhere in the world as they have previously been developed in the United States, Canada and Europe. General estimates of remaining viable resources suggest an overall increase in hydroelectric generation by a factor of 2–3 might be reasonable. Based on future requirements as suggested by equation (1.14), hydroelectricity might continue to contribute a maximum of about 5% of world primary power.

Geothermal energy

Basics of geothermal energy

Geothermal energy is based on the phenomenon that the interior of the Earth is hot. The heat inside the Earth comes from three possible sources:

- Residual heat from the initial creation of the Earth by the condensation of material from the early solar system.

- The decay of radioactive materials in the Earth's interior (primarily isotopes of potassium, thorium and uranium).
- Friction in the Earth's liquid outer core due to solar and lunar tidal forces.

While the exact details of the heating of the Earth's interior is not known, it is believed that the decay of radioactive nuclides is the major contributor. On average, the heat flowing outward from the Earth's interior to the surface amounts to 0.087 W m^{-2} and results from an average thermal gradient of 17 °C –30 °C per km. This heat flow is less than 1% of the solar insolation and has a minimal contribution to the average surface temperature of the Earth. However, the thermal energy stored in the Earth's crust forms the basis of the geothermal heat pumps that are a common source of residential heating in many parts of the world. Large-scale use of geothermal energy generally occurs at the boundary between oceanic plates and continental plates where the thermal gradient is much greater than the average value. Here hot regions occur close to the Earth's surface and allow for thermal energy to be extracted from wells. Geothermal energy is used to produce electricity for distribution on the grid and may, in some cases, provide sufficient thermal energy for direct municipal use.

Geothermal energy resources fall into several categories, in addition to the natural thermal gradient that is used for geothermal heat pumps. These categories include:

- Hot rock. In this case the extraction of geothermal energy typically involves injecting water into a well and recovering heat from the hot water that is pumped back to the Earth's surface.
- Reservoirs of hot water. Typically, these are regions of hot rock that have been permeated by surface water that has penetrated the rock through cracks. These regions may be used as a source of direct thermal energy for space heating, bathing, aquaculture, etc. Hot water reservoirs under pressure are also utilized in flash steam generating stations to produce electricity.
- Reservoirs of steam. These are the most useful deposits for electricity generation as the elevated temperature of the steam makes conversion of thermal energy into mechanical energy to drive a generator through the use of a heat engine (e.g. turbine) suitably efficient.

Below we consider the various applications of geothermal energy.

Applications of geothermal energy

Electricity is generated from fluids extracted from geothermal deposits using conventional turbines connected to generators. While deposits of steam are the most appropriate for this use, they are also somewhat limited. Hot water under pressure is more commonly found in geothermal deposits and these are also useful for producing electricity in so-called 'flash steam plants'. When hot water under pressure is extracted from a geothermal well it vaporizes due to the drop in pressure that occurs. This process produces steam, which is then used to operate a turbine. A typical geothermal area containing an electricity generating station is illustrated in figure 1.28.

Figure 1.28. Geothermal area in Krafla, Iceland. The generating station with a total capacity of 60 MW is in the grey building on the right of the image. Two of the boreholes are covered by the brown domes on the lower left of the image. Insulated pipes carrying the geothermal fluid can be seen connecting the boreholes to the generator building. (Image: Richard A Dunlap.)

In a number of facilities geothermal electricity generation and direct thermal use are combined. Geothermal cogeneration (often called 'combined heat and power' or CHP) is a useful method of efficiently utilizing geothermal energy. In a flash steam plant (for example) steam may be separated from hot water using a steam separator as shown in figure 1.29. The hot water is then used for space heating or other applications as described above. Iceland has been very active in the use of geothermal cogeneration and currently 30% of its electricity comes from geothermal resources and more than 90% of its population obtains heat from various district geothermal heating systems.

Geothermal energy resources

Currently, worldwide capacity for geothermal electricity generation is about 13 GW while worldwide capacity for thermal energy from geothermal deposits is about 70 GW. As the capacity factor of geothermal energy use is fairly high, we can see that the two applications of geothermal energy (~0.08 TW) account for about 0.4% of current world primary power of about 20 TW. Estimates of the total available geothermal power optimistically estimates that this resource amounts to about 2 TW worldwide. From a practical standpoint, probably less than 1 TW of this would be viable from an economic or environmental standpoint. With reference to equation (1.14), geothermal energy is not likely to contribute more than about 2% of our future energy.

Biofuels

Basics of biofuels

Biofuels have been used by humans since antiquity. Energy is extracted from these materials by the combustion of their carbon and hydrogen content. Biofuels fall into

Figure 1.29. Steam separators at the Hellisheiði Geothermal Power Station in Hveragerði, Iceland. (Image: Richard A Dunlap.)

several broad categories. They may consist of solid as-grown plant-based material such as wood. They may consist of gaseous or liquid by-products of plant or animal-based material such as ethanol, biodiesel or biogas. Finally, they may consist of solid waste products of human society such as municipal solid waste. In the present section a few of the most commonly utilized biofuels are reviewed.

Wood

Wood was the first source of energy in nature to be harvested for use in heating and cooking. The use of wood as a fuel grew worldwide as the world population and the demand for energy increased. For the most part, the early use of wood as a fuel was not undertaken in a sustainable manner and forests were not replanted in a systematic way as wood resources were consumed. In industrialized countries the use of wood as an energy source peaked in around 1870. As the industrial revolution put greater and greater demands on energy production, wood was replaced by coal as the principal energy source due to coal's greater availability, energy content and ease of use. Today wood remains an important source of energy for about one third of the world population living in developing countries.

Wood provides energy primarily by the combustion of its carbon content as given in equation (1.2). Wood is quite variable in its energy content but a typical value (for maple with 20% moisture content) is 14 MJ kg^{-1}. This is about one third of the energy content of gasoline per unit mass. The heat produced by the combustion of wood can be used directly or as a means of generating steam to operate a turbine/generator to produce electricity. This latter approach has been used in recent years by coal-fired generating stations that have been converted to burn wood. If wood is used in a renewable manner, then energy is produced with net zero carbon emissions as the carbon dioxide produced during the combustion of wood is sequestered by the growth of new trees.

An example of the approach described above is the Thunder Bay Generating station, as shown in figure 1.30. This generating station first became operational in 1963 as a coal-fired facility. In 2015 it was converted to run on wood pellets (sometimes referred to as advanced biomass). It was the last coal-fired generating station in Ontario and was part of the province's mandate to phase out the use of coal for electricity generation. The plant has been closed since 2018 due to a leak in the boiler. Future plans for the facility are unclear at present.

Ethanol

Ethanol is a light alcohol with the chemical formula C_2H_5OH. Some of its basic properties are summarized in table 1.9. About 95% of ethanol produced worldwide is derived from plant-based material, the remainder is extracted from fossil fuels. Most ethanol that is currently produced is used as a source of energy, as described below, the remainder is used for industrial processes, preparation of chemicals and as alcoholic beverages. Ethanol is an attractive source of energy. As a liquid fuel, it represents a more or less direct replacement of gasoline as a transportation fuel. However, it is important to note that the energy content of ethanol is somewhat lower than that of gasoline; 23.5 MJ L^{-1} compared to 38.5 MJ L^{-1}.

The energy content of ethanol is a manifestation of solar energy that has been converted to chemical energy by photosynthesis in growing plants. Photosynthesis is an endothermic process that uses solar energy to combine carbon dioxide and water to produce glucose by the reaction

$$6CO_2 + 6H_2O \rightarrow C_6H_{12}O_6 + 6O_2 \tag{1.31}$$

The fermentation of glucose produces ethanol and carbon dioxide by the reaction

$$C_6H_{12}O_6 \rightarrow 2C_2H_5OH + 2CO_2 \tag{1.32}$$

The combustion of ethanol is an exothermic process that produces energy along with the by-products carbon dioxide and water;

$$C_2H_5OH + 3O_2 \rightarrow 2CO_2 + 3H_2O \tag{1.33}$$

These three reactions can be seen to be carbon neutral and there is no net carbon dioxide emission.

World ethanol production in 2016 was 9.88×10^{10} L. Of this more than 85% was produced in the United States (5.80×10^{10} L) or Brazil (2.76×10^{10} L). Ethanol produced in the United States is made primarily by the fermentation of corn which yields about 4×10^5 L of ethanol per km^2 of land area. In Brazil, which has a more tropical climate than the United States, ethanol is produced primarily by the fermentation of sugar cane, which produces about 8×10^5 L of ethanol per km^2 of land area.

Figure 1.30. Thunder Bay Generating Station on Mission Island in Thunder Bay, Ontario, Canada. This Thunder Bay Generating Station image has been obtained by the author from the Wikimedia website https://commons.wikimedia.org/wiki/File:ThunderBayCoalGenStn.jpg where it was made available by Loimere under a CC BY 2.0 license. It is included within this book on that basis. It is attributed to Loimere.

Most fuel ethanol is used as an additive to gasoline for light duty vehicles. Annual ethanol production as given above represents about 8% of the global gasoline production of about 1.3×10^{12} L per year by volume or about 5% by energy content. Conventional gasoline automobile engines can operate on mixtures of gasoline

Table 1.9. Some basic properties of ethanol.

Property	Value
Chemical formula	C_2H_5OH
Molecular mass	$46.07\ g\ mol^{-1}$
Density (STP)	$0.789\ g\ cm^{-3}$
Melting point	−114.1 °C
Boiling point	78.4 °C
Heat of combustion	$23.5\ MJ\ L^{-1}$

containing up to 10% ethanol (designated E10) without modification. With minor modification to the fuel system, the so-called 'Flexfuel' vehicles can operate with mixtures of gasoline containing up to 85% ethanol (designated E85). Most light duty vehicles sold in Brazil are designed to operate on any mixture from 100% gasoline to 100% ethanol (E100).

Biodiesel

While ethanol may serve as an attractive replacement for gasoline in light duty vehicles, biodiesel serves a similar purpose with respect to diesel fuel used for trucking. Biodiesel is produced by the transesterification of vegetable oil or animal fat. Estimates of global annual biodiesel production are in the range of 2×10^{10} L. This may be related to annual petroleum-based diesel fuel production which is around 1.8×10^{12} L.

Biodiesel may be produced from recycled used materials, such as sewage sludge or used vegetable oil, or it may be produced from designated crops. Some common crops that are used to produce biodiesel include sunflower and soy, which typically yield around $8 \times 10^4\ L\ km^{-2}$. This is also typical of most plants that grow in temperate climates. Tropical plants such as palm can yield up to $4.7 \times 10^5\ L\ km^{-2}$ and algae can provide up to about $1.7 \times 10^6\ L\ km^{-2}$. In addition to the larger yield, an additional advantage of algae is that it can be grown in ponds and shallow coastal waters that may not have other potential uses.

Most traditional diesel engines will run any mixtures of petroleum derived diesel and biodiesel with minor modifications. These fuels are designated by the percentage of biodiesel, e.g. B99 (for 99% biodiesel, 1% petroleum derived diesel, etc). This is a common mixture, where the 1% petroleum derived diesel is added to inhibit the growth of mold.

Biogas

Biogas is a mixture of combustible gases that is produced by the anaerobic decomposition of organic material. Feedstock for biogas production may be organic municipal waste (as described below), sewage, agricultural waste or manure. The gas that is produced during the anaerobic decomposition of these materials is primarily methane but can also include some amount of carbon dioxide, nitrogen, hydrogen, oxygen and/or hydrogen sulfide. Processing of raw biogas can increase its methane content and make biogas compatible with most uses of fossil fuel sourced natural gas.

The development of biogas as a fuel is at the relatively early stage. Germany has the most active biogas production program where feedstock typically consists of agricultural waste and manure. China and India are also actively involved in the development of biogas resources.

Municipal solid waste

Municipal solid waste refers to the solid component of the waste material produced by humans with the exclusion of hazardous materials and sewage. It contains materials that can be recycled, such as metals, glass and plastics, and also biodegradable materials that can be used for producing compost. The combustible component of municipal solid waste represents a potential source of energy. At present most municipal solid waste is merely discarded, typically in landfills. A breakdown (for the United States) of municipal solid waste disposal methods is shown in table 1.10. Energy may be recovered from municipal solid waste by combustion to produce heat or electricity through thermal generation. The waste may undergo combustion directly or it may be used to produce fuel pellets (referred to as refuse derived fuel, RFD) or gaseous fuel through decomposition. Gaseous fuel produced by the decomposition of municipal solid waste may be most appropriately described by the techniques in the Biogas section above.

The French Island Generating Plant (see figure 1.31) located on French Island in La Crosse, Wisconsin is an example of the commercial utilization of energy produced from municipal solid waste. This facility consists of four units. Units 1 and 2, with generating capacities of 14.3 and 14.0 MW, respectively, were constructed in the 1940s to burn coal. They were converted in the early 1970s to burn oil. In the 1980s both units were converted to burn a 50/50 mixture of municipal solid waste and waste wood. Units 3 and 4 are fuel oil fired combustion turbines.

The utilization of municipal solid waste for energy production is an example of a beneficial product that results from a material that needs to be disposed of in any case. It is also noteworthy that while the combustion of municipal solid waste produces a greenhouse gas (carbon dioxide, CO_2) it eliminates the emission of a much more potent greenhouse gas (methane, CH_4) from organic material that is disposed of in landfills.

Table 1.10. Breakdown of disposal methods for municipal solid waste in the United States (2015). Data from United States Environmental Protection Agency.

Method	Percentage
Discarded (e.g. landfill)	52.5
Recycled (e.g. metals, glass, plastics, composting)	34.7
Energy production	12.8

Figure 1.31. French Island Generating Plant located on French Island in La Crosse, Wisconsin. This French Island Generating Plant image has been obtained by the author from the Wikimedia website https://commons.wikimedia.org/wiki/File:French_Island_Generating_Plant.jpg, where it is stated to have been released into the public domain. It is included within this book on that basis.

There is considerable variability in the amount of municipal solid waste produced per capita in different countries. While industrialized countries tend to produce more municipal solid waste, this is not always the case. Typically, in North America and Europe, the municipal solid waste production amounts to 2–3 kg per capita per day. As the energy content of municipal solid waste is of the order of 10 MJ kg^{-1}, it typically represents a maximum contribution to our primary energy use of about 25 MJ per capita per day. This is about 2% of the energy consumption in industrialized countries.

Biofuel resources

Estimates have suggested that biofuels could supply up to about 6 TW of power, that is about 10% of our future needs. As noted above, many types of biofuels have the potential advantage that they have high energy density and can replace conventional fossil fuels in ways that other renewable sources cannot. This applies largely to energy requirements for transportation.

There are, however, potential adverse effects of biofuel use. Significant land area is needed to grow plant material for wood-fired generating stations, ethanol production and biodiesel production. Much of this land and the materials that are produced (e.g. corn) compete with agricultural requirements for food production for humans and domestic animals. Therefore, it is likely that substantially less than the 6 TW estimate given above would be actually viable.

There are a number of initiatives concerning the development of biofuel resources that may, in the relatively near future, significantly improve the utilization of this renewable energy source. At present, ethanol is produced by the fermentation of glucose. A large component of most plant matter is cellulose.

Cellulose is a polysaccharide with the chemical formula $(C_6H_{10}O_5)_n$ which consists of a linear chain of hundreds to thousands of linked glucose molecules. The development of efficient technologies for the cellulosic production of ethanol would have two significant features. Firstly, the ethanol yield from feedstock such as corn could be significantly improved and, secondly, plant material that is largely cellulose could be used as ethanol feedstock. Perhaps the most relevant of the agricultural material that would fall in this latter category would be switchgrass (*Panicum virgatum*). This is a fast-growing plant that grows well in a variety of habitats, many of which are not suitable for food production. Thus, further research may help to integrate crops for ethanol production into agricultural programs in such a manner that would not directly compete with food production.

The Indian government has undertaken a program to produce biodiesel using oil from the jatropha plant (*Jatropha curcas*). This plant is being grown in waste areas adjacent to the railway tracks between Delhi and Mumbai. The biodiesel produced from these jatropha plants is used as a component of the fuel to operate the train on this railway line.

Algae has the highest yield of any plant grown for biodiesel production. As well, algae grown in ponds or, particularly, in shallow coastal waters would not compete with viable agricultural land. This, however, is not a trivial task. Diesel use in the United States is a good example. The current annual use of diesel fuel (for transportation and heating) in the United States is 1.8×10^{11} L. If the biodiesel yield from algae is 1.7×10^6 L km^{-2} then an area of about 10^5 km^2 would be needed to satisfy diesel fuel needs. As the United States coastline is of the order of 1.6×10^4 km, then utilizing the entire coastline from the shore to a distance of 0.5 km offshore, would have an area of 8×10^3 km^2. This would be able to produce only 8% of the current diesel fuel requirements. The large-scale utilization of biofuels must, therefore, be considered very carefully in the context of other agricultural needs.

1.7 Future energy sources II—nuclear energy

Energy may be obtained from the nucleus by two different methods: nuclear fission and nuclear fusion. The present section provides a brief introduction to these energy sources. Details of the physics of nuclear energy are presented in chapter 2 while the various approaches to fusion energy are the subject of the remainder of this book. Thus, this section provides just a brief overview of how these two aspects of nuclear energy may contribute to fulfilling our energy needs.

Nuclear fission energy

In 1942 the world's first controlled nuclear fission reactor was constructed under the direction of Enrico Fermi. This reactor was designed for the purpose of producing fissile materials from which to construct nuclear weapons for the United States military. The first commercial nuclear fission reactor designed for the purpose of producing electricity was commissioned in 1956 in Sellafield, United Kingdom. During the 1960s and 1970s the number of nuclear power reactors worldwide increased substantially. In the 1980s construction of new nuclear power reactors

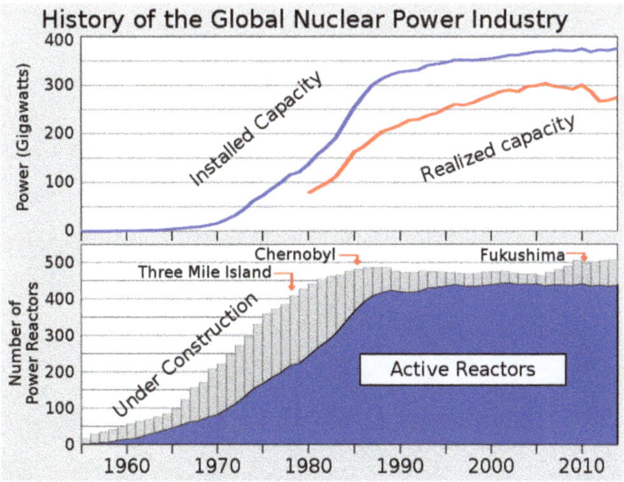

Figure 1.32. Number of nuclear fission reactors worldwide and electricity generated as a function of year. This Nuclear Power History image has been obtained by the author from the Wikimedia website https://commons.wikimedia.org/wiki/File:Nuclear_power_history.svg, where it is stated to have been released into the public domain. It is included within this book on that basis.

decreased substantially. This change in the development of nuclear reactors was the result of several notable nuclear accidents, firstly, Three Mile Island in 1979 and, more seriously, Chernobyl in 1986. In addition to growing public concern over the safety of nuclear reactors, there was also growing concern over the environmental aspects of waste disposal and the security of possible weapons grade radioactive material. These trends in nuclear reactor use are illustrated in figure 1.32. After around 2000 there was some renewed interest in the development of new generation nuclear reactors which were expected to have a lower accident risk. However, since the accident in Fukushima in 2011, the future of nuclear power has become less certain. In the present section we provide a very brief description of the operation of nuclear fission reactors for power production. We begin with the most common design, the traditional thermal neutron reactor. This is followed by a brief account of the development and possible future of fast neutron reactors. The details of the nuclear physics describing reactions that take place in nuclear reactors are considered in the next chapter.

Thermal neutron reactors

Commercial nuclear power reactors (see figure 1.33) utilize the fission of (primarily) uranium to produce thermal energy which is converted into electricity by a traditional steam turbine/generator. Naturally occurring uranium consists of two isotopes; ^{235}U (99.28% naturally abundant) and ^{235}U (0.72% naturally abundant). As described in the next chapter, energy is released when a uranium nucleus undergoes fission and splits into two lighter nuclei and some additional neutrons. Although uranium nuclei undergo spontaneous fission, as this is energetically favorable (i.e. it is exothermic), this process is extremely slow and is not suitable for a practical power reactor. It is the

Figure 1.33. Seabrook Nuclear Generating Station in Seabrook, New Hampshire. This station is a 1244 MW pressurized water fission reactor. (Image: Richard A Dunlap.)

extra neutrons, as described in chapter 2, that induce additional fissions in ^{235}U (which is referred to as fissile). In a thermal neutron reactor. the neutrons do not induce fission in ^{238}U (which is referred to as non-fissile). So, it is only the energy that can be extracted from the less than 1% of the uranium that is ^{235}U that can be recovered by a reactor of this type.

Fast neutron reactors

Fast neutron reactors (often referred to as breeder reactors) are designed so as to make use of the energy content of the non-fissile ^{238}U. This is done by using high energy neutrons to induce reactions in ^{238}U that produce fissile ^{239}Pu. A ^{239}Pu nucleus will undergo fission when a neutron is incident upon it and, in this way, will allow the energy content of the ^{238}U to be realized.

The first fast breeder was constructed in the United States in 1946, although this was not designed to produce power. Since then, a number of fast breeder reactors have been built worldwide. The United States, the United Kingdom, France, Germany, Japan, Russia (and previously the Soviet Union) and India, have at various times had fast breeder reactor programs. At present Russia has two prototype commercial fast breeder reactors that are in operation. Japan and India each have one test reactor of this type. In recent years India has probably been the most active in developing fast breeder reactors. The main reason for this program is to eventually make use of thorium as a reactor fuel. Thorium exists primarily as ^{232}Th (99.98% naturally abundant). ^{232}Th is non-fissile but in a fast breeder reactor neutron reactions with thorium can produce fissile ^{233}U which can be used as a reactor fuel. However, sufficient fissile fuel must be bred using neutrons supplied by induced fission in ^{235}U before a self-sustaining thorium reactor can be operational. It is anticipated that a dedicated effort to produce a self-sustaining thorium reactor would require 30–40 years. Thorium (^{232}Th) has a longer half life (1.4×10^{10} y) than ^{238}U (4.5×10^{9} y) and ^{235}U (7.04×10^{8} y) and is, therefore, more abundant on Earth. Thorium is about three times as abundant in the Earth's crust as uranium, and therefore, in principle, represents a larger resource for energy production. India has the largest known thorium reserves, and this has been their main motivation for pursuing this aspect of nuclear power.

Fission energy resources
The current use of nuclear energy relies almost entirely on the natural existence of ^{235}U as a fuel for thermal neutron reactors. At present nuclear power accounts for about 11% of electricity worldwide. This amounts to about 4% of total global power consumption. The longevity of nuclear fission energy depends on a number of important factors. Specifically, the world uranium resources, the economic viability of these resources and the amount of power nuclear reactors supply. It should be noted that the uranium that is processed each year is not necessarily the same as that which is consumed by power reactors. There are several reasons for this difference. Firstly, some countries have significant stores of uranium which can contribute to uranium for reactor use. Secondly, about 10% of the uranium that is consumed is used for military purposes and not electricity generation. Finally, additional uranium becomes available from reprocessed reactor fuel and from decommissioned nuclear weapons. Overall, however, on the average, the annual uranium use compared with the known resources is a good measure of the longevity of this energy source for thermal reactors. As of 2014, the proven uranium resources worldwide which are recoverable at less than $130 US per kg (which is a current benchmark for economic viability) amounted to about 3.7×10^{9} kg. The annual uranium use (as of 2014) was 7.0×10^{7} kg. Thus, the production to reserves ratio as discussed above for fossil fuels would give an estimated lifetime of 3.7×10^{9} kg/7.0×10^{7} kg y^{-1} = 53 y. Noting that this represents only 4% of global energy use, we see that increasing the number and output of reactors will proportionately decrease the longevity.

The development and implementation of fast breeder reactors utilizing ^{238}U or ^{232}Th would greatly increase the longevity of nuclear fission energy. A rough estimate based on current nuclear energy production would give a longevity of 20 000–30 000 years for ^{238}U and ^{232}Th combined. While this could be even greater if technologies for utilizing lower grade ore were developed, increasing nuclear fission capacity would decrease longevity.

Nuclear fusion energy

As noted above, the utilization of energy from nuclear fission is not without its drawbacks. Certainly, safety, security of fissile material and waste disposal are the major concerns for continued or even increased use of energy from nuclear fission. Improved technology could reduce the risk of nuclear accidents, but increased use of fast breeder reactors could increase security concerns due to the substantially greater amount of fissile material used.

The development of fusion reactors could very successfully address the concerns over nuclear fission energy. Reactors are inherently safe, as the lack of proper control will cause the reaction to die out rather than go out of control. The radioactive materials used in fusion reactors are not suitable for nuclear weapons and security is, therefore, not a concern. Finally, in principle, radioactive waste should not be produced in a nuclear fusion reaction. However, this latter point needs careful consideration in the context of reactor operation. These points are considered in detail throughout the remainder of this book.

It is important to consider the way in which fusion energy could contribute to our future energy needs. The energy released by nuclear fusion is, like the energy released by nuclear fission, in the form of the kinetic energy of the by-products of the nuclear reaction. This kinetic energy is converted into heat and in the simplest scenario, can be used to generate electricity using a steam turbine and a generator. The conversion efficiency of heat from the fusion reaction to electricity is limited, as in the case of thermal generation by burning coal to about 40%. Other possibilities for obtaining electricity from fusion are discussed in chapter 7.

The basics of nuclear fusion reactions are considered in the next chapter, while the remainder of the book deals with the details of specific fusion reactor designs and their viability as a source of future energy.

1.8 Analysis of future energy options

As has been seen in the above discussion, there are a variety of options for future energy that could replace fossil fuels. No doubt, our future energy will come from a mixture of different sources, just as our energy today comes from different sources and, as described in detail below, different sources of energy may be amenable to different applications. The suitability of possible methods of producing carbon-free energy in the future depends on a number of factors. These factors include:

- energy form;
- longevity;
- available power;
- consistency of power;
- cost;
- environmental impact and risk.

Each of these factors is discussed below in the context of the energy options described above in sections 1.6 and 1.7.

Energy form

A breakdown of our energy use is shown in table 1.2. For an analysis of carbon-free energy options, it may be more convenient to categorize the form of our energy use as electrical energy, thermal energy or chemical energy. In the first case, electricity is produced by many forms of carbon-free energy sources. These include wind energy, solar energy (from thermal generation and photovoltaics), hydroelectricity and generating stations based on geothermal energy, biomass energy and nuclear energy. In the second case, direct use of thermal energy may utilize solar energy, geothermal energy or combustion of biofuels. Finally, chemical energy from liquid or gaseous biofuels may directly replace traditional fossil fuels, such as gasoline and diesel fuel, used for transportation. These different approaches to carbon-free energy serve to satisfy different aspects of our end energy utilization.

Longevity

Although in principle, we would expect the longevity of renewable energy sources to be infinite, for all practical purposes, there are factors that need to be considered in making this assessment. This is particularly the case for hydroelectric energy and geothermal energy. In the case of hydroelectric energy, particularly high head facilities, the accumulation of sedimentation upstream from a dam will decrease the efficiency of electricity production, thereby potentially limiting the lifetime of installation. In the case of geothermal energy, the total thermal energy available in the Earth is not unlimited but is large compared to our foreseeable future energy requirements. However, for a particular geothermal plant, it is important not to deplete the local thermal energy faster than it can be replenished from its surroundings. Solar, wind and biofuels do not have any foreseeable limitations.

Available power

The power that is available from each of the renewable sources above have been discussed in some detail. The anticipated maximum fraction of expected future power needs is summarized in table 1.11 for each of these renewable sources. Other renewable sources, such as wave energy and tidal energy, may also contribute a small amount to our future energy mix.

It is clear from the above tabulated information that only solar energy can, by itself, satisfy our future power requirements. It is also clear that the other renewable energy sources discussed above, along with others such as wave and tidal energy,

Table 1.11. Fraction of total future power requirements (e.g. ~54 TW) that can be satisfied by the various renewable sources discussed in detail above.

Source	Fraction of total
Solar	1.0
Wind	0.1
Hydroelectric	0.05
Geothermal	0.05
Biofuels	0.12

could satisfy, perhaps a third to half of future needs. While a mix of the above energy sources could satisfy all of our needs, with a substantial fraction coming from solar, there are other considerations, as discussed below that need to be included to determine if this is, in fact, a viable approach.

Consistency of power

It is important to note that the nameplate capacity used for renewable energy generating stations is the maximum electrical output that the facility can produce. For example, the maximum output for a photoelectric solar facility is the output that that facility would produce when the incident sunlight is at maximum. The capacity factor is the ratio of the actual output to the nameplate capacity. It is obvious that renewable resources such as solar energy will show daily variations. Although not as systematic, wind energy typically exhibits variations that relate to the time of day.

The capacity factor for most types of facilities also shows seasonal variations. These variations are shown in figure 1.34. Solar energy system capacity factors show clear maxima during the summer months (in the Northern Hemisphere) and minima during the winter months. This variation is a direct result of the maximum elevation of the Sun in the sky. Wind energy system capacity factors are somewhat less systematic but typically show maxima in the winter when weather patterns produce the greatest winds. Hydroelectric facilities also show seasonal variations which correspond to fluctuations in the average flow rate of rivers. Capacity factors of municipal solid waste and geothermal energy production show relatively little seasonal variation.

It is clear that the implementation of renewable energy sources must take into account both daily and seasonal fluctuations in capacity factor. Depending on the particular mix of energy sources, the efficient utilization of renewable energy must incorporate appropriate methods for energy storage.

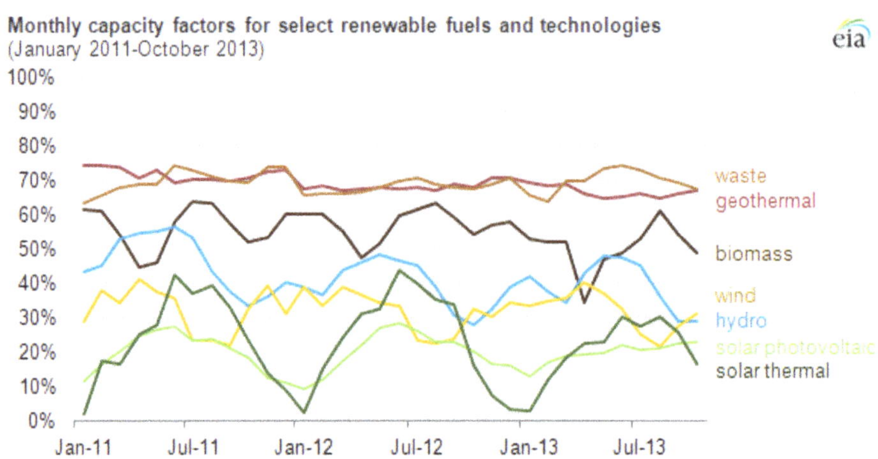

Figure 1.34. Seasonal variations in the capacity factor for the renewable energy sources described in this chapter (https://www.eia.gov/todayinenergy/detail.php?id=14611). Source: U.S. Energy Information Administration (January 2014).

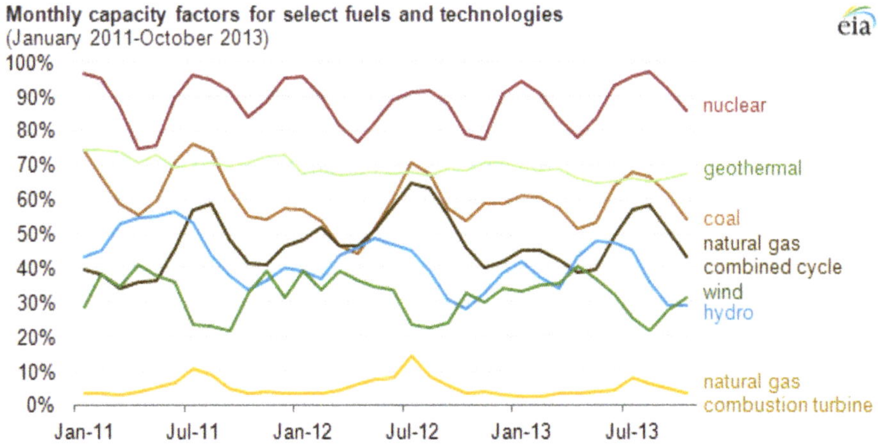

Figure 1.35. Seasonal variations in the capacity factor for nuclear fission energy compared with some renewable energy sources and some fossil fuel sources (https://www.eia.gov/todayinenergy/detail.php?id=14611). Source: U.S. Energy Information Administration (January 2014).

A comparison of capacity factors for nuclear fission energy and some renewable energy sources and some fossil fuel sources is shown in figure 1.35. Hourly and daily fluctuations in the capacity factor for nuclear and fossil fuel energy are typically minimal. There are often some seasonal variations for these energy production methods. These may arise for one or two reasons, variations in thermodynamic efficiency and variations in demand. For nuclear and fossil fuel energy, electricity is generated by conversion of thermal energy produced by nuclear or chemical reactions, respectively, to mechanical energy by means of a heat engine. The efficiency of this process is limited by the thermodynamic efficiency of the heat engine. This is, for example, the Carnot efficiency, η, expressed as a fraction,

$$\eta = \frac{T_{\text{hot}} - T_{\text{cold}}}{T_{\text{hot}}} \tag{1.34}$$

where T_{hot} and T_{cold} are the temperatures of the hot and cold reservoirs, respectively. As the cold reservoir is either water (e.g. river, lake or ocean) or the atmosphere (through the use of cooling towers), T_{cold} will have seasonal variations that will affect the generating station output. More importantly, as seen for the capacity factor for nuclear energy, there are peaks in the capacity factor in winter and summer, corresponding to peaks in demand for heating and cooling, respectively.

The above data show that electricity generated by combustion of fossil fuels or by nuclear energy is more constant in time and better able to meet baseload demands than most types of renewable energy. The substantial reliability on many renewable energy sources, particularly solar and wind, would require a consideration of capacity factor in establishing a reasonable level of installed capacity and would also require the effective implementation of energy storage methods. Current advances in the development of energy storage methods, particularly pumped hydroelectric and batteries, are aimed at dealing with these concerns.

Cost

Cost is a very difficult parameter to assess. An important factor in determining cost results from the maturity of technology. The cost of photovoltaics is an excellent example of the effects of technology on the cost of energy production. In the mid-1970s the infrastructure cost of photovoltaic installations was \$76 US per installed watt. By the mid-2010s the cost had dropped to about \$0.30 US per installed watt. Certainly, a technology that was prohibitively expensive for commercial applications 40 years ago is now commercially viable. Estimates from the United States Energy Information Administration (EIA) for grid-connected installations entering service in the next few years put solar photovoltaics in the same range as hydroelectricity and nuclear (fission) energy per MWh produced. While research and development costs for new technologies may be high, they should not necessarily be indicative of long-term economic viability.

Environmental impact and risk

It is expected that the utilization of renewable energy sources or nuclear energy would eliminate carbon emissions, however, this is not entirely true. The low energy density of many renewables, in particular, and the substantial manufacturing necessary to produce suitable infrastructure is a major factor in the environmental impact of these non-fossil fuel energy sources. This is the case at present, to a large extent, because primarily fossil fuel energy is used to produce renewable energy infrastructure. This situation will be alleviated to a large extent at some point in the future if fossil fuel energy sources are effectively eliminated. This not to say that non-fossil fuel sources are environmentally benign. As well, there is some risk associated with many methods of energy production. Risks may be occupational hazards for workers associated with materials processing, manufacturing and construction, as might be the case for solar photovoltaics, or they may be for the general public, as is the possibility of hydroelectric dam failure. Table 1.12 summarizes some of the more significant environmental concerns and risk factors related to the non-fossil fuel energy sources discussed in this chapter.

As summarized in table 1.12, the overall risk of generating electricity by non-fossil sources is less than that of generating electricity using coal. However, in some cases, e.g. solar and wind, the risk is relatively high because of the low energy density associated with these energy sources and the associated extensive infrastructure required per unit energy capacity. It is also seen for the energy sources described in the table that, although carbon emissions are minimal compared to the use of coal, some environmental concerns exist for any energy production method.

Summary of future energy options

The above discussion illustrates that, of all of the non-fossil fuel energy production methods current in use, only solar energy, with sufficient future development, has the potential to satisfy all of our energy requirements for an extended period of time. This approach is not necessarily desirable for several reasons, as follows:

Table 1.12. Summary of principal concerns related to non-fossil fuel energy sources. The risk factor is normalized to the risk factor for the use of coal (coal risk = 1.0) in terms of risk to both occupational workers and the general public per unit electrical energy generated. Adapted from data presented in Inhaber (1979) (N/A = not assessed).

Energy source	Risk	Environmental concerns
Solar photovoltaics	0.23	• occupies large land area • little possibility of dual land use
Wind	0.33	• occupies large land area
Hydroelectric	0.015	• large reservoir formation • loss of wildlife habitat • reduced fish mobility • potential dam failure • silt transport disruption
Geothermal	N/A	• possible release of underground toxins • possible introduction of geological instability
Biofuels	0.10	• competes with agriculture
Nuclear fission	0.003	• waste disposal • radioactive material security • potential reactor accidents

- The large utilization of land area is not necessarily desirable.
- The intermittent nature of this energy source, both on daily and longer time scales, makes it difficult to accommodate baseload requirements and requires the implementation of extensive energy storage technologies.
- Latitude and local climate requirements make its implementation of minimal viability in some parts of the world.
- Human risk associated with its low energy density is not desirable.
- Availability of rare or expensive materials used for construction may limit future use.

It is certainly most reasonable to utilize any viable energy source that is available, and which minimizes adverse effects on the environment and society. This means that a mix of available energy sources will best fulfill our needs. Nuclear fusion is an energy source that has been under development for a number of years, but which has not reached a point of being a viable commercial energy source. However, nuclear fusion potentially has much to offer our possible future energy mix. Possible advantages of energy from nuclear fusion include:

- Plentiful fuel source.

- High energy density which minimizes land use and potential environmental and societal risks.
- Net zero carbon emissions.
- Improved reactor safety, security concerns and waste disposal problem compared to nuclear fission.

The analysis of the viability of nuclear fusion energy and its potential hazards are complex issues. Chapter 2 provides a brief introduction to the physics of nuclei and the means by which energy may be produced by nuclear reactions. The remainder of this book deals with nuclear fusion and the prospects that it may contribute to our future energy sources.

References and suggestions for further reading

Aubrecht G J 2006 *Energy: Physical, Environmental and Social Impact* 3rd edn (Upper Saddle River, NJ: Pearson Prentice Hall)

Boyle G (ed) 2012 *Renewable Energy: Power for a Sustainable Future* 3rd edn (Oxford: Oxford University Press)

Dunlap R A 2019 *Sustainable Energy* 2nd edn (Boston, MA: Cengage)

Dunlap R A 2020 *Renewable Energy – Volume 1: Requirements and Sources* (San Rafael, CA: Morgan & Claypool)

Fay J A and Golomb D S 2012 *Energy and the Environment, Scientific and Technological Principles* 2nd edn (Oxford: Oxford University Press)

Inhaber H 1979 Risk with energy from conventional and nonconventional sources *Science* **203** 718–23

Kaltschmitt M, Streicher W and Wiese A 2007 *Renewable Energy: Technology, Economics and Environment* (Berlin: Springer)

Maggio G and Cacciola G 2012 When will oil, natural gas and coal peak? *Fuel* **98** 111–23

Ouahabi M H, Benabdelouahab F and Khamlichi A 2017 Analyzing wind speed data and wind power density of Tetouan city in Morocco by adjustment to Weibull and Rayleigh distribution functions *Wind Eng.* **41** 174–84

Patel M R 2006 *Wind and Solar Power Systems: Design, Analysis and Operation* 2nd edn (Boca Raton, FL: CRC Press)

Tester J W, Drake E M, Driscoll M J, Golay M W and Peters W A 2012 *Sustainable Energy: Choosing Among Options* 2nd edn (Cambridge, MA: MIT Press)

Timilsina G R 2013 Biofuels in the long-run global energy supply mix for transportation *Phil. Trans. Royal Soc* **A372** 20120323

Vanek F M, Albright L D and Angenent L T 2016 *Energy Systems Engineering: Evaluation and Implementation* 3rd edn (New York: McGraw-Hill)

IOP Publishing

Energy from Nuclear Fusion

Richard A Dunlap

Chapter 2

Energy from the nucleus

2.1 Introduction

Energy is produced by a nucleus by the conversion of mass, m, to energy, E, according to Einstein's mass–energy relation,

$$E = \Delta mc^2 \tag{2.1}$$

where Δm is the change in mass and c is the speed of light. Changes in nuclear mass can occur during transitions between excited nuclear states and, most importantly for the present discussion, during fission, that is the splitting of heavy nuclei into two lighter nuclei, or fusion, that is, the fusing together of two light nuclei into one heavier nucleus. The reasons for changes in nuclear mass are related to changes in the nuclear binding energy. Before we discuss the details of nuclear binding energy, we begin in the next section with an introduction to basic nuclear structure.

2.2 Nuclear composition

The nucleus of an atom is a bound system of neutrons and protons. The number of protons, Z, determines the identity of the element, and the number of neutrons, N, combined with the number of protons determines the mass number, A, as

$$A = N + Z \tag{2.2}$$

Neutrons and protons are often referred to collectively as nucleons. The term nuclide refers to a specific nuclear species as determined by N and Z. A nuclide is designated as ${}^{A}E$, where E is the name of the element. Since E is determined by the value of Z, then the number of neutrons may be determined for the nuclide using equation (2.2). While the term isotope is sometimes used interchangeably with nuclide, it more specifically refers to the family of nuclides with the same value of Z, that is, of the same element.

Certain combinations of neutrons and protons may form stable nuclei or unstable nuclei while other combinations may not bind together, even for short

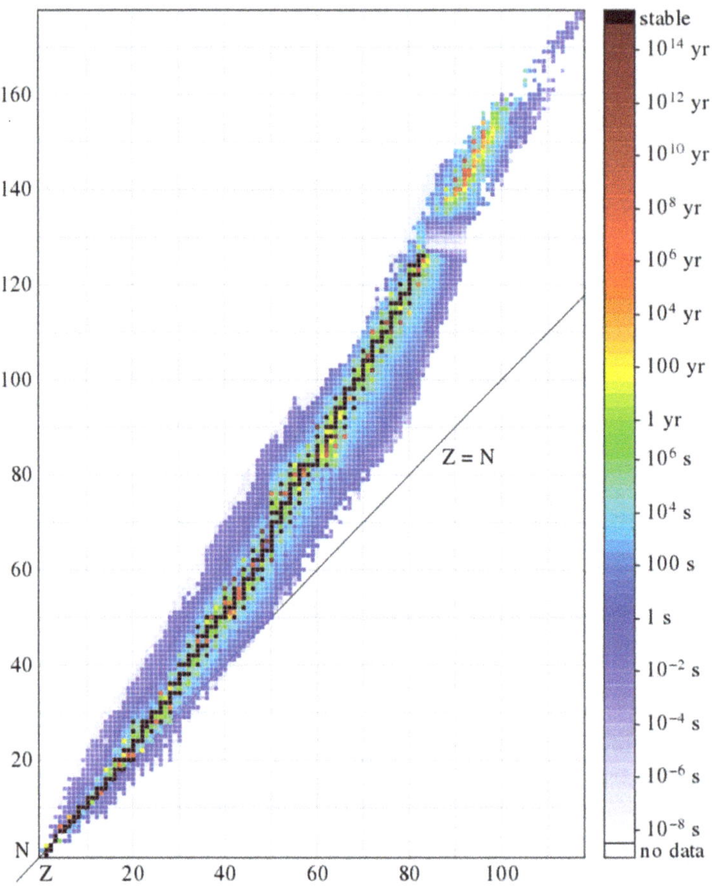

Figure 2.1. Segrè plot showing the number of protons (horizontal axis) and neutrons (vertical axis) for stable and unstable nuclides. This Segrè Plot has been obtained by the author from the Wikimedia website https://commons.wikimedia.org/wiki/File:Isotopes_and_half-life.svg, where it is stated to have been released into the public domain. It is included within this book on that basis.

periods of time. We can describe this behavior graphically in the form of a Segrè plot as shown in figure 2.1. This plot is named after Emilio Segrè, who was awarded the Nobel Prize in Physics in 1959 for his discovery of the antiproton, and shows the composition of stable and unstable nuclei as a function of the neutron and proton numbers. An important observation that can be made concerning the Segrè plot is that light nuclei have roughly the same number of neutrons and protons while heavy nuclei have significantly more neutrons than protons. This is discussed further below.

2.3 Nuclear binding energy

In order to understand how the nucleons in the nucleus are bound together, we begin by looking at some of the physical properties of the nucleus. Specifically, it is important to consider the size of the nucleus. While an atom, which is defined by the extent of the distribution of localized electrons, is of the order of 10^{-10} m in diameter, the nucleus has been determined on the basis of scattering experiments to be of the order of 10^{-15} m in diameter.

It is important to consider how interactions between the neutrons and protons hold the nucleus together. There are actually four known interactions in nature, the gravitational interaction, the electromagnetic interaction, the strong interaction and the weak interaction. The properties of these four interactions are summarized in table 2.1. The gravitational and electromagnetic interactions are long-range interactions with a potential that decreases as the inverse of distance ($1/r$) while the strong and weak interactions are short range and do not extend beyond the limits of the nucleus.

Since the Coulomb repulsion of the positively charged protons in the nucleus could not bind the neutrons and protons together, it is the attractive strong interaction, which acts on these particles on a distance scale of about 10^{-15} m, that is most important in binding the particles of the nucleus.

A general understanding of the interactions that are important for nuclear binding, allows us to accurately describe the behavior of the nuclear binding energy. The binding energy, B, is defined as a positive quantity which is a measure of the energy that would be required to completely separate the particles in a bound system. Because of the equivalence of mass and energy, the binding energy influences the mass of a system. We can understand this in the following way. The mass of a system of isolated (i.e. non-interacting) particles is merely the sum of their individual masses, $\sum m_i$. The mass of a bound system of particles is the sum of the individual masses minus the mass equivalent of the binding energy, B/c^2, that is

$$m = \sum m_i - B/c^2 \tag{2.3}$$

The fact that the binding energy reduces the mass of a system of particles can be easily explained by the observation that one must add energy (and equivalently mass) to the bound system to separate the particles far enough apart so that they are non-interacting. The more tightly bound the system of particles, the greater the binding energy. It follows, therefore, that the more tightly bound the system of particles, the smaller the mass.

We can understand the binding energy, and hence the mass, of various nuclei as a function of the number of neutrons and protons on the basis of a simple

Table 2.1. The four interactions in nature and their basic properties.

Interaction	Acts on	Range (m)	Strength
Gravitational	Masses	$\sim\infty$	10^{-39}
Electromagnetic	Charges	$\sim\infty$	10^{-2}
Strong	Neutrons, protons and similar particles	10^{-15}	1
Weak	Neutrons, protons, electrons and most other subatomic particles	10^{-18}	10^{-5}

semiempirical model. This is the liquid drop model and is the basis for the derivation of the so-called semiempirical mass formula. The liquid drop model is based, in part, on an analogy between the forces binding the nucleons in a nucleus and the forces binding the molecules in a drop of liquid. The interactions between the neutral molecules in a liquid can be expressed in terms of the Lennard-Jones potential (named after the British theoretical physicist John Lennard-Jones) of the form

$$V(r) = -\frac{a}{r^6} + \frac{b}{r^{12}} \tag{2.4}$$

where the ar^{-6} is an attractive term that describes the van der Walls interactions and the br^{-12} term is a repulsive term that results from the Pauli exclusion principle. Although the form of the Lennard-Jones potential is not the same as that for the strong nuclear interaction, they are similar in that both are very short-ranged interactions. This is seen for the Lennard-Jones potential by a comparison of the r^{-6} dependence of the attractive term compared to the r^{-1} dependence of the electromagnetic interaction.

In a classical analog, because of the short-range nature of the strong interaction, each nucleon interacts only with its nearest neighbors and not with all of the nucleons in the nucleus. Thus, to first order, the total binding energy associated with a collection of A nucleons will be directly proportional to A. Therefore, we may write the binding energy per nucleon as

$$B_V/A = a_V \tag{2.5}$$

where a_V is a constant representing the so-called volume term. This may be compared with the Coulomb energy associated with a system of Z charges, which is proportional to $Z(Z-1)$.

This simple first order approximation is an overestimate of the binding energy per nucleon for several reasons and can be improved by including additional correction terms. The first correction term comes about because, in this classical picture, the nucleons on the surface of the nucleus are not as tightly bound as those in the interior because they have fewer neighbors. Therefore, we can subtract a term which is proportional to the surface area of the nucleus. If we assume the nucleus to be spherical with a radius R, then the surface area will be proportional to R^2. If each nucleon occupies the same volume, then the radius may be related to $A^{1/3}$. An empirical expression based on experimental data is

$$R = 1.2A^{1/3} \tag{2.6}$$

where the radius is expressed in fm or 10^{-15} m. (This unit is commonly called a fermi.) Thus, the surface term may be expressed in terms of the constant a_S as

$$B_S/A = -a_S A^{2/3} \tag{2.7}$$

Another factor that reduces the nuclear binding energy is the Coulomb repulsion between the positively charged protons. A classical derivation of the Coulomb energy associated with a spherically symmetric distribution of charges of radius R

shows that this energy is proportional to $Z(Z-1)/R$. Using the above relationship between nuclear radius and mass number we write the Coulomb term in the semiempirical mass formula as

$$B_C/A = -a_C Z(Z-1)A^{-4/3} \tag{2.8}$$

where the constant a_C can be determined empirically from nuclear mass data but can also be approximated using classical *ab initio* calculations. Since the nuclear configuration with the greatest binding energy is the most stable than a simple inspection of equation (2.8) would suggest that it would be advantageous to minimize the number of protons in a nucleus. That is, we might expect large nuclei comprised only of bound neutrons. This is obviously not the case, although figure 2.1 would suggest that heavier nuclei benefit from having more neutrons than protons. In order to account for the features shown in figure 2.1, an additional term, referred to as the symmetry term is included in the semiempirical mass formula such that there is an advantage to having roughly equal numbers of neutrons and protons in a nucleus but that this advantage becomes less important for heavier nuclei. The usual form that is used for this term is

$$B_{sym}/A = -a_{sym}(A-2Z)^2A^{-2} \tag{2.9}$$

Although the inclusion of the symmetry term in the semiempirical mass formula is entirely empirical, the neutron-to-proton ratio as a function of nuclear size as shown in figure 2.1, is also predicted on the basis of quantum mechanical models of nuclear structure. Combining the above equations gives the total binding energy per nucleon as

$$\begin{aligned} B/A &= B_V/A + B_S/A + B_C/A + B_{sym}/A \\ &= a_V - a_S A^{2/3} - a_C Z(Z-1)A^{-4/3} - a_{sym}(A-2Z)^2A^{-2} \end{aligned} \tag{2.10}$$

The coefficients, a_V, a_S, a_C and a_{sym}, can be obtained by fitting equation (2.10) to experimentally measured values of the nuclear binding energy of various nuclei. The total nuclear binding energy is obtained from measured nuclear mass, m_{nuc}, according to equation (2.3) as

$$B = [Nm_n + Zm_p - m_{nuc}]c^2 \tag{2.11}$$

where m_n and m_p are the masses of the free neutron and proton, respectively. Generally, atomic masses (of neutral atoms) rather than nuclear masses are measured. In this case, the expression for the nuclear binding energy will include electron masses, m_e, and total electronic binding energy, b, as

$$B = [Nm_n + Z(m_p + m_e) - m_{nuc} - b/c^2]c^2 \tag{2.12}$$

Figure 2.2 shows the relative importance of the various terms in the semiempirical mass formula as a function of the total number of nucleons. Relative neutron and proton numbers can be estimated as a function of mass number from the region of stable nuclei, as shown in figure 2.1.

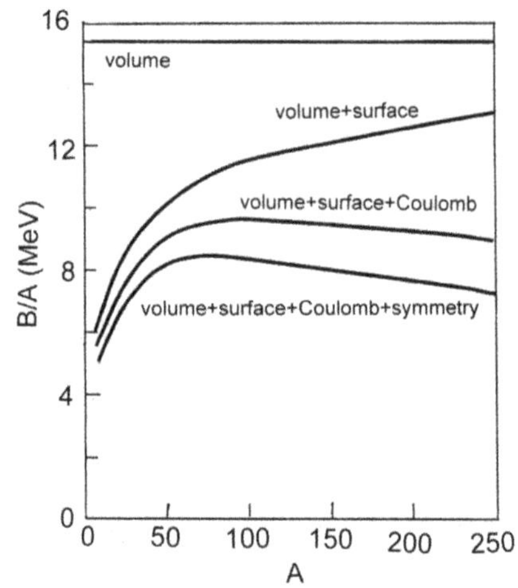

Figure 2.2. Relative importance of the various terms in the semiempirical mass formula as a function of mass number.

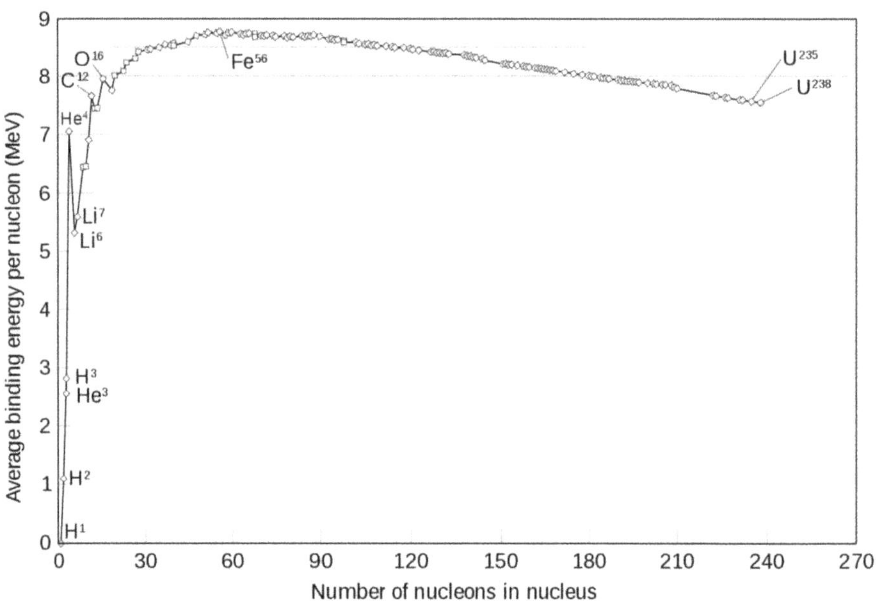

Figure 2.3. Binding energy per nucleon as a function of nucleon number for stable and long-lived nuclides. This Binding Energy image has been obtained by the author from the Wikimedia website https://commons.wikimedia.org/wiki/File:Binding_energy_curve_-_common_isotopes.svg, where it is stated to have been released into the public domain. It is included within this book on that basis.

The experimentally determined nuclear binding energies are illustrated in figure 2.3. The figure shows that the general form of the binding energy curve is well described by the semiempirical mass formula. It is clear, however, that there are some anomalies for lighter nuclei, where binding energies are greater than that predicated by the smooth curve corresponding to the semiempirical mass formula. The details of the binding energy for light nuclei are illustrated in figure 2.4, and a careful examination of this figure provides some important insight into this behavior.

Figure 2.4 shows that some nuclei, notably ^{4}He, ^{8}Be and ^{12}C, have binding energies per nucleon that fall above the general trend exhibited by the data. Others, notably ^{8}B, fall below the general trend line. We can understand the reasons for this behavior by looking at the number of neutrons and protons in these nuclei as given in table 2.2. It is obvious that nuclei with even numbers of like nucleons have anomalously high binding energies while nuclei with odd numbers of like nuclei have anomalously low binding energies. This behavior is the result of a strong pairing tendency of like nucleons. That is, neutrons in nuclei with an even number of

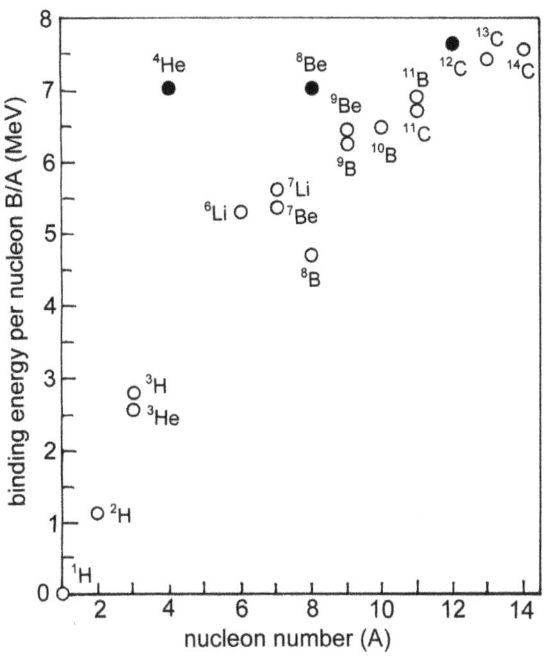

Figure 2.4. Binding energy per nucleon for light nuclei as a function of nucleon number. Even–even nuclei as described in the text are shown by the solid data points.

Table 2.2. Number of neutrons and protons in some nuclei with anomalous binding energies as seen in figure 2.4.

Nuclide	Mass number (A)	Neutrons (N)	Protons (Z)
^{4}He	4	2	2
^{8}Be	8	4	4
^{8}B	8	3	5
^{12}C	12	6	6

neutrons will form pairs with their quantum mechanical spins anti-aligned, and this will lead to increased nuclear stability (and correspondingly high binding energy). The same feature occurs for protons. Thus, in the liquid drop model, so-called even–even nuclei (i.e. even N, even Z) have increased binding energy while odd–odd (odd N, odd Z) nuclei have decreased binding energy and even–odd or odd–even nuclei will fall somewhere in between. This may be expressed quantitatively as an additional (pairing) term in the binding energy of equation (2.10) of the form

$$B_{\mathrm{p}}/A = -a_{\mathrm{p}}A^{-7/4} \tag{2.13}$$

where a_{p} is a parameter which is negative for even–even nuclei, zero for even–odd or odd–even nuclei and positive for odd–odd nuclei. As shown in figure 2.3 and as can be deduced from the functional form of the pairing term shown in equation (2.13), the pairing term becomes relatively less important for heavy nuclei. The significance of the pairing energy for the fusion of light nuclei will become apparent below.

2.4 Energy from nuclear fission

Thermal neutron reactors

The features shown in figure 2.4 provide a simple illustration of how energy can be obtained from nuclear processes. Two possible processes are obvious from the figure, fission (i.e. splitting) of a heavy nucleus into two lighter nuclei and fusion (i.e. combining) of two light nuclei into one heavy nucleus. In both cases it is seen that the initial configuration has a lower binding energy per nucleon than the final configuration. This means that since the total number of nucleons does not change, then the total binding energy after the process is greater than it was before. According to the equivalence of mass and energy, this increase in total binding energy results in a decrease in total mass and this lost mass is given off as energy (in the form of kinetic energy of the reaction products).

Fission reactions are the basis of energy production in all existing nuclear reactors. Figure 2.3 provides a simple means of estimating the energy liberated in a typical fission process. Consider a heavy nucleus with (e.g.) A~236 which undergoes fission and splits symmetrically into two lighter nuclei with A~118. Figure 2.3 shows that the initial heavy nucleus has a nuclear binding energy of a nucleus with A~236 is 7.56 MeV per nucleon. This gives a total binding energy for the nucleus of 236 × 7.56 MeV = 1.78 GeV. The graph shows that each of the fission fragments has a binding energy of 8.51 MeV per nucleon for a total binding energy of 236 × 8.51 MeV = 2.00 GeV. This gives an energy released from the fission of 2.00 GeV − 1.78 GeV = 220 MeV.

This description is a bit overly simplistic for a couple of reasons. Firstly, in addition to the fission fragments there are generally neutrons left over after the fission and secondly when fission occurs the two fission fragments are generally not the same size, that is, the fission is not symmetric. Therefore, the fission process may be written as

$${}^{A}X \rightarrow {}^{B}Y + {}^{A-B-\alpha}Z + \alpha n \tag{2.14}$$

where n is a neutron. The number of left-over neutrons, α, is typically 2 or 3 per fission. These residual neutrons are the result of the nuclear characteristics shown in figure 2.1 and are crucial for the creation of the fission chain reaction that is necessary for the operation of a nuclear fission reactor, as discussed below. Since heavy nuclei have a larger neutron to proton ratio than light nuclei, then the two fission fragments do not need all the neutrons available from the heavier nucleus. Even so, the fission fragments typically have more neutrons than are needed to produce stable nuclei. Subsequently, they will typically undergo β^- decay, thereby converting a neutron to a proton, p, and emitting an electron, e^- and an electron antineutrino, $\bar{\nu}_e$, via the weak interaction;

$$n \rightarrow p + e^- + \bar{\nu}_e \tag{2.15}$$

While it is obvious from the brief discussion above concerning binding energies that the fission of any heavy nucleus will be an exothermic process. However, spontaneous fission is actually a very slow process in most nuclei. The reason for this has to do with the nuclear potential well that results from the strong interaction and which binds the nucleons together in the nucleus. Figure 2.5 shows a typical example of the nuclear potential well holding two nuclear fission fragments, with an exothermic fission energy of E_b, together in the parent nucleus before a fission reaction occurs. In order escape from the strong nuclear potential well, the fission fragments must tunnel through the Coulomb barrier.

We can look at this process a little more quantitatively by plotting the Coulomb barrier as a function of the total number of nucleons in a nucleus, as shown in figure 2.6. It is seen in the figure that for nuclei with more than about 300 nucleons, the Coulomb barrier goes to zero and spontaneous fission will occur on a very short time scale. This type of analysis places a maximum size (around $A = 300$) for which nuclei can exist. We can also see from the figure that for nuclei such as uranium (with $A = 235$–238) that the Coulomb barrier is around 5–6 MeV. This value is too large

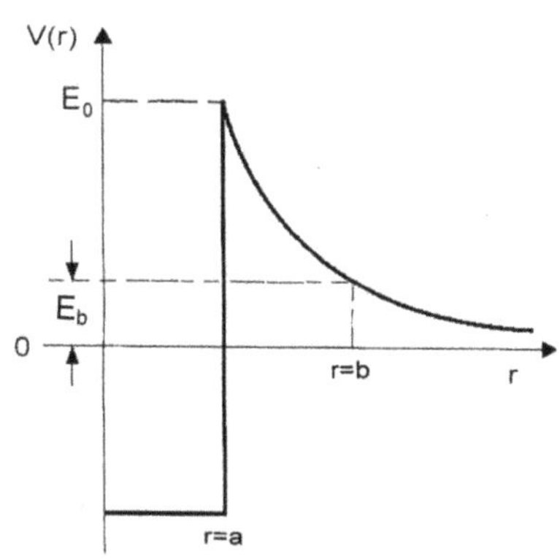

Figure 2.5. Potential barrier associated with two positively charged nuclei as a function of the distance between their centers. The more or less spherical square well for small values of distance, r, is the result of the strong interaction. For larger radii, a repulsive Coulomb interaction exists between the positively charged fission fragments.

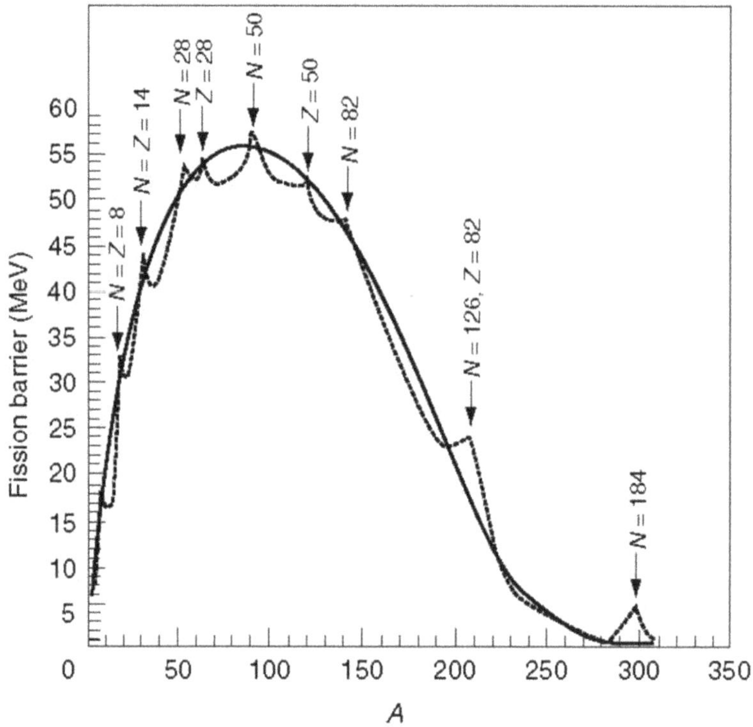

Figure 2.6. Coulomb barrier for spontaneous fission of heavy nuclei as a function of the total number of nucleons. Reprinted from Myers and Swiatecki (1966). Copyright (1966), with permission from Elsevier.

for spontaneous fission to occur on a time scale that would be compatible with the construction of a viable fission power reactor.

In order to understand fission reactions in practical terms, we need to go back and consider the extra neutrons that are emitted in the fission reaction shown in equation (2.14). The neutrons that are produced in such fission reactions are typically at an energy of about 2 MeV. Thermal fission reactors utilize neutrons that are at thermal energies, typically <1 eV. The high energy (fast) neutrons are thermalized during their passage through a moderating material which is typically water, heavy water or graphite. In a mixture of ^{235}U and ^{238}U reactor fuel, as described in the previous chapter, thermal neutrons may cause the following reactions

$$n + {}^{235}U \rightarrow {}^{236}U + 6.46\ \text{MeV} \tag{2.16}$$

and

$$n + {}^{238}U \rightarrow {}^{239}U + 4.78\ \text{MeV} \tag{2.17}$$

where the excess energy on the right-hand side of the equations is the energy that is released for a low energy (thermal) neutron. It is important to note that the thermal neutron cross section is about four orders of magnitude greater for ^{235}U nuclei than for ^{238}U nuclei. So even though the natural abundance of ^{235}U is much less than ^{238}U, in a naturally occurring mixture of uranium isotopes most of the thermal neutrons will react with ^{235}U nuclei. The energy in equations (2.16) and (2.17) raises the energy of the progeny nucleus on the right-hand side of the equation in the Coulomb well shown in figure 2.5. A comparison of equations (2.16) and (2.17) and the details of figure 2.6, shows that the excess energy in equation (2.16) is greater than the Coulomb barrier while the excess energy in equation (2.17) is less than the Coulomb barrier. Therefore, in equation (2.16), the ^{236}U nucleus formed by the capture of a thermal neutron by a ^{235}U nucleus can undergo so-called 'induced fission' without the need to quantum tunnel though the Coulomb barrier, while the ^{239}U nucleus formed by the capture of a thermal neutron by a ^{238}U nucleus cannot. Because of this behavior, a ^{235}U nucleus is referred to as 'fissile' while a ^{238}U nucleus is referred to as 'non-fissile'. These characteristics of neutron reactions in a mixture of ^{235}U and ^{238}U nuclei are the reason why thermal neutron reactors basically make use of only the fission energy associated with the ^{235}U nuclei. This concept of fissile and non-fissile nuclei can be extended to other heavy nuclei. A detailed inspection of neutron reactions in heavy nuclei generally shows that nuclei with an even number of protons and an odd number of neutrons, such as ^{235}U ($Z = 92$, $N = 143$), or vice versa, are fissile while nuclei with an even number of protons and an even number of neutrons, such as ^{238}U ($Z = 92$, $N = 146$), are non-fissile. This feature is related to the nucleon pairing tendency, as illustrated in figure 2.4, that shows that nuclei such as ^{235}U have a lower binding energy per nucleon than nuclei such as ^{238}U, and are, therefore, inherently less stable and will more readily undergo fission.

A typical distribution of fission fragment sizes from the fission of some common heavy nuclei is shown in figure 2.7. This distribution is referred to as the fission yield. We see from the figure that the fission of a ^{235}U nucleus will, on average, produce a

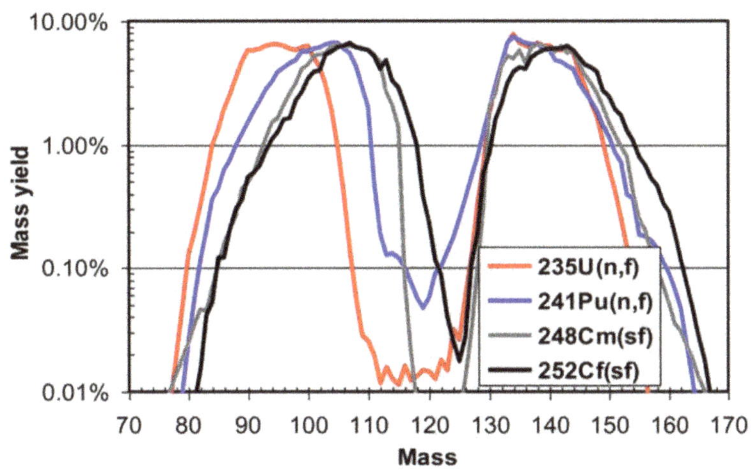

Figure 2.7. Fission yield for fission of ^{235}U, ^{241}Pu, ^{248}Cm and ^{252}Cf. Reproduced from de France *et al* (2018). CC BY 4.0.

lighter fission fragment with a mass number around 95 and a heavier fission fragment with has a mass number around 140. Thus, two typical induced fission reactions for ^{235}U might be

$$n + {}^{235}\mathrm{U} \rightarrow {}^{236}\mathrm{U} \rightarrow {}^{141}\mathrm{Ba} + {}^{93}\mathrm{Kr} + 2n \tag{2.18}$$

and

$$n + {}^{235}\mathrm{U} \rightarrow {}^{236}\mathrm{U} \rightarrow {}^{137}\mathrm{I} + {}^{96}\mathrm{Y} + 3n \tag{2.19}$$

showing that typically two or three neutrons are left over and these can then go on to induce further fissions in ^{235}U. The above description overviews the basic reactions that take place in a thermal neutron fission reactor. The manner in which a controlled nuclear fission reaction is established and maintained in a commercial fission power reactor is described in most introductory nuclear physics texts. *Introductory Nuclear Physics* (Krane 1988) is an excellent resource for further information about reactor design and control.

Based on a detailed analysis of the fission process in ^{235}U, which is the principal fuel in thermal neutron power reactors, the most common type of commercial nuclear reactor worldwide, about 80% of the fission energy or about 170 MeV per fission, becomes available as kinetic energy of the fission fragments. This kinetic energy is manifested as heat which can be used to generate electricity. These values give rise to the substantial amount of energy which can be obtained from nuclear fission, as discussed in chapter 1.

Fast breeder reactors

As noted in chapter 1, fast breeder reactors are able to utilize a substantial fraction of the nuclear energy that is available from the non-fissile ^{238}U component of natural uranium. In more recent years the idea that this approach may also be used to obtain useful energy from non-fissile naturally occurring ^{232}Th has become prominent. The following brief overview of breeder reactions and their applications considered both fast breeder ^{238}U reactors and potential future ^{232}Th reactors.

The fast breeder reactors which have been constructed thus far utilize the interaction of fast neutrons on ^{238}U nuclei. At energies around 2 MeV, as is typical for neutrons emitted in uranium fission processes, the cross sections for neutron absorption by ^{235}U and ^{238}U nuclei are similar. This is in contrast to the situation for neutrons with thermal energies where, as stated above, the neutron cross section is substantially larger for ^{235}U nuclei. Thus, in a mixture of uranium isotopes, fast neutron absorption by a ^{238}U nucleus, as shown in equation (2.17) is a likely process. This reaction produces ^{239}U, which is unstable and decays by β^- decay (with a half-life of about 23 min) to ^{239}Np, which is also unstable and decays by β^- decay (with a half-life of about 2 days) to ^{239}Pu. Overall, fast neutrons absorbed by a ^{238}U nucleus will cause the reaction

$$n + {}^{238}\mathrm{U} \rightarrow {}^{239}\mathrm{U} + \gamma \rightarrow {}^{239}\mathrm{Np} + e^- + \bar{\nu}_e \rightarrow {}^{239}\mathrm{Pu} + e^- + \bar{\nu}_e \tag{2.20}$$

where γ is a gamma ray. ^{239}Pu is β-stable, although it will spontaneously undergo α-decay with a half-life of about 24 000 years. It is also fissile and a ^{239}Pu nucleus can undergo induced fission when a thermal neutron is incident upon it;

$$n + {}^{239}Pu \rightarrow {}^{240}Pu \rightarrow \text{fission} \tag{2.21}$$

Excess neutrons from this induced fission reaction can then be used to breed more fissile fuel according to equation (2.20). It is important to note, however, that the neutrons on the left-hand side of equation (2.20) come, initially, from the fission of fissile ^{235}U in the reactor. However, since more than one excess neutron is emitted in the process in equation (2.21), more than one fissile ^{239}U nucleus can be bred for each induced ^{239}U fission. The ratio of fissile nuclei created to induced fissions is referred to as the 'breeding ratio'. Obviously, the above description indicates that a breeding ratio greater than unity for these processes will lead to an increase in the fissile material present in the reactor. A self-sustaining condition results when sufficient fissile material is generated. A measure of the time required to achieve this condition is the doubling time, that is the time required for the number of fissile nuclei to double. This time depends on the nuclides involved and also, to a lesser extent, on the chemical form of the reactor fuel. For the $^{238}U \rightarrow {}^{239}Pu$ cycle typical doubling times are in the range of 10–20 years.

The concept of creating a fission reactor that utilizes the potential of the substantial thorium resources on earth has been presented in the previous chapter. Naturally occurring thorium is virtually all ^{232}Th, which is non-fissile (having 142 neutrons and 90 protons) for the reasons described above. Thus, thorium has no naturally occurring fissile isotope. This possibility of constructing a nuclear reactor that utilizes thorium must breed fissile fuel from non-fissile ^{232}Th using fast neutrons. Neutrons incident on ^{232}Th nuclei are captured to produce ^{233}Th by the process

$$n + {}^{232}Th \rightarrow {}^{233}Th + \gamma \tag{2.22}$$

The ^{233}Th thus produced decays by β^- decay with a half-life of 22 min;

$${}^{233}Th \rightarrow {}^{233}U + e^- + \bar{\nu}_e \tag{2.23}$$

The ^{233}U nuclei are stable against β^- decay but decay by α decay with a half-life of 160 000 years. These ^{233}U nuclei are fissile and serve as a fission reactor fuel, undergoing induced fission by thermal neutron absorption according to the reaction

$$n + {}^{233}U \rightarrow {}^{234}U \rightarrow \text{fission} \tag{2.24}$$

Additional natural thorium may be input into the reactor and additional fissile ^{233}U is then bred according to equations (2.22) and (2.23) using excess neutrons from the induced fission reaction in equation (2.24). As above for the $^{238}U \rightarrow {}^{239}Pu$ cycle, the breeding ratio is greater than unity, leading to an increase in the number of fissile nuclei in the reactor. Typical doubling times for $^{238}U \rightarrow {}^{239}Pu$ cycle reactors are in the range of 70–100 years. This indicates that a prolonged and sustained effort to breed fissile fuel is necessary for the sustainable implementation of thorium fission reactors.

2.5 Energy from nuclear fusion

From figures 2.3 and 2.4 it is readily seen that combining very light nuclei will, in general lead to an increase in binding energy per nucleon and is a potential source of energy production. In fact, as described in more detail in the next chapter, fusion is the means by which the Sun and other stars produce energy. The sun primarily produces energy by fusing four ^{1}H nuclei together to form one ^{4}He nucleus. This fusion process takes place in a series of steps. The most probable sequence of reactions that produces energy in the Sun begins with the fusing of two ^{1}H nuclei (i.e. protons) by the process

$$p + p \rightarrow d + e^{+} + \nu_e \qquad (2.25)$$

where d is the deuteron, i.e. the ^{2}H nucleus, e^{+} is a positron and ν_e is an electron neutrino. As a bound pair of two protons is not stable, the fusing of two protons is accompanied by the conversion of one of the protons into a neutron by a β^{+} decay;

$$p \rightarrow n + e^{+} + \nu_e \qquad (2.26)$$

As noted above for β^{-} decay (see equation 2.15), β^{+} decay is the result of the weak interaction. Since the Sun is, at this stage of its life, comprised primarily of ^{1}H, the most likely process involving the deuteron produced in the reaction in equation (2.25) will be the fusion reaction

$$p + d \rightarrow {}^{3}He + \gamma \qquad (2.27)$$

^{3}He nuclei produced by this reaction can then fuse to produce ^{4}H by the process.

$${}^{3}He + {}^{3}He \rightarrow {}^{4}He + 2p \qquad (2.28)$$

The above description describes how the Sun produces the enormous amount of power that it radiates into space. It is tempting to speculate that if we could create a reactor here on earth with the same conditions that exist inside the Sun, then we would have a viable source of energy. There are a couple of simple facts that indicate that this is not the case.

Firstly, the Sun is about 4.6 billion years old and it will continue to produce energy (at a slightly increasing rate) for another 5 billion years or so, after which it will become a red giant and will eventually end up as a planetary nebula. Roughly speaking this means hydrogen fusion will serve as a source of energy for the Sun for about 10 billion years, more or less. Obviously, if we provide fuel to a reactor here on earth, we will not want to wait 10 billion years to extract its energy. Basically, the Sun consumes its fuel too slowly to be a viable power reactor. The main reason for this is the existence of the p–p fusion reaction in equation (2.25) which is the first step in the fusion of hydrogen into helium. As noted above this is a very slow process because of the necessity of the weak interaction (as described in table 2.1) to convert a proton to a neutron.

Secondly, we can actually look at how much energy the Sun produces. The power output from the Sun is about 3.86×10^{26} W. This is in comparison to the average energy use of humanity of about 1.9×10^{13} W. The sun produces a lot of power, but the Sun is also very large. Most of the energy produced by the Sun is produced near its core. The standard solar model estimates the power density at the center of the Sun to be 276.5 W m^{-3}. We can compare this to the power produced by a human body. Consider a human that consumes food with an energy content of 2000 Calories per day. Converting 2000 Cal to Joules gives (2000 Cal) × (4184 J Cal^{-1}) = 8.37×10^6 J. For a period of one day (86 400 s) the average power output will be about (8.37×10^6 J)/(8.44×10^4 s) = 97 W. That is, on the average a human body produces about as much power as a 100 W light bulb. Since a person's density is about the same as water (1000 kg m^{-3}), a typical 75 kg human will have a volume of (75 kg)/(1000 kg m^{-3}) = 0.075 m^3. This volume of the Sun will produce (0.075 m^3) × (276.5 W m^{-3}) = 21 W, only about 20% the power output of a human body. Obviously, this would not be a sufficient power density for a viable reactor.

The conclusion from the above discussion is that the Sun actually produces energy very slowly. It only produces a lot of power because it is so large. In order to understand how we might do things differently here on earth to produce a useful fusion power reactor, it is important to consider which fusion reactions, as described in the next section, are viable candidates.

Fusion reactions for power reactors

In order to be a viable fusion reaction for a power reactor the fusion process must fulfill several criteria as follows:

1. Produce net energy. This means the reaction must be exothermic, meaning that the product nuclei must have a greater binding energy per nucleon than the reactant nuclei.
2. Have only two reactants. Processes involving three reactants, as occur in the interior of some very hot stars, such as the so-called triple alpha process ($3\,^4He \rightarrow {}^{12}C$), require too high temperatures to be practical in a reactor.
3. Conserve both the number of neutrons and the number of protons. If neutrons are converted to protons (or vice versa) as in equations (2.15) or (2.26) then the weak interaction is involved, and the reaction will proceed too slowly.
4. Have two or more massive particles as products. Reactions such as in equation (2.27) in which there is only one massive product and a photon are not practical as they require the electromagnetic interaction to conserve both energy and momentum.

Table 2.3 lists the possible fusion reactions that satisfy these general criteria. There are several considerations for choosing a suitable reaction for a commercial power reactor. Some factors important to consider are as follows:

1. The ability to create conditions amenable to a sustained fusion reaction.
2. A large energy output per fusion reaction.
3. The lack of a free neutron as a by-product of the fusion reaction.

Unfortunately, no reaction ideally satisfies all of these criteria. Details of each of these factors are discussed briefly below.

Table 2.3. Exothermic fusion reactions that satisfy the criteria listed in the text as possible reactions for power reactors.

No.	Reactions	Energy (MeV)
1	$d + d \rightarrow t + p$	4.03
2	$d + d \rightarrow {}^3He + n$	3.27
3	$d + t \rightarrow {}^4He + n$	17.6
4	$d + {}^3He \rightarrow {}^4He + p$	18.3
5	$t + t \rightarrow {}^4He + 2n$	11.3
6	$t + {}^3He \rightarrow {}^4He + n + p$	12.1
7	$t + {}^3He \rightarrow {}^4He + d$	14.3
8	${}^3He + {}^3He \rightarrow {}^4He + 2p$	12.9
9	$p + {}^6Li \rightarrow {}^4He + {}^3He$	4.0
10	$d + {}^6Li \rightarrow 2{}^4He$	22.4
11	$d + {}^6Li \rightarrow {}^3He + {}^4He + n$	2.56
12	$d + {}^6Li \rightarrow {}^7Be + n$	3.4
13	$d + {}^6Li \rightarrow {}^7Li + p$	5.0
14	${}^3He + {}^6Li \rightarrow 2{}^4He + p$	16.9
15	$p + {}^{11}B \rightarrow 3{}^4He$	8.7

The ability to create conditions for a sustained fusion reaction is a necessary condition to the construction of a viable fusion power reactor. Creating appropriate conditions to maintain a fusion reaction is quite different than that needed to maintain a fission reaction. The reason for this difference is a consequence of the Coulomb interaction. If it is energetically favorable (that is, the reaction is exothermic) then once the fission fragments separate, they will be pushed apart by the repulsive Coulomb interaction between the positively charged protons in each fragment. For fusion to occur, even when the reaction is exothermic, the two reactants must overcome the repulsive Coulomb interaction between their positive charges to allow the attractive strong interaction to bind them together. The potential energy associated with two nuclei approaching each other with a kinetic energy E_b, is illustrated in figure 2.5. The figure shows that at a distance, b in the figure, that the Coulomb repulsion energy becomes greater than the nuclei's kinetic energy and the nuclei will scatter from one another. Classically, it is only if the kinetic energy of the nuclei is greater than E_0 in the figure that the nuclei will come in contact with each other and the attractive strong interaction will cause them to fuse together. Since E_0 is greater than the energy that could realistically be achieved in a reactor, fusion is the result of quantum mechanical tunneling through the Coulomb barrier when two nuclei with sufficient energy are in sufficiently close proximity for a sufficiently long time. Thus, the relevant parameters for the construction of a viable fusion reactor are particle energy (or equivalently speaking, temperature), particle density and confinement time. These factors will be discussed in detail in the later chapters of this book for different experimental approaches to fusion.

It is clear from the above discussion that minimizing the Coulomb barrier is beneficial in achieving fusion. This is accomplished by minimizing the nuclear charges. For this reason, laboratory fusion of hydrogen isotopes should generally be more easily accomplished than the fusion of the heavier nuclei in table 2.3 which have a greater positive charge. The fusion rate, f, is given in terms of the reactivity and the number densities of the two species of nuclei as,

$$f = n_1 n_2 \langle \sigma v \rangle \tag{2.29}$$

where the reactivity is defined as the average of the product of the fusion cross section, σ, and the relative velocity of the two nuclei, v, that is, $\langle \sigma v \rangle$, and is a quantitative measure of the probability of fusion. n_1 and n_2 are the number densities of the two reaction species. Note that for fusion processes such as d–d fusion or t–t fusion where there is only one species of nuclei, reaction rates benefit by an additional factor of two because each nucleus can react will all other nuclei and not just half of them (for the case where there are equal numbers of two different nuclei, as in d–t fusion).

Figure 2.8 shows the reactivity for some potentially useful fusion reactions. It is clear that for temperatures up to about 10^9 K (or 100 keV), the d–t fusion is one to two orders of magnitude more likely than d–d fusion. This is as expected. The Coulomb barrier for d–d and d–t fusion is the same, but the cross section for d–t fusion will be greater because of the additional nucleon (neutron) associated with the tritium nucleus compared to the deuterium nucleus. Fusion reactions involving more massive (and more highly charged) nuclei are typically less probable for most energies than those involving only isotopes of hydrogen due to the increased Coulomb barrier.

For reasons as described above, conditions for d–t fusion should be the least demanding. As well, the large energy release per fusion should be advantageous. Ultimately, however, other fusion processes may also be important for commercial energy production because of the d–t reaction's high neutronicity

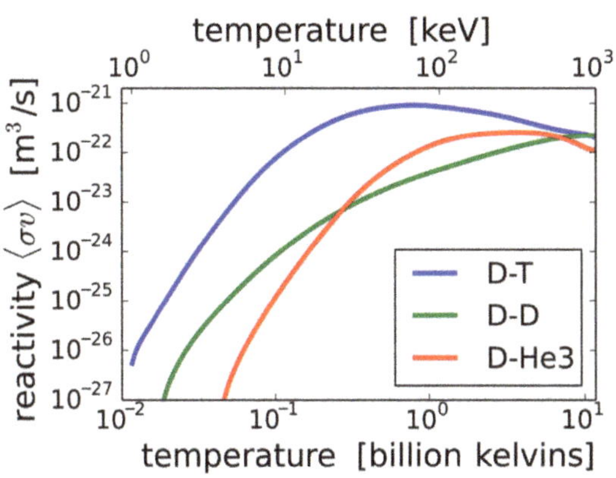

Figure 2.8. Reactivities of some fusion reactions as a function of plasma temperature. Note 1 K = 8.3×10^{-5} eV. This Reactivities of Some Fusion Reactions image has been obtained by the author from the Wikimedia website https://commons.wikimedia.org/wiki/File:Fusion_rxnrate.svg where it was made available by Dstrozzi under a CC BY 2.5 license. It is included within this book on that basis. It is attributed to Dstrozzi.

(as described below) and potentially limited resource availability as discussed later in the book.

As seen in table 2.3, many fusion processes, including d–d fusion and d–t fusion have a neutron as one of the final products and this situation has a number of drawbacks. In such cases, some of the energy released during the fusion appears as kinetic energy of the neutron. This energy is difficult to recover because the uncharged neutron (unlike charged protons and nuclei) does not readily interact with charges in the reactor fuel and does not significantly contribute to heating. The neutronicity of a fusion reaction is the fraction of energy released in the processes that the neutron acquires. Conservation of energy and momentum requires that for a reaction that has two by-products, with masses m_1 and m_2, then the kinetic energy is distributed in the proportions $m_2/(m_1+m_2)$ to mass m_1 and $m_1/(m_1+m_2)$ to mass m_2. A simple calculation for d–t fusion shows that about 80% of the kinetic energy is given to the neutron. That is, the neutronicity is 0.8. Thus, only 20% of the 17.6 MeV of energy per fusion (or about 3.5 MeV) is readily available for power production.

Other considerations for fusion reactions with a high neutronicity, include radiation damage, neutron reactions and neutron activation of reactor structural materials. Radiation damage occurs when neutron displace atoms in the crystalline structure of a material leaving a vacancy behind. Over time, this process can lead to the weakening of the material and eventually to the need to replace reactor components.

A neutron can also interact with a nucleus in a reactor material. A variety of different reactions can occur, but one example is

$$\mathrm{n} + {}^{56}\mathrm{Fe} \rightarrow {}^{56}\mathrm{Mn} + \mathrm{p} \tag{2.30}$$

where the iron nucleus absorbs a neutron and emits a proton. Since ^{56}Fe is 92% naturally abundant, this reaction will occur in any iron containing materials. The liberated proton will attract an electron in the material and will form a hydrogen atom. Hydrogen atoms, thus formed, will coalesce and form gas bubbles which will reduce the structural integrity of the metal.

Neutron activation, as discussed in further detail in chapter 7, results from the reaction of neutrons with nuclei leading to the production of radioactive materials. Many transition elements, such as Cr, Mn, Fe Co, Ni and Co, are common components of structural steels (e.g. stainless steel) and are susceptible to neutron activation. Cobalt is an interesting example of this process. ^{59}Co is 100% naturally abundant and has a large neutron absorption cross section. Neutrons incident on cobalt produce ^{60}Co by the reaction

$$\mathrm{n} + {}^{59}\mathrm{Co} \rightarrow {}^{60}\mathrm{Co} + \gamma \tag{2.31}$$

^{60}Co is radioactive with a half life of 5.2 years and decays by β^- decay

$${}^{60}\mathrm{Co} \rightarrow {}^{60}\mathrm{Ni}^{*} + \mathrm{e}^{-} + \bar{\nu}_\mathrm{e} \tag{2.32}$$

where $\bar{\nu}_\mathrm{e}$ is an electron antineutrino and $^{60}\mathrm{Ni}^*$ is an excited state of ^{60}Ni (indicated by the asterisk) that decays by emitting gamma rays.

The discussion above shows that there are certain advantages to fusion reactions with a low neutronicity; that is ability to utilize a greater fraction of the fusion energy and reduced concern over radiation damage and production of radioactive by-products. Reactions such as d–^{3}He fusion and p–^{11}B fusion are referred to as aneutronic as, in principle, they produce no neutrons. These two reactions are the most likely aneutronic fusion reactions to be achieved in a controlled reactor and are discussed further in chapters 6 and 7.

It is interesting to note that even in a reactor that utilizes an aneutronic reaction, some neutrons (although a very small number) may be produced. p–^{11}B fusion is an interesting example. The primary reaction between the fuel components is

$$p + {}^{11}B \rightarrow 3\alpha \qquad (8.7\,\text{MeV}) \tag{2.33}$$

In addition to the primary reaction in equation (2.33) a minor branching ratio for p–^{11}B fusion is the reaction

$$p + {}^{11}B \rightarrow {}^{11}C + n \qquad (-2.8\,\text{MeV}) \tag{2.34}$$

As this reaction is endothermic and therefore occurs rarely and only in cases where the proton carries enough kinetic energy to make the reaction energetically favorable. However, it is a minor source of neutron emission from an aneutronic p–^{11}B fusion reactor. Another contribution to neutron emission in a p–^{11}B reactor can occur if the hydrogen fuel (which provides the protons) is not isotopically pure and contains a small fraction of deuterium. If this is the case, neutrons can be produced by the fusion reaction

$$d + {}^{11}B \rightarrow {}^{12}C + n \qquad (13.7\,\text{MeV}) \tag{2.35}$$

This is highly exothermic and produces neutrons with kinetic energy in excess of 12.6 MeV. A final possibility exists if a ^{11}B nucleus reacts with the by-product of a previous p–^{11}B fusion, that is an alpha particle, according to the reaction

$$\alpha + {}^{11}B \rightarrow {}^{14}N + n \qquad (0.157\,\text{MeV}) \tag{2.36}$$

This reaction is slightly exothermic where the neutron acquires about 0.15 MeV of energy from the fusion reaction. However, the neutron also acquires most of the kinetic energy associated with the incident alpha particle. As each alpha particle in equation (2.33) acquires one third of the fusion energy (or about 2.9 MeV) the resulting neutron in equation (2.36) typically has an energy of around 3 MeV. A detailed analysis of fusion reactions in an aneutronic d–^{3}He reactor would also show the emission of a small number of neutrons. In the simple case for this reaction, d–d fusion will occur (leading to neutron emission) as well as d–^{3}He fusion.

The neutron flux from an aneutronic reaction such as p–^{11}B fusion typically accounts for less than 1 percent of the fusion reactions. Thus, although there is some energy lost as kinetic energy of emitted neutrons and some activation of reactor materials, both of these factors are much less than in reactors that utilize d–d or d–t fusion reactions.

References and suggestions for further reading

Cohen B L 1983 Breeder reactors: a renewable energy source *Am. J. Phys.* **51** 75

Cohen B L 1990 *The Nuclear Energy Option—An Alternative for the 90s* (New York: Plenum)

Cottingham W N and Greenwood D 1986 *An Introduction to Nuclear Physics* (Cambridge: Cambridge University Press)

de France G, Blanc M, Jentschel U, Köster U, Leoni S, Mutti P, Simpson G, Soldner T and Urban W 2018 EXOGAM at the ILL: the EXILL campaign *EPJ Web Conf.* **178** 01004

Dunlap R A 2004 *An Introduction to the Physics of Nuclei and Particles* (Belmont, CA: Brooks/Cole)

Dunlap R A 2013 A macroscopic analogue of the nuclear pairing potential *Eur. J. Phys. Educ.* **4** 1–6

Dunlap R A 2018 *Particle Physics* (San Rafael, CA: Morgan & Claypool)

Enge H A 1966 *Introduction to Nuclear Physics* (Reading, MA: Addison-Wesley)

Inglis D R 1973 Nuclear Energy – Its Physics and Its Social Challenge (Reading, MA: Addison-Wesley)

Krane K S 1988 *Introductory Nuclear Physics* (New York: Wiley)

McCracken G and Stott P 2013 *Fusion: The Energy of the Universe* 2nd edn (New York: Academic)

Morse E 2018 *Nuclear Fusion* (Cham: Springer)

Myers W D and Swiatecki W J 1966 Nuclear masses and deformations *Nucl. Phys.* **8** 1–60

Chapter 3

The early history of nuclear fusion

3.1 Introduction

Throughout the 19th century our knowledge of the properties of the Sun increased. This resulted from experimental observations that led to an understanding of the Sun's size and composition. However, throughout that century the mechanism by which the Sun produced energy remained a mystery. Classical theories of solar energy production such as chemical reactions or gravitational energy were not adequate to explain the known longevity of the Sun. In the early part of the 20th century the introduction of Einstein's mass–energy relation and the development of experimental techniques for the accurate measurement of nuclear masses laid the groundwork for our understanding of how nuclear fusion could give rise to the production of substantial energy. It was in the 1930s, however, that the discovery of the neutron and the development of quantum mechanics provided a complete picture of the physical processes involved in nuclear fusion.

The present chapter overviews some of the important developments in our understanding of nuclear fusion up to the 1960s and how fusion can be used for the production of useful energy. This overview includes the development of our understanding concerning stellar fusion processes, the understanding of basic nuclear structure, the development of laboratory techniques for the artificial creation of nuclear fusion reactions and the events leading to the creation of the first thermonuclear weapons.

3.2 An historical overview of solar properties

Stars, including our Sun, produce energy through nuclear fusion reactions. We owe our very existence to the presence of these solar reactions. However, it was only about seventy years ago that we developed a basic understanding of energy production in stars and, specifically, in our Sun. We begin the story of how we came to understand the fusion processes in the Sun by looking at some of the most significant scientific investigations of the Sun prior to around the middle of the 19th century. Two solar

doi:10.1088/978-0-7503-3307-8ch3

properties that are necessary for the description of the mechanism by which the Sun produces energy are the solar mass and the solar composition.

The mass of the Sun

A simple method of determining the mass of the Sun follows from the laws of motion and the theory of gravitational interactions developed by Isaac Newton (ca. 1643–1727). Newton's law of gravitational attraction states that the force, F, between two masses, M and m, separated by a distance, r, may be expressed as

$$F = \frac{GMm}{r^2} \tag{3.1}$$

where G is the gravitational constant ($6.674\,30(15) \times 10^{-11}$ m^3 kg^{-1} s^{-2}). The gravitational constant was first estimated by Newton to be about 7×10^{-11} m^3 kg^{-1} s^{-2}. This estimate is based on a measurement of the gravitational acceleration, g, and the radius of the Earth, R_e. From Newton's law, $F = ma$, the gravitational constant may be written as

$$G = \frac{gR_e{}^2}{M_e} \tag{3.2}$$

where M_e is the mass of the Earth and R_e is the Earth's radius. As the mass of the Earth was not known at the time of Newton, it was more convenient to rewrite equation (3.2) in terms of the mean density of the Earth, ρ, as

$$G = \frac{3g}{4\pi R_e\rho} \tag{3.3}$$

Newton's estimate of the gravitational constant was made on the basis of equation (3.3) and the assumption that the mean density of the Earth was around 5×10^3 kg m^{-3} (i.e. 5 times the density of water). The first direct laboratory measurement of the gravitational constant was made by the English scientist Henry Cavendish (1731–1810) in 1798 using masses suspended on a torsion pendulum. Using the measured value of the gravitational constant allows for a determination of the actual mean density of the Earth as well as its total mass. Modern measurements give a mean density of 5.51×10^{-3} kg m^{-3} and a total mass of 5.972×10^{24} kg.

Once the gravitational constant is known, the mass of the Sun, M_s, may be determined from equation (3.2) by writing the force between the Sun and the Earth as

$$F = M_e\frac{v^2}{R} = \frac{GM_eM_s}{R^2} \tag{3.4}$$

where v is the velocity of the Earth in its orbit around the Sun and R is the radius of the orbit. The term v^2/R is the centripetal acceleration of the Earth. Writing the velocity in terms of the period of the orbit ($T = 1$ year $= 3.154 \times 10^7$ s)

$$v = \frac{2\pi R}{T} \tag{3.5}$$

and solving equation (3.4) for the Sun's mass gives

$$M_s = \frac{4\pi^2 R^3}{GT^2} \tag{3.6}$$

Determining the solar mass amounts to measuring the distance between the Earth and the Sun. This was not known with any degree of certainty during Newton's lifetime, it was determined within about 2% of the currently accepted value in the 1760s using observations of the transit of Venus and geometric methods proposed in 1716 by the English astronomer Edmond Halley (1656–1742). The currently accepted value of the solar mass is 1.989×10^{30} kg.

The composition of the Sun

Our knowledge of the composition of the Sun comes from the study of its atomic spectra. The earliest observation of optical spectra that exhibited features related to the atoms that gave rise to the emission of the radiation was made in the very early years of the 19th century. William Hyde Wollaston (1766–1828) was an English scientist who made important discoveries in a number of fields of chemistry, physics and astronomy. He is, perhaps, best known as the discoverer of the elements palladium and rhodium. Wollaston also made significant observations of electromagnetic induction that preceded the work of Michael Faraday (1791–1867) on the same subject by a decade. What is of most relevance to the present discussion was his observation of optical solar spectra. In 1802, Wollaston (1802) observed the presence of about half a dozen dark lines in the continuum spectrum of the optical radiation from the Sun.

A little more than a decade after Wollaston's discovery, the Bavarian physicist Joseph Ritter von Fraunhofer (1787–1826) expanded on this early work on solar spectroscopy. Fraunhofer invented the spectroscope, with which he observed bright emission lines in the optical spectra of hot gases in the laboratory. Fraunhofer also observed dark absorption lines in the optical spectrum from the Sun. The greater sensitivity of his spectrometer, compared to that of Wollaston, allowed him to observe 574 dark absorption lines in the solar spectrum and also to observe absorption lines in the spectra of some bright stars. His observations showed that the absorption lines of the stars he studied differed from each other and also from those associated with the Sun.

Perhaps the next significant development of relevance was made by the French physicist Jean Bernard Léon Foucault (1819–68). Foucault is best known for the pendulum that bears his name with which he demonstrated the effects of the rotation of the Earth. As well, in 1849, Foucault demonstrated that emission and absorption lines that appear at the same wavelength in optical spectra of different sources are due to the presence of the same material in the source.

The German physicist Gustav Robert Kirchhoff (1824–87) is best known for the law that bears his name and describes the properties of currents in electric circuits. Kirchhoff also made important contributions to the field of optical spectroscopy and is responsible for what are now known as Kirchhoff's three laws of spectroscopy.

These three laws summarized the knowledge of atomic spectroscopy at that point in time and may be expressed as:

1. A solid material, a liquid or a gas under high pressure which is heated to high temperature will emit a continuum spectrum.
2. A gas under low pressure which is heated to high temperature will emit a spectrum consisting of bright emission lines.
3. A continuum spectrum, as produced in accordance with the first law, will exhibit dark absorption lines when viewed through a region of cool gas.

The laws stated by Kirchhoff are empirical in nature and were based on experimental observation. Based on a description of atomic structure, which developed in the early part of the 20th century, we now have a detailed understanding of how atomic spectra are related to the basic properties of the atom.

Of particular relevance to our understanding of the properties of the Sun was the work in the early 1860s by the Swedish physicist Anders Jonas Ångström (1814–74) who is the namesake of the unit equal to 10^{-10} m. Ångström's work on solar spectroscopy established that the Sun is composed mostly of hydrogen and this eventually led to the description of the nuclear fusion processes that are responsible for its energy production.

3.3 Darwin's dilemma

In 1859 Charles Robert Darwin (1809–82) published *On the Origin of the Species* (Darwin 1859) which presented his theory of evolution. Within a decade or so of the publication of *On the Origin of the Species*, the basic principles Darwin's theory were generally accepted within the scientific community. However, the time scale that Darwin envisioned for the evolutionary process was at odds with the ages of the Earth and Sun that we generally accepted at the time. The scientific reasons for the accepted values of the Earth's and Sun's ages are reviewed below.

The age of the Earth

Darwin was aware that evolution based on natural selection would be a very slow process and would require that the age of the Earth was sufficiently long. Common religious beliefs in Europe in the mid-19th century indicated that the Earth was about 6000 years old. This value was clearly inconsistent with the requirements of the theory of evolution. As evidence for the time that the Earth had been in existence, Darwin used his observations of geological features. Specifically, Darwin studied the Weald, which is an area in South East England in the counties of Hampshire, Surrey, Sussex and Kent. The Weald is a region that was created by erosion and is bordered by the parallel chalk escarpments known as the North and South Downs. Based on his assumptions concerning erosion rates, Darwin estimated that it took 3×10^8 years for the Weald to be formed and he used this as an estimate of the minimum age of the Earth in the first edition of his book. Modern studies show that the Weald is actually about 1.3×10^8 years old and is much younger than the actual age of the Earth.

In 1862 William Thomson (later known as Lord Kelvin) (1824–1907) published his studies concerning the age of the Earth which were based on thermodynamic arguments. Kelvin assumed that the Earth formed as a molten object and radiated heat into space as it cooled to its present temperature. He calculated the length of time required for the near-surface temperature gradient of the Earth to decrease to its currently measured value. From these studies he estimated that the most likely age of the Earth was around 1×10^8 years. Based on later analyses (published in 1897) he revised his estimate for the age of the Earth to around 2×10^7 years. While scientifically convincing, Kelvin's analysis yielded incorrect results because they omitted two very important factors that influence the temperature within the Earth's interior; firstly, the heat that is produced in the Earth's interior by the decay of radioactive nuclides, most notable isotopes of uranium, thorium and potassium, and secondly, heat produced by friction in the liquid outer core. Both these factors contribute to a much slower decrease in the surface temperature gradient than Kelvin estimated. However, Kelvin's work on the age of the Earth was a major source of controversary with regard to Darwin's estimate of the age of the Weald and the time required for evolution to occur. In fact, the discrepancy between Kelvin's estimate of the Earth's age and his own was a major cause of concern for Darwin and by the third edition of *On the Origin of the Species* he had completely removed arguments based on estimated age of the Weald from his book.

It was well into the 20th century when the classical approach to determining the age of the Earth that was used in Darwin's time, was replaced by methods based on the isotopic abundance of radioactive nuclides. In 1929 Francis William Aston (1877–1945) used mass spectroscopy to observe the relative natural abundances of different isotopes of lead and uranium. Using Aston's results, along with a knowledge of the half lives of radioactive nuclides, Ernest Rutherford (1871–1937) calculated the age of the Earth as 3.4×10^9 years. This approach has become the standard method for determining the age of the Earth and current measurements indicate an age of 4.527×10^9 years. As a point of reference, current fossil evidence indicates that the total time required for evolution from the simplest life forms to the present was 3.5×10^9 to 3.8×10^9 years.

The age of the Sun

The burning gas theory

An even greater discrepancy in time scales and source for concern with regard to Darwin's theory of evolution dealt with the age of the Sun. For many years it had been assumed that the Sun produced energy by chemical combustion. A simple calculation based on this hypothesis shows that the maximum lifetime of the Sun, t, would be

$$t = \frac{E}{P} \tag{3.7}$$

where E is the total energy available from combustion and P is the power irradiated (i.e. energy per unit time). The total energy available from combustion is given by

$$E = QM_s \tag{3.8}$$

where Q is the heat of combustion per unit mass of the solar material. This gives

$$t = \frac{QM_s}{P} \tag{3.9}$$

This simple approach assumes that the power output of the Sun remains constant throughout its lifetime. This is probably not exactly true, but this simple approach will provide an approximate value of the lifetime of the Sun. Using estimates of the quantities in equation (3.9) that were available in the mid-19th century, Kelvin estimated the lifetime of the Sun on the basis of chemical reactions to be 3000 years, even less than the Biblical estimate of the age of the Earth. Using modern values of the quantities in equation (3.9) ($Q = 1.42 \times 10^8$ J/kg for hydrogen, $M_s = 1.989 \times 10^{30}$ kg and $P = 3.846 \times 10^{26}$ W) gives the lifetime of the Sun as 23 400 years.

The gravitational energy theory

Certainly, by the mid-1800s it was known that chemical combustion was not a viable method for energy production in the Sun. Ideas concerning a possible alternative mechanism for energy production in the Sun developed in the mid-1800s, largely as a result of work by Lord Kelvin and Hermann Ludwig Ferdinand von Helmholtz (1821–94), a German physician and physicist. The so-called Kelvin–Helmholtz mechanism is a means by which the gravitational potential energy of the Sun can be converted into radiation. The basic process by which this occurs is described below.

We begin with the Sun, i.e. a spherically symmetric gravitationally bound collection of gas, in hydrostatic equilibrium. When the system is in hydrostatic equilibrium the outward force from the pressure of the gas is balanced by the inward gravitational force. As the Sun radiates energy, it loses energy and cools. Cooling reduces the internal pressure, and gravity causes the Sun to contract to maintain hydrostatic equilibrium. Thus, the contraction of the Sun causes the decrease in gravitational energy, or at least part of it, to be carried away as radiation.

The potential energy, U, associated with the gravitational interaction between two masses follows from the force given in equation (3.1);

$$U = -\frac{GMm}{r} \tag{3.10}$$

To determine the total potential energy associated with a spherical mass, we integrate the potential between a differential mass shell of thickness dr and the mass inside that radius, $m(r)$, from 0 to the total radius of the sphere (Sun), R_s, as

$$U = -G\int_0^{R_s} \frac{m(r)4\pi r^2\rho(r)}{r}\mathrm{d}r \tag{3.11}$$

where $\rho(r)$ is the radially dependent density. As the radial dependence of the density is not known *a priori*, we make the simple assumption that the density is constant. Therefore, we can write

$$m(r) = \frac{4}{3}\pi r^3\rho \tag{3.12}$$

and equation (3.11) can be integrated to give

$$U = -\frac{16}{15} G\pi^2 \rho^2 R_s^{\,5} \tag{3.13}$$

Writing the total mass as

$$M_s = \frac{4}{3}\pi R_s^{\,3} \tag{3.14}$$

and substituting into equation (3.13) gives

$$U = -\frac{3}{5}\frac{GM_s^{\,2}}{R_s} \tag{3.15}$$

As the Sun cools and collapses, it radiates away some of its gravitational potential energy in the form of photons. The virial theorem tells us that half of the change in gravitational potential energy becomes radiation. We can show that this is the case by calculating the kinetic (i.e. thermal) energy of the Sun. The kinetic energy of each gas particle is given as $\frac{3}{2}k_B T$ where k_B is the Boltzmann constant and T is the absolute temperature. The total kinetic energy is given by the integral over the Sun's volume, V,

$$K = \int \frac{3}{2} k_B T n(r) \mathrm{d}V \tag{3.16}$$

where $n(r)$ is the number density of gas particles. Using the ideal gas law $PV = nk_B T$, where P is pressure, in equation (3.16), gives

$$K = \frac{3}{2}\int_0^{R_s} 4\pi r^2 P(r)\mathrm{d}r = [2\pi P(r) r^3]_0^{R_s} - 2\pi \int_0^{R_s} r^3 \frac{\mathrm{d}P(r)}{\mathrm{d}r}\mathrm{d}r \tag{3.17}$$

where the form on the right-hand side of the equation is obtained by integration by parts. The first term on the right is zero, since $P(R_s) = 0$. The second term on the right can be evaluated using the condition for hydrostatic equilibrium

$$\frac{\mathrm{d}P(r)}{\mathrm{d}r} = -\frac{Gm(r)\rho(r)}{r^2} \tag{3.18}$$

Equation (3.17), therefore gives

$$K = \frac{G}{2}\int_0^{R_s} \frac{m(r) 4\pi r^2 \rho(r)}{r}\mathrm{d}r \tag{3.19}$$

A comparison of equation (3.19) with equation (3.11) readily shows that

$$K = -\frac{U}{2} \tag{3.20}$$

or

$$\Delta K = -\frac{\Delta U}{2} \tag{3.21}$$

This means that as U decreases, K increases by $\frac{\Delta U}{2}$, and conservation of energy requires that radiation with energy $\frac{\Delta U}{2}$ is emitted. The radiated energy available from the gravitational interaction is, therefore,

$$E = -\frac{U}{2} = \frac{3}{10}\frac{GM_s^2}{R_s} \tag{3.22}$$

Kelvin's 1862 estimate of the lifetime of the Sun, based on gravitation as its source of energy, was 2×10^7 years, consistent with his later estimates of the age of the Earth. Using contemporary values for the parameters in equation (3.22) and the assumption that the energy output of the Sun is constant, equation (3.7) gives a value of the Sun's lifetime of 9.5×10^6 years. Obviously, classical arguments based on combustion and gravitational interactions are not sufficient to explain the longevity of the Sun as a source of radiation.

In 1920 Arthur Stanley Eddington (1882–1944) utilized Einstein's mass–energy relation and Aston's nuclear mass measurements to hypothesize that the Sun produced energy by fusing four hydrogen nuclei together to form one helium nucleus (Eddington 1920). It was this work that finally led to a calculated lifetime of the Sun that was compatible with the age of the Earth and with the time scale required by Darwin's theory of evolution. At that time, the details of nuclear structure were not well known. Nor was it obvious how four positively charged hydrogen nuclei (with a total charge of +4e) could fuse together to form a helium nucleus with a net positive charge of +2e. During the latter part of the 1920s and the early 1930s, our knowledge of the quantum mechanical nature of subatomic particles and nuclear structure began to be pieced together. In the next section some of the most significant developments in these two areas are summarized.

3.4 Early developments in nuclear physics

The development of our current ideas on the properties of the nucleus began in the latter part of the 19th century. The earliest work related to nuclear properties dealt with the observation and characterization of natural decay processes. The present section overviews some of the more significant scientific developments during the final years of the 19th century and the first 35 years or so of the 20th century that led up to our understanding of nuclear fusion.

Radioactive decay

The first experimental evidence for nuclear processes came in 1896 with the discovery of radioactivity by the French physicist Antoine Henri Becquerel (1852–1908). Becquerel's earliest scientific interests were in the field of phosphorescence and absorption of light by crystals. In particular, at that time it was known that phosphorescent materials could emit light of one color after being exposed to light of a different color. In 1896 Wilhelm Conrad Röntgen (1845–1923) reported that a Crookes tube was capable of emitting a previously unknown type of penetrating radiation, which became known as x-rays. Becquerel was investigating the possibility that phosphorescent materials might emit similar penetrating radiation after exposure to light. Becquerel was working with some uranium salts that were known to be phosphorescent and found that they were capable of emitting penetrating radiation. He soon discovered, as well, that non-phosphorescent uranium salts also emitted penetrating radiation. He concluded, therefore, that the uranium salts were emitting the radiation without the need of being previously exposed to light and that the radiation originated from the uranium salt itself. Nearly forty years earlier (in 1857) the French photographic inventor Claude Félix Abel Niépce de Saint-Victor (1805–70) observed that photographic plates could be darkened when placed in the proximity of uranium salts. While Abel Niépce de Saint-Victor concluded that the plates were darkened by an 'invisible' form of radiation, it was Becquerel's observations that highlighted the scientific significance of this phenomenon. His early work, along with that of Rutherford, also showed that the radiation produced by different radioactive materials was differently affected by the presence of a magnetic field and therefore had different charges. Perhaps, most significantly, around 1900 Becquerel measured the mass-to-charge ratio of negatively charged particles from nuclear decay using the method developed by J J (Joseph John) Thomson (1856–1940) three years earlier to show that the so-called cathode rays were negatively charged particles which he referred to as 'corpuscles'. Thomson's 'corpuscles' were later renamed 'electrons', a name suggested earlier by George Johnstone Stoney (1826–1911). Becquerel discovered that the negatively charged decay particles had the same mass-to-charge ratio as Thompson's electrons and suggested that these particles were, in fact, the same.

Following Becquerel's work, there was intense scientific interest in radioactivity. Significant early work was done by Marie Skłodowska Curie (1867–1934) and Pierre Curie (1859–1906) who, in 1903, shared the Nobel Prize with Becquerel for work on radiation. Some important findings from research on radioactivity during this period included the discovery of new radioactive elements, polonium and radium, Pierre Curie's observation that radium continuously gave off heat and the clarification of three distinct types of radiation based on their charge—positive, negative and electrically neutral. It was Pierre Curie's observation of heat emission by a radioactive material that can be viewed as the first step toward the future development of nuclear power.

Einstein's mass–energy relation

The description of nuclear processes relies on the relationship between mass and energy. The simple principle of conservation of energy is sometimes considered to be a direct consequence of Newton's laws but remained a somewhat ambiguous concept well into the 19th century. The main difficulty in clarifying the concept of conservation of energy came from an incomplete understanding of the relationship between different forms of energy. In 1844 the Welsh scientist William Robert Grove (1811–96) hypothesized that a single 'force' as he called it (in modern terminology his term 'force' refers to 'energy') could be manifested in different forms such as heat, light, electricity, magnetism and mechanical energy. The term 'conservation of energy' is derived from Grove's work and was first used by William Rankine (1820–72) in 1850. The first *ab initio* proof of the law of conservation of energy, known as Noether's first theorem, was developed in 1915 by the German mathematician Amalie Emmy Noether (1882–1935).

In 1905 Albert Einstein (1879–1955) postulated the equivalence of mass and energy and the concept of 'the conservation of energy' became 'the conservation of mass–energy' (Einstein 1905). It is interesting to note that in Einstein's (1905) paper he did not present the well-known equation, $E = mc^2$, rather he expressed this concept in words as (English translation): *If a body gives off energy, L, in the form of radiation, then its mass will decrease by* L/c^2. It was only through subsequent work by Einstein and many others, including Max Planck (1858–1947), Johannes Stark (1874–1957), Gilbert Newton Lewis (1875–1946), Richard Chace Tolman (1881–1948), Hendrik Antoon Lorentz (1853–1928) and Max von Laue (1879–1960), that the current form of the mass–energy equation was adopted along with our current understanding of its meaning. It is the concept of mass–energy equivalence as stated by Einstein that forms the basis of our understanding of energy release by nuclear processes.

Aston's nuclear mass measurements

The application of Einstein's mass–energy relation to nuclear processes, for example, fission and fusion reactions, as well as the decay processes, requires a very accurate knowledge of nuclear masses or, as described in chapter 2, atomic masses. Perhaps the earliest experimental measurements of atomic masses were made by the German physicist Wilhelm Wien (1864–1928) beginning around the very end of the 19th century. Wien developed the so-called 'Kanalstrahlen' which was an early version of the mass spectrometer which utilized electric and magnetic fields to separate particles on the basis of their mass-to-charge ratio. During the early 20th century, Francis William Aston greatly improved the mass spectrometer for the purpose of separating the isotopes of chemical elements. In 1920 Aston showed that the mass of four ^1H nuclei was greater than the mass of one ^4He nucleus and this observation led, during the next two decades, to the development of a detailed theory to explain energy production in the Sun.

Development of quantum theory

The Austrian physicist Erwin Schrödinger (1887–1961) laid the foundations for quantum mechanics in his 1926 paper 'Quantisierung als Eigenwertproblem' (Quantization as an Eigenvalue Problem) in which he developed what is now known as the Schrödinger equation (Schrödinger 1926). Shortly thereafter, in 1927, the German physicist Friedrich Hermann Hund (1896–1997) discovered quantum tunneling while calculating the ground state of the double-well potential. Hund is best known for developing a set of rules (i.e. Hund's rules) for determining the quantum numbers of electrons in atomic ground states. The following year the Soviet-American physicist George Gamow (born Georgiy Antonovich Gamov,

1904–68) used the concept of tunneling along with the Schrödinger equation to describe alpha decay. Alpha decay was known to occur in some heavy nuclei where a particle of charge +2e was emitted. Gamow applied the Schrödinger equation to a potential along the lines of that shown in figure 2.5 to show how a positively charged alpha particle could tunnel through the Coulomb barrier and escape from a nucleus. The same principles were later applied to show how two positively charged nuclei could tunnel through the Coulomb barrier and fuse together.

Early nuclear theories

Until the mid-1920s it was commonly believed that a nucleus was comprised of a number of protons, to provide the observed mass, and a number of electrons, to electrically cancel some portion of the protons in order to provide the correct atomic number. The introduction of Bose–Einstein statistics (named for Satyendra Nath Bose (1894–1974) and Albert Einstein) in 1924–25 and Fermi–Dirac statistics (named for Enrico Fermi (1901–54) and Paul Dirac (1902–84)) in 1926 created difficulties in describing the properties of many nuclei using the hypothesis that the nucleus contained electrons. Shortly thereafter, the Swedish physicist Oskar Benjamin Klein (1894–1977) showed that from a quantum mechanical standpoint it would not be possible to confine an electron inside a nucleus. A simple argument that illustrates this point is based on the de Broglie wavelength of an electron. In 1924 Louis de Broglie (1892–1987) postulated that the wavelength, λ, associated with a particle of, m, is given in terms of the particle's momentum, p, as

$$\lambda = \frac{h}{p} \tag{3.23}$$

where h is the Planck constant. For relativistic particles Einstein's relation gives

$$E^2 = m_0{}^2c^4 + p^2c^2 \tag{3.24}$$

where m_0 is the particles rest mass and c is the speed of light. If the particle's kinetic energy is much greater than its rest mass energy, then equation (3.24) reduces to

$$E = pc \tag{3.25}$$

and solving for energy in equation (3.23) gives

$$E = \frac{hc}{\lambda} \tag{3.26}$$

The maximum energy available to bind an electron within a nucleus is given by Coulomb's law as

$$E = \frac{Ze^2}{4\pi\varepsilon_0 r} \tag{3.27}$$

For an electron to be bound inside a nucleus the energy in equation (3.27) should be greater than the energy in equation (3.23) where λ is the nuclear radius. In the most advantageous case, i.e. for a large nucleus such as ^{238}U, the electron kinetic energy would be ~170 MeV compared to a Coulomb energy of ~18 MeV. For small nuclei such as ^{2}H the comparison would be between a kinetic energy of ~800 MeV and a Coulomb energy of ~1 MeV. In the above calculations it is convenient to use the constants $hc = 1240$ MeV fm and $\frac{e^2}{4\pi\varepsilon_0} = 1.44$ MeV fm, and the empirical relation for the nuclear radius (in fm) $r = 1.2\ A^{1/3}$. This simple calculation clearly shows that it is not possible to confine an electron within the dimensions of a nucleus by means of the Coulomb interaction.

Discovery of the neutron

The true nature of the nuclear composition became known in 1932 with the discovery of the neutron by James Chadwick (1891–1974) (Chadwick 1932). Chadwick's work followed directly from the work of German physicist Walther Bothe (1891–1957) and his student, Herbert Becker. Bothe and Becker used alpha particles that were produced by the decay of polonium,

$$^{210}\text{Po} \rightarrow {}^{206}\text{Pb} + \alpha \tag{3.28}$$

to irradiate beryllium. This caused the reaction

$$\alpha + {}^{9}\text{Be} \rightarrow {}^{12}\text{C} + \text{n} \tag{3.29}$$

Bothe and Becker observed that the emitted particle was electrically neutral and, as at that time the neutron was not known to exist, assumed that it was a gamma ray. Shortly after Bothe and Becker's work, Irène Joliot-Curie (1897–1956) and Jean Frédéric Joliot-Curie (1900–58) continued experimental work along these lines. They used the particles produced by the reaction in equation (3.29) to irradiate a sample of paraffin. Paraffin is a hydrocarbon with a high hydrogen content and, therefore, provides a target that is rich in protons. The neutrons scatter elastically from the protons and thereby impart energy and momentum to the protons. The Joliot-Curies observed the emission of protons with an energy of 5.3 MeV from this experiment. A simple apparatus to perform experiments as described by the Joliot-Curies is shown in figure 3.1. An analysis based on the conservation of energy and momentum indicated that the neutral particle in equation (3.28) could not be a massless photon but was, instead, a neutral particle with mass.

Chadwick continued the study of the interactions of neutrons with nuclei using targets containing a variety of different elements including nitrogen, oxygen, helium and argon and showed that these results were consistent with the existence of a neutral massive particle. In order to determine the mass of this particle he conducted an experiment where the beryllium target was replaced by a boron target to produce neutrons by the reaction

$$\alpha + {}^{11}\text{B} \rightarrow {}^{14}\text{N} + \text{n} \tag{3.30}$$

Based on the conservation of energy (i.e. rest mass energy plus kinetic energy) the reaction in equation (3.30) can be analyzed as

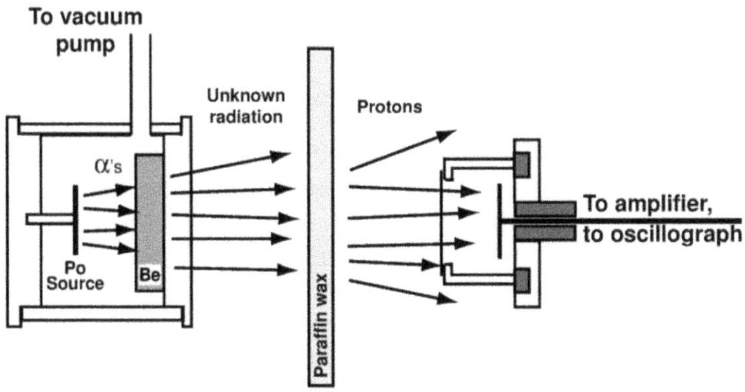

Figure 3.1. Schematic of an apparatus to observe the reaction shown in equation (3.29) and the subsequent elastic scattering of neutrons and protons. This Chadwick Neutron Discovery image has been obtained by the author from the Wikimedia website https://commons.wikimedia.org/wiki/File:Chadwick_Neutron_Discovery.png where it was made available by Bdushaw under a CC BY-SA 4.0 license. It is included within this book on that basis. It is attributed to Bdushaw.

$$\frac{1}{2}m_\alpha {v_\alpha}^2 + m_\alpha c^2 + m_\mathrm{B} c^2 = \frac{1}{2}m_\mathrm{N} {v_\mathrm{N}}^2 + m_\mathrm{N} c^2 + \frac{1}{2}m_\mathrm{n} {v_\mathrm{n}}^2 + m_\mathrm{n} c^2 \tag{3.31}$$

where m and v are mass and velocity, respectively, and the subscripts refer to alpha particle (α), neutron (n), boron nucleus (B) and nitrogen nucleus (N). Equation (3.31) may be solved for the mass of the neutron as

$$m_\mathrm{n} = \frac{\frac{1}{2}m_\alpha {v_\alpha}^2 + m_\alpha c^2 + m_\mathrm{B} c^2 - \frac{1}{2}m_\mathrm{N} {v_\mathrm{N}}^2 - m_\mathrm{N} c^2}{\left(c^2 + \frac{{v_\mathrm{n}}^2}{2}\right)} \tag{3.32}$$

All of the terms on the right-hand side of this equation were known to Chadwick or could be measured experimentally except the velocity of the neutron, v_n, and the velocity of the nitrogen nucleus, v_N. In order to determine the velocity of the neutron, which Chadwick assumed had a mass that was similar to the proton mass, he bombarded a sample containing protons (i.e. hydrogen atoms), as in figure 3.1, with the neutrons from the reaction in equation (3.30). The neutrons and protons scattered elastically and the velocity of the protons, which assumed the velocity of the neutrons before the collision, could be readily measured. Once the velocity of the neutron was known, the velocity of the nitrogen nucleus could be determined using conservation of momentum. From this work, Chadwick obtained a neutron mass of 938±1.8 MeV/c^2. This may be compared with the currently accepted value of 939.57 MeV/c^2.

For some time, the exact nature of the neutron was not known. A simple model based on the knowledge at the time viewed the neutron as a tightly bound composite particle consisting of a proton and an electron. While this model correctly predicted the neutron charge, it failed in other respects. The necessary binding energy was larger than the observed electron energies from beta-decay, as described below. At the time, however, assumptions concerning the identity of the neutron were based on the neutron's mass relative to the proton's mass (938.27 MeV/c^2). If the neutron consisted of a bound proton–electron pair, then its mass would be less than the mass of the proton. The reason for this is that the binding energy, as discussed in the previous chapter, would reduce the mass of the system by more than the mass of the electron. Chadwick's initial neutron mass measurements, as given above, were not sufficiently accurate to make this determination. However, Chadwick's subsequent measurements were able to show that the neutron mass was, in fact, greater than the proton mass, indicating that the neutron was not a composite particle. It would, however, be another three decades before the exact relationship between the proton and neutron would be understood on the basis of quark theory.

Beta decay

The final point that needed to be resolved in order to implement Eddington's suggestion for nuclear fusion as a source of energy in the Sun dealt with the identity of the proton and the neutron, as the suggested fusion process required the conversion of two protons into two neutrons.

Beta particles, that is, electrons, were discovered in 1897 by Thomson and were known since around 1900 from the work by Rutherford and Becquerel to be produced in certain radioactive decay processes. In 1901 Rutherford and Frederick Soddy (1877–1956) showed that beta-decay resulted in a transmutation of one element into another. This work was extended by Soddy and (independently) by Kazimierz Fajans (1887–1975) and in 1913 the so-called radioactive displacement law was proposed which states that beta-decay (now known as β^- decay) from one element produces another element which is one position to the right of it in the periodic table.

With Chadwick's discovery of the neutron, the details of the β^- decay process became clear. A neutron in the nucleus was changed into a proton and an electron was emitted. Following from Chadwick's determination that a neutron was not a proton–electron bound pair, it also became obvious that β^- decay was a process that dealt with subatomic particles on a very fundamental level.

In 1911 Lise Meitner (1878–1968) and Otto Hahn (1879–1968) measured the energy spectrum of electrons emitted during the β^- decay and showed that these electrons had a continuous energy distribution. The continuum spectrum observed for electrons from β^- decay was in sharp contrast to the discrete line spectrum observed for ^{4}He nuclei emitted during an alpha decay process. As it was known that the transitions involved in the β^- decay process were quantized, it was assumed that the electrons that were produced would be of a single energy. For nearly two decades it was believed that the electrons that were emitted during β^- decay were losing energy somehow before they were being detected.

An additional inconsistency concerning β^- decay that remained unresolved at the time dealt with the conservation of angular momentum. This became obvious with the introduction of Bose–Einstein statistics and Fermi–Dirac statistics in the mid-1920s. The mass number of a nucleus was not changed during β^- decay, but an electron of spin 1/2 was emitted and this process would violate conservation of angular momentum.

In 1930 Wolfgang Pauli (1900–58) suggested that the nucleus contained a light uncharged particle that was being emitted along with the electron during the β^- decay process and that the transition energy was being shared between these two particles during the decay. Pauli called the uncharged particle the 'neutron'. However, as Chadwick subsequently discovered the heavy uncharged particle which became known as the neutron, Fermi suggested that Pauli's light uncharged particle should be named the 'neutrino'. If the neutrino was a spin 1/2 particle (i.e. a fermion) then Pauli's suggestion would also resolve the apparent violation of angular momentum conservation. The basic process of β^- decay could now be written as

$$\mathrm{n} \rightarrow \mathrm{p} + \mathrm{e}^- + \bar{\nu}_\mathrm{e} \tag{3.33}$$

where $\bar{\nu}_\mathrm{e}$ is an electron antineutrino. This process is energetically favorable and will occur spontaneously with a half-life of about 10.3 min This means that a free neutron is not stable but neutrons which are bound in a nuclear potential can be stabilized.

An understanding of β^- decay, in itself, did not fill in all of the gaps in Eddington's proposal concerning solar energy production by hydrogen fusion. Eddington's hypothesis required that two protons would be changed into two neutrons during the fusion, the opposite of equation (3.33). The first step in describing the details of this process occurred in 1929. In that year, the Soviet physicist Dmitri Skobeltsyn (1892–1990) was using a cloud chamber to investigate cosmic rays. During his studies he observed particles that behaved like electrons but curved in the opposite direction in the presence of a magnetic field, indicating that they had a charge opposite to that of the electron. However, at the time, the significance of this observation was not appreciated.

In 1932 Carl David Anderson (1905–91) was conducting cloud chamber experiments on cosmic rays similar to those of Skobeltsyn and again observed electron-like particles with a positive charge. Anderson's careful measurements showed that this positively charged particle, which became known as the positron, had the identical mass-to-charge ratio as the electron. This new particle provided the first experimental evidence for the existence of antimatter and was consistent with predictions made in 1928 by Dirac that both negatively charged and positively charged versions of the electron could exist.

While Anderson's work showed the existence of the positron in cosmic rays, it was the 1934 experiments of Irène and Frédéric Joliot-Curie that showed the positron's relationship to beta-decay. The Joliot-Curies bombarded aluminum with alpha particles and observed the reaction

$$^{4}\mathrm{He} + {}^{27}\mathrm{Al} \rightarrow {}^{30}\mathrm{P} + \mathrm{n} \tag{3.34}$$

The nuclide ^{30}P is unstable and decays with a half-life of about 2.5 min by the process

$$^{30}\mathrm{P} \rightarrow + \mathrm{e}^+ + \nu_\mathrm{e} \tag{3.35}$$

where ν_e is the electron neutrino. This decay, known as β^+ decay, corresponds to the conversion of a proton into a neutron along with the emission of a positron and a neutrino. The basic process is, therefore,

$$p \to n + e^+ + \nu_e \tag{3.36}$$

This process is not energetically favorable and will not occur spontaneously for a free proton, meaning that the proton is a stable particle. However, protons in nuclei can be destabilized by their interactions with the other nucleons.

In addition to β^- decay and β^+ decay, a third beta-decay process, which is equivalent to β^+ decay is also possible. This decay process, known as electron capture, was first hypothesized in 1934 by Gian Carlo Wick (1909–92) and corresponds to the reaction

$$p + e^- \to n + \nu_e \tag{3.37}$$

We can see that this process will not be energetically favorable for a free proton, as the mass of the neutron is greater than the combined mass of the proton and electron. However, when the proton is bound in a nucleus the nuclear binding energy can make this process favorable. We can see how electron capture in equation (3.37) relates to β^+ decay in equation (3.36) by adding an electron to each side of equation (3.36). The left-hand side of the equation will become the same as the left-hand side of equation (3.37). The right-hand side will be the same as the right-hand side of equation (3.37) except it will contain an additional e^+e^- pair. This electron–positron pair will annihilate and produce gamma rays with a total energy of $2m_ec^2$. There is also a small additional energy term related to the electronic binding energy of the captured electron. Thus, when the proton is bound in a nucleus, electron capture can be energetically favorable when β^+ decay is not. The first experimental observation of electron capture was made in 1937 by Luis Alvarez (1911–88) for the decay process

$$^{48}V + e^- \to {}^{48}Ti + \nu_e \tag{3.38}$$

From the above discussion, we can see that by 1937 all of the fundamental nuclear physics that was required to understand fusion processes in the Sun and other stars was known. The first detailed descriptions of stellar fusion came in the late 1930s, as described in the next section.

3.5 Stellar fusion processes

Around 1938 Hans Bethe (1906–2005) and Carl Friedrich von Weizsäcker (1912–2007) independently proposed the CNO cycle or carbon–nitrogen–oxygen cycle as a mechanism for fusing four hydrogen nuclei into one helium nucleus as a means of energy production in stars (Bethe 1939, von Weizsäcker 1937, 1938). This fusion cycle is sometimes referred to as the Bethe cycle or Bethe–Weizsäcker cycle. A ^{12}C nucleus is used as a catalyst for this process which includes four fusion reactions between protons and heavier nuclei along with two β^+ decays to convert two of the protons to neutrons. The CNO cycle is described as follows:

$$p + {}^{12}C \to {}^{13}N + \gamma \qquad 1.95\ \text{MeV} \tag{3.39}$$

$$^{13}N \rightarrow {}^{13}C + e^{+} + \nu_e \text{ (half-life 598 s)} \qquad 1.20 \text{ MeV} \tag{3.40}$$

$$p + {}^{13}C \rightarrow {}^{14}N + \gamma \qquad 7.54 \text{ MeV} \tag{3.41}$$

$$p + {}^{14}N \rightarrow {}^{15}O + \gamma \qquad 7.35 \text{ MeV} \tag{3.42}$$

$$^{15}O \rightarrow {}^{15}N + e^{+} + \nu_e \text{ (half-life 122 s)} \qquad 1.73 \text{ MeV} \tag{3.43}$$

$$p + {}^{15}N \rightarrow {}^{12}C + {}^{4}He \qquad 4.96 \text{ MeV} \tag{3.44}$$

The energy associated with each process is shown, along with the half-lives of the two β^+ decays. This cycle is shown graphically in figure 3.2.

The energy from the CNO cycle is released in the form of kinetic energy of the reaction by-products, for all processes, and in the form of γ-ray energy from nuclear de-excitations, such as in equations (3.39), (3.41) and (3.42). The total energy release for the cycle is about 24.7 MeV. The two positrons created in the β^+ decays in equations (3.40) and (3.43) will annihilate with electrons in the stellar environment by the process

$$e^{+} + e^{-} \rightarrow 2\gamma \tag{3.45}$$

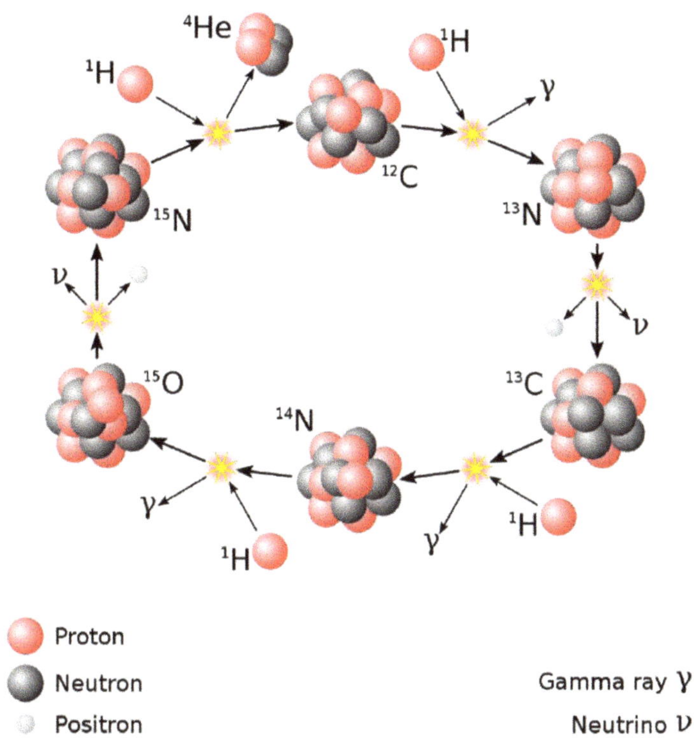

Figure 3.2. Graphical representation of the CNO cycle. This CNO Cycle image has been obtained by the author from the Wikimedia website https://commons.wikimedia.org/wiki/File:CNO_Cycle.svg where it is stated to have been released into the public domain. It is included within this book on that basis.

releasing an additional $4m_ec^2 = 2.04$ MeV of energy, giving a total energy release of about 26.7 MeV per CNO cycle. Some of this energy (about 1.7 MeV per cycle) is carried away as kinetic energy of the electron neutrinos in equations (3.40) and (3.43). Since the neutrinos have an extremely low interaction cross section, this energy is lost, giving a net energy of about 25 MeV per CNO cycle that can contribute to the stellar luminosity.

At the time the CNO cycle was proposed, it was believed that the Sun consisted of about 10% nitrogen. Thus, Bethe suggested that this cycle would be the principal source of energy in the Sun. The total nitrogen concentration in the Sun is actually less than half of one percent meaning that, although the CNO process can occur in the Sun, it is not the dominant mechanism for energy production. It is now known that the CNO cycle is the dominant process for energy production in stars that are more massive than about 1.3 solar masses.

The mechanism by which the Sun produces most of its energy was discovered by Bethe in 1939 (Bethe 1939). This reaction, referred to as the proton–proton reaction (or p–p reaction) begins with the fusing of two protons along with an accompanying β^+ decay to produce deuterium. The complete reaction to produce ^{4}He is as follows:

$$\mathrm{p} + \mathrm{p} \rightarrow \mathrm{d} + \mathrm{e}^+ + \nu_\mathrm{e} \qquad 0.42\ \mathrm{MeV} \tag{3.46}$$

$$\mathrm{p} + \mathrm{d} \rightarrow {}^3\mathrm{He} + \gamma \qquad 5.49\ \mathrm{MeV} \tag{3.47}$$

$$^3\mathrm{He} + {}^3\mathrm{He} \rightarrow {}^4\mathrm{He} + 2\mathrm{p} \qquad 12.86\ \mathrm{MeV} \tag{3.48}$$

where d is the deuteron, i.e. the ^{2}H nucleus. This process is the simplest and also the dominant process by which hydrogen is fused into helium in the Sun. It is, therefore, referred to as the p–p I cycle. The energy associated with each process is shown in the equation. The first two processes occur twice for each proton–proton cycle and the last reaction occurs once. This gives a total energy for the cycle of 24.7 MeV. The positrons in equation (3.46) will annihilate with electrons to give an additional 2.04 MeV of energy. Thus, the total energy release is 26.7 MeV. This is the same as the total energy release for the CNO cycle and shows that the carbon in the CNO cycle merely acts as a catalyst for the fusing of hydrogen into helium. The energy release in the processes in equation (3.46) is about 15% of the total energy release in equations (3.40) and (3.43). Since p–p fusion occurs twice in the p–p I cycle, the energy lost to neutrinos is only about 0.5 MeV for the p–p I reaction compared to 1.7 MeV for the CNO cycle. Thus, the p–p I reaction provides about 26.2 MeV to the Sun's luminosity. Figure 3.3 shows a graphical representation of the p–p I reaction.

The p–p I cycle is responsible for about 85% of the energy produced by the proton–proton reaction in the Sun. Bethe also described the other possible process that begin with equation (3.46) that occur in the Sun. These proton–proton processes are shown in figure 3.4, along with their respective branching ratios. Most all of the energy produced by the proton–proton cycle that is not the result of the p–p I cycle is

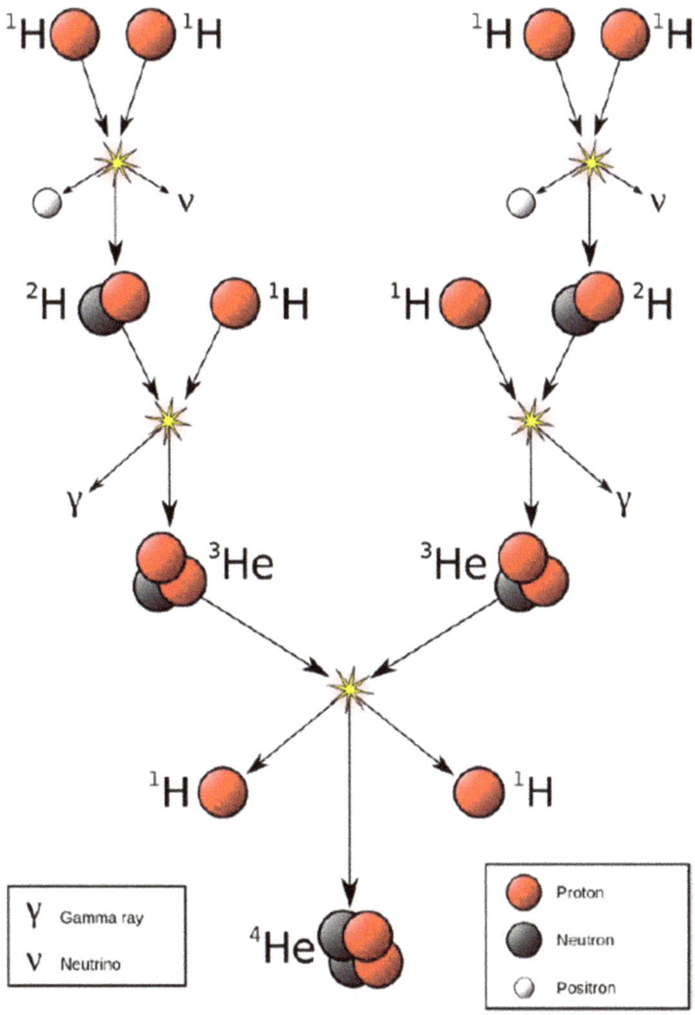

Figure 3.3. Graphical representation of the proton–proton (p–p I) cycle. This p–p I Cycle image has been obtained by the author from the Wikimedia website https://commons.wikimedia.org/wiki/File:Fusion_in_the_Sun.svg where it is stated to have been released into the public domain. It is included within this book on that basis.

produced by the p–p II cycle, as shown in the figure. Energy production by other routes is minimal. In all cases, the total energy produced in one proton–proton cycle is the same, 26.7 MeV, although the amount lost to neutrinos varies between the different cycles because of the energy associated with the relevant β^+ decays. Neutrino energy loss is about 1.1 MeV per cycle for the p–p II reaction and about 7.5 MeV per cycle for the p–p III reaction.

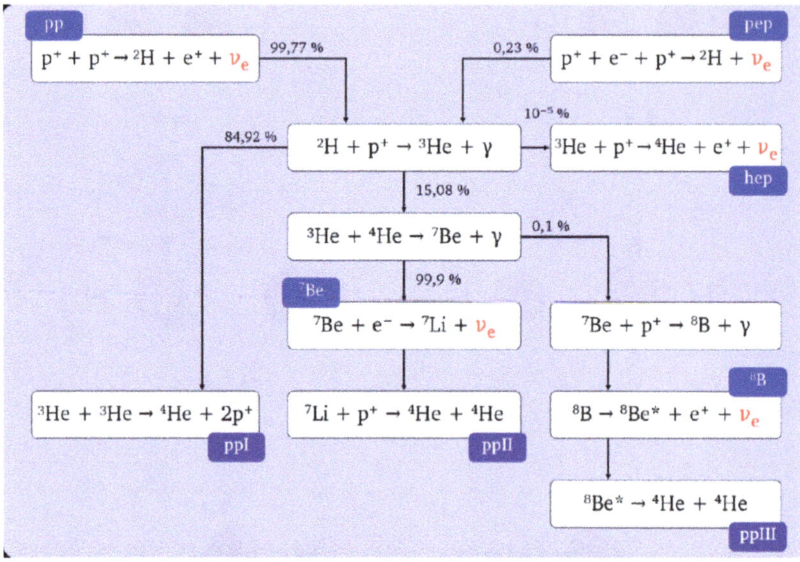

Figure 3.4. The various branches of the proton–proton cycle. This proton–proton cycle image has been obtained by the author from the Wikimedia website https://commons.wikimedia.org/wiki/File:Proton_proton_cycle.svg where it was made available by Dorottya Szam under a CC BY-SA 2.5 license. It is included within this book on that basis. It is attributed to Dorottya Szam.

3.6 Laboratory fusion

The discovery of laboratory fusion

The utilization of nuclear fusion as a source of energy requires the creation of appropriate fusion conditions here on Earth. In the early 1930s there was a concentrated effort directed by Ernest Rutherford at the Cavendish Laboratory to advance the understanding of the nucleus. The researchers involved in this work included Patrick Blackett (1897–1974), James Chadwick, John Cockcroft (1897–1967), Pyotr Kapitsa (1894–1984), Mark Oliphant (1901–2000) and Ernest Walton (1903–95). During this era Oliphant constructed a particle accelerator capable of accelerating protons to an energy of 600 keV. He first used this device to confirm results of Cockcroft and Walton from a year or so earlier on the transmutation of elements by high energy proton irradiation.

Around 1934, Oliphant conducted a series of experiments involving the bombardment of various nuclei with high energy deuterons produced in his particle accelerator (Oliphant *et al* 1935). During these experiments he discovered that deuterons reacted with other deuterons, as well as ^{3}H nuclei and ^{3}He nuclei, exothermically, thereby releasing excess energy in the process. The reasons for these exothermic reactions can be seen by an inspection of figures 2.3 and 2.4 where the fusing of deuterons with other light nuclei releases excess binding energy.

Early concepts in controlled nuclear fusion

In the mid-1940s interest began in the use of nuclear fusion as a source of energy. The major task in implementing such an approach is to raise the temperature of the fusion fuel to a point where nuclei have a sufficiently high probability of tunneling through the Coulomb barrier and fusing. In order to achieve such a condition, the fuel is in the form of a plasma. The method of containment of the plasma becomes a critical issue. The 1934 work on plasma physics by Willard Harrison Bennett (1903–87) formed an important component of early work on the design of a suitable fusion reactor. In that year Bennett published a theoretical paper (Bennett 1934) describing what he referred to as 'self-focusing' which is now referred to as the pinch effect (often referred to as the 'Bennett pinch'). The basic pinch phenomenon arises as a result of the interaction between an electric current in a plasma and the magnetic field that the current produces. Bennett's work on the pinch effect remained largely unknown until it became an important part of the post-World War II interest in controlled nuclear fusion. It was realized that the pinch effect could be created by running a current through the plasma and this would not only be effective at increasing plasma density and temperature but would also serve to isolate the walls of the chamber from the extreme temperatures necessary for fusion.

An obvious problem with a linear pinch geometry, as suggested by figure 3.5, is that the plasma is unconstrained at the ends of the machine. There are a few possible solutions to this problem. One solution is to enclose the device in a magnetic solenoid which produces an increased magnetic field at the ends of the plasma channel as shown in figure 3.6. In this arrangement the plasma particles will be reflected at the ends of the tube. This type of device is referred to as a magnetic mirror machine.

Another possible solution to the problem of plasma confinement is to bend the chamber into a torus, as shown in figure 3.7. In order to produce a toroidal current in the plasma, poloidal magnetic coils are wrapped around the torus as shown in the figure. This would seem to be an appropriate means of confining the plasma along the toroidal current direction and prevent interactions with the chamber walls. However, there is a major flaw in this design, which was pointed out by Fermi in

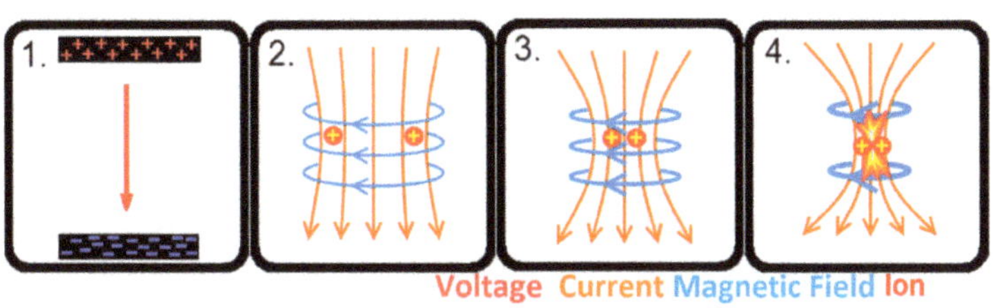

Figure 3.5. Illustration of the pinch effect: (1) a voltage is applied across a region containing a plasma, (2) current flows through the plasma in the axial direction (i.e. the z-direction) thereby creating an azimuthal magnetic field, (3) the Lorentz force between the current and the magnetic field that it produces pushes the lines of current together, thus, pinching the plasma and (4) pinched ions fuse and release energy. This Pinch Effect image has been obtained by the author from the Wikimedia website https://commons.wikimedia.org/wiki/File:Plasma_Pinch_Mechanism.png where it was made available by WikiHelper2134 under a CC BY-SA 3.0 license. It is included within this book on that basis. It is attributed to WikiHelper2134.

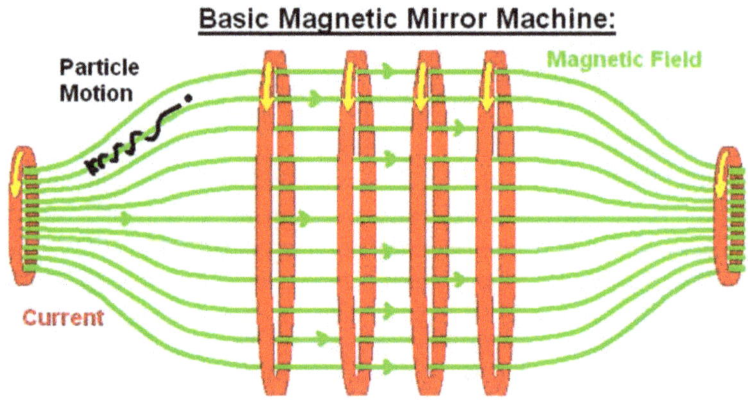

Figure 3.6. Basic geometry of a magnetic mirror machine. This Magnetic Mirror Machine image has been obtained by the author from the Wikimedia website https://commons.wikimedia.org/wiki/File:Basic_Magnetic_Mirror.jpg where it was made available by WikiHelper2134 under a CC BY-SA 3.0 license. It is included within this book on that basis. It is attributed to WikiHelper2134.

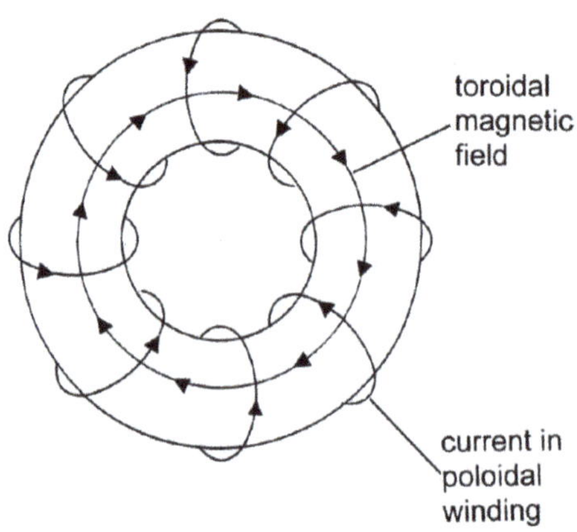

Figure 3.7. Toroidal geometry for a fusion reactor showing currents in the poloidal magnet windings and the resulting toroidal current in the plasma.

around 1950. The magnetic coils, as seen in figure 3.7, are closer together on the inside of the torus than they are on the outside of the torus. This will give rise to magnetic field gradients within the torus which will affect the induced plasma current. Since electrons and ions in the plasma have opposite charge, they will be affected differently by the magnetic field inhomogeneities and this will cause the particles to drift away from the center of the plasma chamber.

In 1951 Lyman Spitzer, Jr (1914–97) proposed a solution to the problem of particle drift in the toroidal fusion chamber. He suggested twisting the torus into a figure-eight shape as shown in figure 3.8. In this arrangement particles which drift in

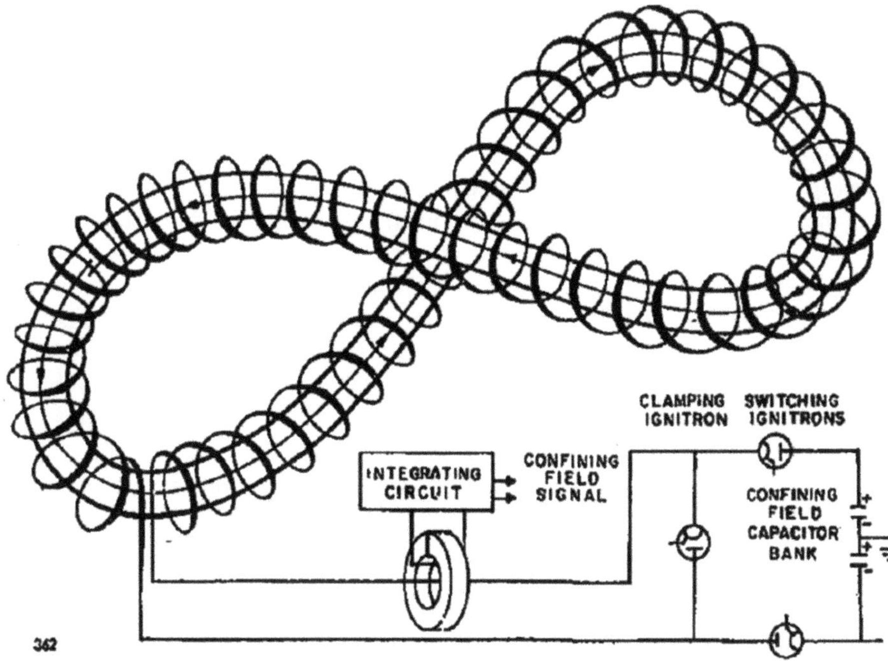

Figure 3.8. Diagram of a simple version of the stellarator proposed by Spitzer. Reproduced from Coor *et al* (1958) with permission of AIP Publishing.

one direction in one half of the figure-eight will drift in the opposite direction in the other half. In this way the drifts will tend to average out over the complete cycle and the charge separation, as occurs in the simple torus, will cancel out. Spitzer referred to this device as a stellarator.

In the early 1950s serious experimental work began in the UK and, shortly thereafter, in the United States to construct a controlled nuclear fusion reactor with the ultimate goal of producing useable energy. Research in this area throughout the 1950s and much of the 1960s concentrated on the three approaches as described above, pinch machines, mirror machines and the stellarator. Much of the early work (up to around 1958) in the United States has been nicely summarized in the book by Amasa Bishop, *Project Sherwood—The U.S. Program in Controlled Fusion* (Bishop 1958).

The early history of experimental nuclear fusion research in the Soviet Union is quite interesting. Much of the early initiative, both in controlled magnetic confinement fusion and in the development of thermonuclear fusion weapons, came from a letter written by Oleg Lavrent'ev (1926–2011) to the Central Committee of the Communist Party of the Soviet Union in 1950. The details of this communication were summarized half a century later by Bondarenko (Bondarenko 2001). Lavrent'ev was, at the time, a sergeant in the Red Army, who was self-educated in physics. The letter, which was directed to Soviet physicist Andrei Sakharov (1921–89) detailed Lavrent'ev's ideas concerning the use of electrostatic fields to confine a plasma in a controlled nuclear fusion reactor, as well as his ideas on the use of a fission bomb to ignite a fusion weapon based on lithium deuteride. While Sakharov felt that electrostatic confinement was not a viable approach to controlled nuclear fusion, he was inspired by Lavrent'ev's ideas and considered the possibility of using magnetic fields to contain a fusion plasma. In 1951, Sakharov, in collaboration with Soviet physicist Igor Tamm (1895–1971) proposed the design of the tokamak, which has since become the most successful approach to fusion energy, as described in detail in the next chapter.

Inertial confinement fusion developed somewhat later than magnetic confinement fusion. Early ideas in the 1950s focused mainly on the possibility of utilizing fusion bombs to generate useful energy, for example, the use of a hydrogen bomb to heat water inside a cavern to produce steam to drive turbine/generators. Work in this area remained very speculative. It was around 1962, shortly after the invention of the laser, that the idea of laser fusion as a means of inertial confinement began to develop. However, at this time, lasers did not have sufficient power output to make this approach practical. It was in the mid- to late-1970s before serious experimental efforts began in this field. The subsequent development of this field of fusion research is discussed in chapter 5.

3.7 Thermonuclear weapons

The development of the atomic bomb, that is a fission weapon, officially began in late 1942 in the United States with the support of Canada and the United Kingdom. The project, which became known as the 'Manhattan Project' developed plutonium and uranium fission weapons. The first detonation of a fission weapon occurred on 16 July 1945 at the Alamogordo Bombing and Gunnery Range in New Mexico. This detonation, known as the 'Trinity Test' utilized an implosion-type plutonium bomb.

In 1941, even before the official beginning of the 'Manhattan Project', discussions between Enrico Fermi and Edward Teller (1908–2003) laid out the basic idea of thermonuclear weapons. These weapons are sometimes known as 'hydrogen bombs', 'H-bombs' or 'fusion weapons'. Fermi suggested to Teller that a small fission bomb could be used as a trigger for a larger fusion bomb. The design of the first operational device was developed by Teller in collaboration with the Polish physicist Stanisław Ulam (1909–84). The thermonuclear weapon consists of two stages. The first stage is the fission weapon which is detonated and subsequently ignites the second stage consisting of a fusion weapon. As deuterium–tritium has a larger fusion cross section than deuterium–deuterium, it is the choice of fusion fuels for such devices, although the earliest experimental tests of thermonuclear weapons used pure deuterium in order to simplify construction. Current weapons typically use solid lithium deuteride, typically enriched in ^{6}Li, as a fusion fuel. Neutrons from the detonation of the fission primary stage are incident on the lithium in the lithium deuteride and produce tritium, ^{3}H, by the reaction

$$n + {}^{6}Li \rightarrow {}^{3}H + {}^{4}He \tag{3.49}$$

A common design of a modern Teller–Ulam fusion weapon is illustrated in figure 3.9. The diagram shows that the primary stage consists of a conventional high explosive outer lens that is used to compress the fissile core into supercriticality. The fissile hollow core also includes a tritium or deuterium–tritium fusion booster inner core. This booster undergoes fusion when compressed by the fissile hollow core and produces extra neutrons to facilitate fission reactions. The secondary stage consists of a plutonium 'sparkplug' surrounded by the lithium deuteride fusion fuel. The sequence of events that occurs when the thermonuclear weapon is detonated is described in figure 3.10.

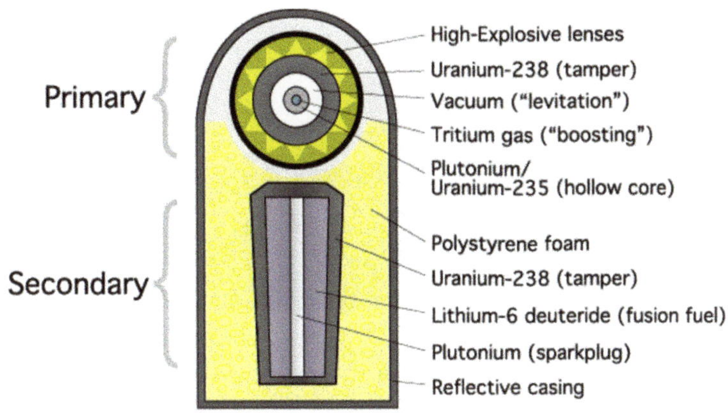

Figure 3.9. Basic design of a Teller–Ulam fusion weapon. This Teller–Ulam Fusion Weapon image has been obtained by the author from the Wikimedia website https://commons.wikimedia.org/wiki/File:Teller-Ulam_device.png where it is stated to have been released into the public domain. It is included within this book on that basis.

Figure 3.10. Detonation sequence of a thermonuclear weapon. (a) Weapon before detonation; components as described in figure 3.9. (b) Detonation of high explosive compresses fissile plutonium/uranium hollow core into supercriticality and inducing a fission chain reaction. (c) X-rays from primary fission stage reflect from casing and irradiate polystyrene foam. (d) Polystyrene foam becomes a plasma thereby compressing the lithium deuteride fusion fuel and plutonium sparkplug. Plutonium sparkplug begins fissioning. (e) Neutrons from plutonium fission create tritium from ^{6}Li according to equation (3.49) and the resulting deuterium–tritium fuel undergoes fusion. This Detonation Sequence image has been obtained by the author from the Wikimedia website https://commons.wikimedia.org/wiki/File:Teller-Ulam_device_firing_sequence.png, where it is stated to have been released into the public domain. It is included within this book on that basis.

Figure 3.11. The Ivy Mike explosion on 1 November 1952. This Ivy Mike image has been obtained by the author from the Wikimedia website https://commons.wikimedia.org/wiki/File:%22Ivy_Mike%22_atmospheric_nuclear_test_-_November_1952_-_Flickr_-_The_Official_CTBTO_Photostream.jpg where it was made available by The Official CTBTO Photostream under a CC BY 2.0 license. It is included within this book on that basis. It is attributed to The Official CTBTO Photostream.

The first operational thermonuclear weapon was detonated by the United States on 1 November 1952 on the island of Elugelab in Enewetak Atoll. This atoll is now part of the Marshall Islands, an independent island nation. This test device was named Ivy Mike and consisted of a plutonium fission primary stage and a liquefied deuterium fusion secondary stage. Ivy Mike was a test of concept device and was not intended as a deployable weapon as it weighed 74 tons and its dimensions were 27 m × 14 m × 19 m. A photograph of the Ivy Mike explosion is shown in figure 3.11. The history of the development of the hydrogen bomb is described in detail in the book by Richard Rhodes, *Dark Sun—The Making of the Hydrogen Bomb* (Rhodes 1995).

References and suggestions for further reading

Bennett W H 1934 Magnetically self-focussing streams *Phys. Rev.* **45** 890–97

Bethe H A 1939 Energy production in stars *Phys. Rev.* **55** 434–56

Bishop A S 1958 *Project Sherwood—The U.S. Program in Controlled Fusion* (Boston, MA: Addison-Wesley)

Bondarenko B D 2001 Role played by O A Lavrent'ev in the formulation of the problem and the initiation of research into controlled nuclear fusion in the USSR *Phys. Usp.* **44** 844–51

Byatt W J, Rogers F T Jr and Waltner A 1949 On the beta-particle spectrum from the decay of tritium *Phys. Rev.* **75** 909

Chadwick J 1932 The existence of a neutron *Proc. R. Soc.* **A136** 692–708

Coor T, Cunningham S P, Ellis R A, Heald M A and Kranz A Z 1958 Experiments on the ohmic heating and confinement of plasma in a stellarator *Phys. Fluids* **1** 411–20

Darwin C R 1859 *On the Origin of Species* (London: John Murray) (Note that the complete title of the first edition was On the Origin of Species by Means of Natural Selection, or the Preservation of Favoured Races in the Struggle for Life)

Eddington A S 1920 The internal constitution of the stars *Sci. Mon.* **11** 297–303

Einstein A 1905 Ist die Trägheit eines Körpers von seinem Energieinhalt abhängig? *Ann. Phys.* **18** 639–43

Hirsch R L 1967 Inertial-electrostatic confinement of ionized fusion gases *J. Appl. Phys.* **38** 4522–34

Oliphant M L E, Kempton A R and Rutherford E 1935 The accurate determination of the energy released in certain nuclear transformations *Proc. R. Soc.* **A149** 406–16

Rhodes R 1995 *Dark Sun—The Making of the Hydrogen Bomb* (New York: Simon & Schuster)

Schrödinger E 1926 Quantisierung als Eigenwertproblem *Ann. Phys.* **384** 273–376

von Weizsäcker C F 1937 Über Elementumwandlungen in Innern der Sterne I *Phys. Z.* **38** 176–91

von Weizsäcker C F 1938 Über Elementumwandlungen in Innern der Sterne II *Phys. Z.* **39** 633–46

Wollaston W H 1802 A method of examining refractive and dispersive powers, by prismatic reflection *Phil. Trans. R. Soc.* **92** 365–80

Chapter 4

Magnetic confinement fusion

4.1 Introduction

As noted in the previous chapter, work up until the 1960s on controlled nuclear fusion for the purpose of providing a commercial source of energy focused on three approaches, pinch machines, mirror machines and the stellarator. Some research has continued to the present day on pinch machines, which are classified as inertial confinement devices. The *z*-pinch machine, at Sandia National Laboratories in Albuquerque, New Mexico is the most notable example and is described in some detail in chapter 6. However, this research is aim mainly at understanding nuclear fusion weapons and not at energy production. Most research activities on mirror machines ended in the 1980s when it became apparent that circular machines such as the tokamak were less susceptible to the instability problems that machines with a linear geometry faced. A brief description of previous mirror machine research has been presented in the previous chapter. While interest in stellarators dwindled after the introduction of the tokamak, there has been renewed activity in this area in recent years. The present chapter overviews the basic physics of controlled nuclear fusion devices which employ magnetic confinement. Following this overview, research on tokamaks is reviewed along with an update on recent stellarator research. Some other, less conventional, approaches to controlled nuclear fusion that employ magnetic fields are discussed in chapter 6.

4.2 Achieving fusion conditions

The goal of creating a useful nuclear fusion reactor is, obviously, to create conditions such that the energy output is more than the energy required to operate the reactor. The economics of a successful power reactor requires that the energy output be substantially greater than the energy input. Several different criteria appear in the literature to describe the conditions of a successful fusion reaction and it is important to understand the distinction between these different approaches.

doi:10.1088/978-0-7503-3307-8ch4

Some important terms that are used to describe conditions that are suitable for a successful fusion reaction are *ignition*, the *Lawson criterion* and *breakeven*, as discussed below.

Ignition

As an example of the different approaches to determining appropriate conditions for a fusion reactor, we consider deuterium–tritium fusion,

$$\mathrm{d} + \mathrm{t} \rightarrow \alpha + \mathrm{n} \quad (17.6\ \mathrm{MeV}) \tag{4.1}$$

When this fusion reaction occurs, the fusion energy is given up as kinetic energy of the alpha particle and the neutron. Conservation of energy and momentum requires that the energy is distributed between the two particles in proportion to their masses. In this case the energies of the alpha particle (E_α) and the neutron (E_n) are given by

$$\frac{E_\alpha}{E_\mathrm{n}} = \frac{m_\mathrm{n}}{m_\alpha} \tag{4.2}$$

or $E_\alpha \sim 3.5$ MeV and $E_\mathrm{n} \sim 14.1$ MeV. In a typical reactor design, the neutrons, or at least some fraction of them, are captured in a blanket around the reactor and are used to breed new tritium fuel for future reactions and, as well, will contribute some of their energy to the total heat production of the reactor. This process is described further in chapter 7. It is, however, the energy associated with the charged alpha particles that is of the most immediate concern. The alpha particles will give up energy through Coulomb interactions to the remaining unfused fuel. This can heat some of the remaining fuel to fusion conditions and these additional fusions will release additional alpha particles. Ignition is the condition where this reaction is self-sustaining, that is, the rate of alpha particle heating is equal to the rate at which the plasma loses energy.

The alpha particle heating rate is calculated in terms of the fusion rate as defined in equation (2.29);

$$f = n_\mathrm{a} n_\mathrm{b} \langle \sigma v \rangle \tag{4.3}$$

where n_a and n_b are the number densities of the two reaction fuels and $\langle \sigma v \rangle$ is the average of the product of the fusion cross section, σ, and the relative velocity of the two nuclei, v, at the temperature of the plasma, T. It is customary to use equal mixtures of the two fuel components, so that $n_\mathrm{a} n_\mathrm{b} = n^2/4$, where n is the total particle number density. The alpha particle heating rate, i.e. heating power, P_α, is, therefore

$$P_\alpha = \frac{n^2}{4} \langle \sigma v \rangle E_\alpha \tag{4.4}$$

where E_α is the alpha particle kinetic energy.

The rate at which the plasma loses energy is calculated from the thermal energy per particle, $3k_\mathrm{B}T/2$, where k_B is the Boltzmann constant. Since there are equal numbers of nuclei and electrons then the energy per unit volume will be $3nk_BT$, where n is defined as above. The rate of energy loss (i.e. power) for the plasma is written as

$$P_{plasma} = \frac{3nk_BT}{\tau} \tag{4.5}$$

where τ is the plasma confinement time. Equating equations (4.4) and (4.5) gives the condition for ignition as

$$n\tau = \frac{12k_B}{E_\alpha}\frac{T}{\langle\sigma v\rangle} \tag{4.6}$$

The temperature dependence of $n\tau$ on the left-hand side will be given by the temperature dependence of $T/\langle\sigma v\rangle$ on the right-hand side. A simple way of analyzing these results is to use the temperature dependence of $T/\langle\sigma v\rangle$ in equation (4.6) to plot $n\tau$ as a function of temperature, as illustrated in figure 4.1. This plot shows that there is a minimum at around a temperature of 26 keV, where the value of $n\tau$ is approximately 1.7×10^{20} m^{-3} s. The minimum condition for ignition is, therefore, expressed as

$$n\tau > 1.7 \times 10^{20}\ \text{m}^{-3}\ \text{s} \tag{4.7}$$

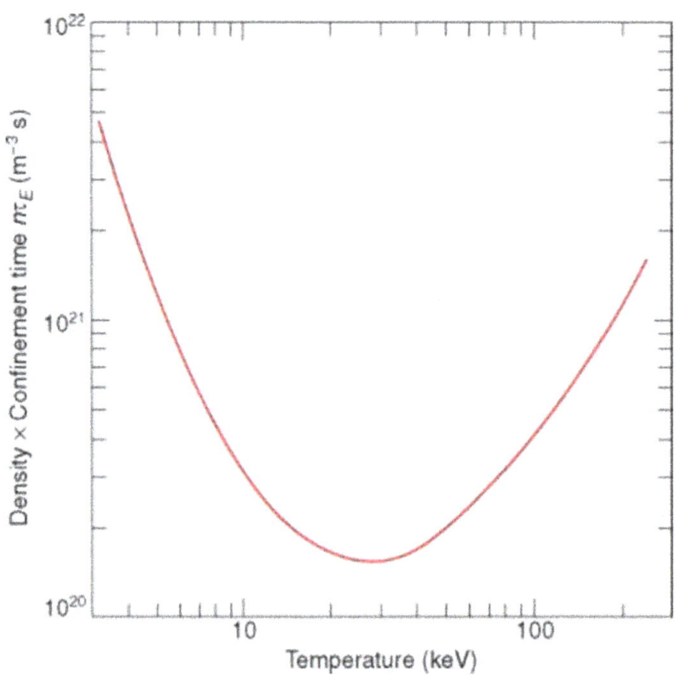

Figure 4.1. Relationship between $n\tau$ (in m^{-3} s) and temperature, T (in keV) for d–t fusion. Reprinted from McCracken and Stott (2013). Copyright (2013), with permission from Elsevier.

at a temperature of 26 keV (or about 3×10^8 K). At higher temperatures or lower temperatures, a greater value of $n\tau$ as shown in figure 4.1 will be necessary to achieve ignition.

The Lawson criterion

The Lawson criterion for achieving a useful nuclear fusion reaction was originally proposed in the mid-1950s by British engineer and physicist John D Lawson (1923–2008). Lawson's derivation follows along the lines of the derivation above for the condition for ignition, except that Lawson assumed that all of the energy from the fusion reaction was recovered as heat and that this was converted into electricity at an efficiency of 33% (appropriate for thermal conversion at the time) and returned to maintain the plasma conditions by ohmic heating at an efficiency of 100%. Thus, the application of the above approach using Lawson's assumptions means that in equation (4.6) the energy is one third of the total fusion energy, i.e. 0.33×17.6 MeV = 5.9 MeV rather than just the 3.5 MeV acquired by the alpha particle. Therefore, the Lawson criterion is a somewhat less stringent condition than the ignition condition given above yielding the inequality

$$n\tau > 1.0 \times 10^{20} \text{ m}^{-3} \text{ s} \tag{4.8}$$

Breakeven

Breakeven is a somewhat ambiguous term. Sometimes it refers to the situation where the fusion energy is equal to the energy input into the fuel. This is often referred to as scientific breakeven and is, in general, a less stringent criterion than the Lawson criterion as it does not involve the efficiency of returning fusion energy to the reactor to facilitate future fusion reactions. The factor Q is sometimes used to denote the reactor gain and is defined as

$$Q = \frac{P_{\text{fusion}}}{P_{\text{external}}} \tag{4.9}$$

where P_{fusion} is the fusion power produced and P_{external} is the external heating power supplied to the plasma. A value of $Q = 1$, therefore, represents scientific breakeven. Breakeven, from a practical standpoint, must take this efficiency into account in order to evaluate the commercial viability of a fusion reactor. While Lawson's approach is aimed at an analysis of magnetic confinement fusion, we can generalize the above discussion to consider d–t fusion in general where the efficiency of returning the 17.6 MeV released in each fusion event to the reactor to drive subsequent fusion reactions. If we define this efficiency as, η, expressed as a fraction, then combining equations (4.4) and (4.5) will yield

$$n\tau > 7 \times 10^{-4} \frac{T}{\eta\langle\sigma v\rangle} \tag{4.10}$$

where the temperature is expressed in keV. This expression is appropriate for all d–t fusion reactions and, as discussed further in chapter 5 for inertial confinement fusion, illustrates the importance of the efficiency of returning fusion energy to the reactor for further fusion reactions in establishing appropriate criteria for a successful commercial reactor.

The triple product

The use of the Lawson criterion as discussed above was an early approach to understanding the progress made in reaching ignition conditions in nuclear fusion reactors. In recent years, another criterion, known as the triple product, $nT\tau$, has been used as a figure of merit for fusion reactors. Multiplying both sides of equation (4.6) by temperature gives

$$nT\tau = \frac{12k_B}{E_\alpha}\frac{T^2}{\langle\sigma v\rangle} \tag{4.11}$$

The minimum in $nT\tau$ for d–t fusion follows from the data in figure 4.1 and occurs at a temperature of around 14 keV. This temperature corresponds to the maximum fusion energy output and represents the most advantageous operating condition for a viable reactor. In the temperature region around 14 keV, i.e. from around 10 keV to 20 keV, the value of the value of $\langle\sigma v\rangle$ can be approximated as const $\times\ T^2$, meaning that the triple product in this region is approximately temperature independent with a value of

$$nT\tau = 3 \times 10^{21}\ \mathrm{m}^{-3}\ \mathrm{keV\ s} \tag{4.12}$$

corresponding to the condition for ignition. This gives an approximate value for minimum plasma conditions for which self-heating will lead to a chain d–t fusion reaction. Typical values of the three factors in the triple product can be assessed for magnetic confinement reactors on the basis of plasma stability. Stability requires a density of less than about $2 \times 10^{20}\ \mathrm{m}^{-3}$. Using the ideal temperature of 14 keV, yields a minimum confinement time to reach ignition of about 1 s.

Reactor energy gain

While the above discussion concerning plasma conditions can be adapted to all types of fusion confinement technologies, magnetic confinement fusion can be considered from a somewhat different approach. When plasma conditions in a magnetic confinement fusion reactor satisfy the ignition criteria, then a self-contained chain reaction will be created. This means that the plasma will be heated through alpha particle heating in order to maintain the fusion reaction without the need for additional external energy input. This is advantageous from an energy standpoint but is not necessarily the most desirable situation from a reactor control standpoint. It may be advantageous for the reactor to require additional energy input (i.e. heating) in order to maintain fusion conditions. If the plasma reaches ignition, then no external heating is required and $Q = \infty$. If the fusion power is equal to the external heating power then $Q = 1$, and this condition is, as discussed above, the situation that we have defined as scientific breakeven. Reaching scientific breakeven is an interesting achievement from a scientific standpoint but is not sufficient to produce an economically viable commercial reactor. As suggested above, scientific breakeven requires less demanding plasma conditions than ignition and is not a sufficient condition for a commercial power reactor. Considering typical energy conversion efficiencies, a value of $Q \approx 5$ is typical of a reactor operation with net energy gain, and a value of Q in the range of 25–30 is generally considered to be appropriate for a power rector. Thus, an evaluation of Q for an experimental reactor, along with an assessment of the triple product (and/or the Lawson criterion) serves to provide an assessment of reactor performance.

4.3 Some basic plasma physics

General characteristics of plasmas

As discussed above, the temperature at which fusion reactions, specifically d–t fusion reactions, occur are in the range of 2×10^8 K (i.e. 20 keV). At these temperatures, hydrogen atoms are fully ionized, thereby creating a plasma. Although plasmas are, overall, electrically neutral, the positive charges (ions) and negative charges (electrons) are not necessarily uniformly distributed in space, giving rise to oscillations and turbulence in the plasma. Plasmas are common outside of the Earth and make up the majority of matter in the Universe (although the possible contribution of dark matter to the mass of the Universe is presently unknown) including stellar interiors and the interstellar medium. On Earth, plasmas may be naturally occurring, as in lightning and upper atmospheric regions, or they may be artificially produced, as in gas discharge tubes (e.g. neon lights) and those produced in the pursuit of controlled nuclear fusion. While many, such as those that exist in stellar interiors, interstellar matter and in nuclear fusion experiments, consist of electrons and fully ionized atoms, others, such as those in gas discharge tubes, consist of electrons and partially ionized atoms. For typical d–d and d–t plasmas that have been the subject of most experimental studies of controlled nuclear fusion reactions, the atoms are fully ionized at the conditions under which the experiments are conducted. As well, since all of the atoms involved are isotopes of hydrogen, the ions are all singly charged and there are equal numbers of ions and electrons in the plasma. In general, plasmas can exist over a wide range of densities and temperatures. Figure 4.2 summarizes typical properties of some plasmas.

Plasma properties

The behavior of a normal gas is governed by short-range interactions between the atoms or molecules. This property is consistent with the general characteristic that normal gases are not electrically conductive, at least not for electric fields that are less than that which causes electrical breakdown. The behavior of a plasma is quite different and this difference stems from the fact that plasmas are electrically conductive. As a result of this electrical conductivity, the behavior of a plasma is influenced by long-range electromagnetic fields. Several parameters are in common use to describe the behavior of a plasma. Three of the most important parameters

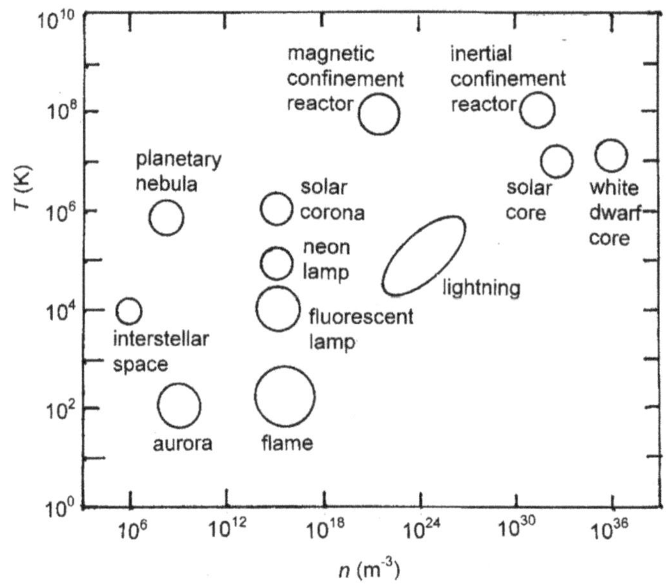

Figure 4.2. Density, n, and temperature, T, for some naturally occurring and artificially produced plasmas. The values for controlled magnetic confinement and inertial confinement reactors are based on hypothetical designs that are expected to fulfill the necessary fusion criteria. Normal materials, i.e. solids, liquids and gases, are located in the lower right of the graph.

that describe a plasma are the plasma frequency, the Debye length and the plasma parameter. These are each described below.

The plasma frequency

The plasma frequency is defined as the natural oscillation frequency of the electrons in a plasma. This oscillatory behavior is the result of the Coulomb interaction between the negatively charged electrons and the positively charged ions. This frequency may be approximated, assuming that the ion mass is infinite, as

$$\omega_{\mathrm{p}} = \left[\frac{n_{\mathrm{e}}e^2}{\varepsilon_0 m_{\mathrm{e}}}\right]^{1/2} \tag{4.13}$$

where n_{e} is the electron number density, e is the electronic charge, ε_0 is the permittivity of free space and m_{e} is the electron mass. When the plasma frequency is much larger than the electron–neutral particle collision frequency then a gas is properly described as a plasma and electrostatic interactions dominate over the classical dynamics of an ordinary gas.

The Debye length

A charged ion, with charge $+Ze$, in a plasma is screened from the electrostatic influence of charged particles outside a sphere by the concentration of charge associated with the electron gas around the particle. The radius of this sphere, λ_{D}, is referred to as the Debye radius or Debye length (sometimes called the Debye screening length). The Debye radius is the effective range of the electrostatic potential around the ion, which is given by the Debye–Hückel potential, $\phi(r)$, with the radial dependence of the form

$$\phi(r) = \frac{Ze}{4\pi\varepsilon_0 r}\exp\left[-\frac{r}{\lambda_D}\right] \tag{4.14}$$

Outside of the Debye radius the electrostatic energy is less than the thermal energy. The Debye length also represents the distance travelled by an electron with the thermal velocity between electron–electron collisions. From the definition of the plasma frequency, the mean time between electron–electron collisions is merely ω_p^{-1} and for a mean electron velocity, v, the Debye radius is given by $\lambda_D = v/\omega_p$. For electrons at a temperature, T_e, this may be written as

$$\lambda_D = \left[\frac{\varepsilon_0 k_B T_e}{n_e e^2}\right]^{1/2} \tag{4.15}$$

When the Debye radius is much smaller than the overall spatial dimensions of the plasma, then edge effects can be ignored, and the plasma can be considered as quasineutral.

The plasma parameter

The plasma parameter, Λ, is a dimensionless constant that represents the ratio of the maximum impact parameter to the distance of closest approach for classical Coulomb scattering. This is written as

$$\Lambda = 4\pi n_e \lambda_D^3 \tag{4.16}$$

Equation (4.16) is considered valid for plasmas in which the electron thermal velocities are much greater than the ion thermal velocities. A simpler description of the plasma parameter is based on the mean number of electrons in a sphere of radius equal to the Debye radius. This is sometimes denoted N_D, to distinguish it from the expression in equation (4.16), and is written as

$$N_D = \frac{4\pi}{3} n_e \lambda_D^3 \tag{4.17}$$

so that $N_D = \Lambda/3$. Substituting for the Debye radius from equation (4.15) gives

$$\Lambda = \frac{4\pi}{n_e^{1/2}}\left[\frac{\varepsilon_0 k_B T_e}{e^2}\right]^{3/2} \tag{4.18}$$

and

$$N_D = \frac{4\pi}{3n_e^{1/2}}\left[\frac{\varepsilon_0 k_B T_e}{e^2}\right]^{3/2} \tag{4.19}$$

Another parameter which is useful in defining the behavior of a plasma, and which is closely related to the plasma parameter, is the coupling parameter. This dimensionless parameter, Γ, is defined as the ratio of the Coulomb energy to the thermal energy, i.e.

$$\Gamma = \frac{E_C}{k_B T_e} \tag{4.20}$$

The Coulomb energy per electron is defined as

$$E_C = \frac{e^2}{4\pi\varepsilon_0 \langle r \rangle} \tag{4.21}$$

where $\langle r \rangle$ is the mean distance between electrons. Typically, $\langle r \rangle$ is taken to be the Wigner–Seitz radius, which, by analogy with the traditional definition used in solid state physics, is the radius of a sphere which contains one particle (in this case, one electron). The Wigner–Seitz radius is expressed as

$$\langle r \rangle = \left[\frac{4\pi}{3} n_e\right]^{-1/3} \tag{4.22}$$

and this gives the coupling constant as

$$\Gamma = \frac{e^2}{4\pi\varepsilon_0 k_B T_e}\left[\frac{4\pi}{3} n_e\right]^{1/3} \tag{4.23}$$

Comparing equation (4.18) for the plasma parameter and equation (4.23) for the coupling constant shows that to within a factor of approximately unity

$$\Gamma \approx \Lambda^{-2/3} \tag{4.24}$$

Classification of plasmas

As noted in figure 4.2, plasmas cover a wide range of different densities and temperatures. The plasma parameters allow for the classification and understanding of plasma properties. In the case where $\Lambda \gg 1$, the system is defined as an ideal plasma. In an ideal plasma, collective electrostatic interactions dominate over individual two-particle interactions and the behavior of the plasma may be described by the interaction of particles with a mean background potential.

An inspection of equations (4.18) and (4.24) provides some insight into the relationship between plasma properties and the value of the plasma parameter and the coupling constant. Clearly from these equations, Λ increases with increasing plasma temperature and decreases with increasing plasma density. Γ shows just the opposite behavior. Table 4.1 summarizes some of the properties of plasmas based on the value of the plasma parameter. Plasma in a magnetic confinement fusion reactor, as discussed in the present chapter, has properties, as illustrated in figure 4.2, that are typical of a weakly coupled (ideal) dilute plasma. Fuel in an inertial confinement

Table 4.1. Plasma properties as a function of plasma parameter.

Plasma parameter	$\Lambda \gg 1$	$\Lambda \ll 1$
Ideal plasma	Yes	No
Coupling constant	$\Gamma \ll 1$	$\Gamma \gg 1$
Coupling	Weak	Strong
Density	Low	High
Temperature	High	Low
Direct electrostatic interactions	Unimportant	Important

fusion reactor, as described in the next chapter, at the time it undergoes fusion, has properties as illustrated in figure 4.2. Although the temperatures for magnetic confinement and inertial confinement fusion are similar, the difference in densities, i.e. typically 10 orders of magnitude, are very significant. Inertial confinement plasmas, therefore, have properties that are typical of strongly coupled plasmas.

Magnetized plasmas

The charged particles that constitute a plasma are, in principle, subject to the influence of externally applied magnetic fields. If the behavior of the charged particles is substantially altered by the presence of a magnetic field, then the plasma is said to be magnetized. The quantitative condition that describes a plasma as being magnetized is the relationship between the frequency of gyration of a charged plasma particle around the magnetic field lines and the frequency of particle–particle collisions. If the particle makes at least one gyration on the average before a collision occurs, then the plasma is considered to be magnetized. For a particle of charge, Ze, and mass, m, in a magnetic field, B, the angular frequency of gyration, ω, often referred to as the gyromagnetic frequency of cyclotron frequency, is

$$\omega = \frac{ZeB}{m} \tag{4.25}$$

Typically for nuclear fusion experiments, as described in the present book, particles are singly charged. Equation (4.25) shows that, because of the significant difference in mass between electrons and ions, there are significant differences in the cyclotron frequency for electrons and the cyclotron frequency for ions. As a result, in some plasmas under certain external magnetic field conditions, the electrons can be magnetized while the ions are not. The spatial path of a charged particle, of charge Ze in a magnetic field can be described by the gyroradius, or Larmor radius, r_L, given as a function of temperature, T,

$$r_L = \left[\frac{2mk_BT}{Z^2e^2B^2}\right]^{1/2} \tag{4.26}$$

This equation shows that the Larmor radius of an electron is much smaller, because of its smaller mass, than the Larmor radius of an ion. If an additional vector force,

$\boldsymbol{F}$, acts on the charged particle, e.g. an electrostatic force, then the particle will drift with a velocity, $\boldsymbol{v}_D$,

$$\boldsymbol{v}_D = \frac{\boldsymbol{F} \times \boldsymbol{B}}{Ze|\boldsymbol{B}^2|} \tag{4.27}$$

In general, this additional force causes the particle to travel in a helical path around the magnetic field lines.

The importance of magnetic fields in the confinement of a plasma may be considered by looking at the balance of forces acting on the particles of the plasma. The total plasma pressure, P_p, pushing outward, is the sum of the partial pressures due to the electrons, e, and the ions, i. This is expressed as

$$P_p = n_e k_B T_e + n_i k_B T_i \tag{4.28}$$

The force acting on the plasma particles as a result of the applied magnetic field can be expressed as a pressure, P_m, of the form

$$P_m = \frac{B^2}{2\mu_0} \tag{4.29}$$

where μ_0 is the permeability of free space. The parameter β for the plasma is expressed as the ratio of the plasma pressure to the magnetic pressure as

$$\beta = \frac{2\mu_0 P_p}{B^2} \tag{4.30}$$

This expression gives β as a fraction, although it is sometimes expressed as a percentage. The ability to operate a magnetic confinement fusion reactor at a higher value of β means that the same performance can be achieved for smaller magnetic fields, leading to a reduction in reactor size and overall infrastructure costs.

4.4 Tokamaks

Early history of the tokamak

As noted in the previous chapter, the earliest work on magnetic confinement fusion reactors dealt with linear magnetic mirror devices. In order to deal with problems of instabilities and plasma loss at the ends of the plasma channel, research moved in the direction of the stellarator (see figure 3.9). The stellarator, which was invented by Lyman Spitzer in the early 1950s, was researched extensively during the 1950s and 1960s. The largest portion of this work was conducted at the Princeton Plasma Physics Laboratory in the United States and at the Max Planck Institute for Plasma Physics in Germany. However, in the 1950s an alternative approach to the stellarator geometry, the tokamak, was being developed in the Soviet Union. The concept of the tokamak was developed by Soviet physicists Andrei Sakharov and Igor Tamm as noted in chapter 3. The word tokamak is an acronym for the Russian name of the device—**to**roidal'naya **ka**mera s **ma**gnitnymi **k**atushkami (meaning, toroidal chamber with magnetic coils). The name was suggested in 1957 by the Soviet physicist Igor Golovin (1913–97), who was the vice-director of the Laboratory of Measuring Apparatus of Academy of Science.

The world's first tokamak, the T-1, was operational at the Kurchatov Institute in Moscow from 1957 to 1959. This device had a major radius of 0.625 m, a minor radius of 0.13 m and a maximum magnetic field of 1 T. Little information was released at the time about the results from this device. In 1962, the world's second tokamak, the T-3, also at the Kurchatov Institute, became operational. This device had a major radius of 1.0 m, a minor radius of 0.12 m and a maximum magnetic field of 2.5 T. In 1968, results from T-3 experiments were published. These showed plasma properties that were much superior to those obtained on current stellarators, and a general shift in magnetic confinement fusion research from stellarators to tokamaks occurred. The first tokamak outside of the Soviet Union, was constructed at the Princeton Plasma Physics Laboratory and operated from 1970 to 1974. This device, referred to as the ST (for symmetric tokamak) was converted from Princeton's Model C stellarator. Since that time more than 50 operational tokamaks have been constructed worldwide and about half of these are still functional. In recent years, renewed interest in stellarators has led to further developments in this field, as discussed in the next section.

Basic physics of the tokamak

As noted by Enrico Fermi in the early 1950s, the simple toroidal geometry as illustrated in figure 3.8 is not appropriate for the magnetic confinement of a plasma. This results from the fact that the poloidal coils are closer together on the inside of the torus than on the outside of the torus.

This behavior can be described more quantitatively by calculating the toroidal magnetic field $B_\phi(r)$, that results from a poloidal current, I_p. This field has a radial dependence that can be expressed as

$$B_\phi(r) = \frac{\mu_0 I_\mathrm{p}}{2\pi r} \tag{4.31}$$

where μ_0 is the permeability of free space and I_p is the total poloidal current. ϕ represents the direction of the magnetic field along the toroidal direction as shown in figure 3.8. The magnetic field gradient, $\nabla \boldsymbol{B}$, that arises from equation (4.31) produces a drift of the negative charges (electrons) in one direction (e.g. down) and positive charges (ions) in the opposite direction (e.g. up). The resulting charge separation produces an electric field, $\boldsymbol{E}$, perpendicular to the magnetic field and this electric field exerts a Lorentz force on the charges in the plasma that is proportional to $\boldsymbol{E} \times \boldsymbol{B}$. The net result of the Lorentz force is that both electrons and ions drift outward and will, eventually, impact the wall of the plasma vessel. This is obviously detrimental as it cools the plasma and degrades the integrity of the vessel wall.

One solution to the drift problem was suggested by Lyman Spitzer and resulted in the figure-eight stellarator design shown in figure 3.9. The stellarator may also use asymmetric poloidal coils to produce a helical magnetic field as described below in section 4.5. The design of the tokamak provides another approach to solving the drift problem. In addition to the toroidal magnetic field that is produced by the poloidal coils, the tokamak incorporates an additional poloidal field that is produced by a toroidal current. The net magnetic field is, therefore, the sum of the toroidal and poloidal fields and has a helical geometry. Thus, as the charged plasma particles follow the magnetic field lines, they alternately move from the inside of the torus to the outside of the torus and the drift is averaged out. Although this cancellation is not perfect it sufficient to maintain suitable plasma conditions if magnetic field conditions are appropriately chosen.

It is easy to imagine the creation of the toroidal magnetic field from currents in external poloidal windings, as illustrated in figure 3.8. However, it is not possible to place a physical toroidal coil in the middle of the plasma vessel to create the poloidal magnetic field. In order to create the poloidal magnetic field, a toroidal current is generated by induction using a time varying current in a set of inner poloidal field coils, often referred to as the central solenoid, as shown in figure 4.3. This central solenoid creates a poloidal field, and the time variation of this field produces an induced current in the toroidal direction. The induction of a toroidal current in the plasma chamber is possible because of the conductive nature of the plasma itself, so this toroidal current is sometimes referred to as the plasma current and it is an

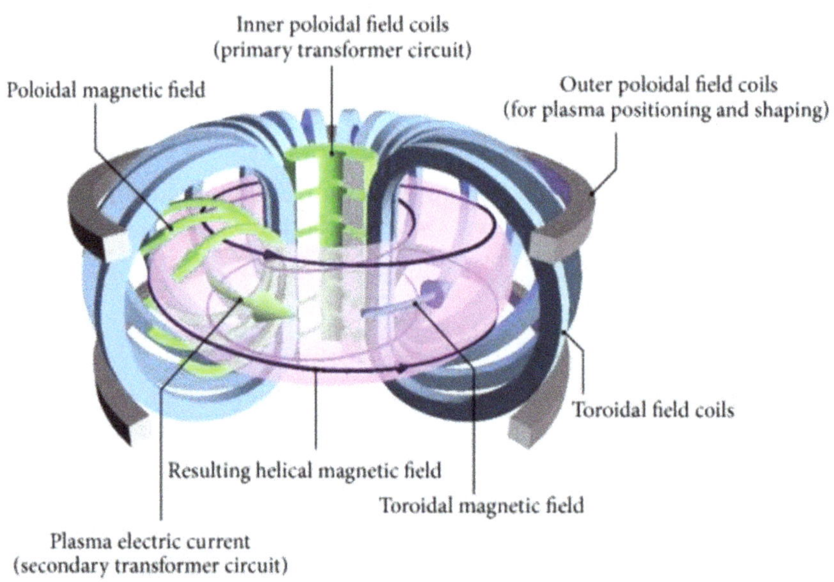

Figure 4.3. Diagram of a generic tokamak showing the geometry of external coils and the corresponding magnetic fields that they produce. The total, helical, magnetic field is illustrated. This Diagram of a Generic Tokamak image has been obtained by the author from the Wikimedia website https://commons.wikimedia.org/wiki/File:Schematic-of-a-tokamak-chamber-and-magnetic-profile.jpg where it was made available by Li S, Jiang H, Ren Z and Xu C under a CC BY 4.0 license. It is included within this book on that basis. It is attributed to Li *et al* (2014).

essential component of the tokamak design. This induced toroidal current, I_t, may be expressed in terms of Faraday's law as

$$I_t = \frac{V_t}{R} = -\frac{1}{R}\frac{d\Phi}{dt} \tag{4.32}$$

where V_t is the induced toroidal EMF, R is the plasma resistance and Φ is the magnetic flux density. The magnetic flux density is proportional to the magnetic field, which in turn, is proportional to the current in the central solenoid, as well as the magnet geometry.

The steps in the creation of the helical magnetic field in the plasma chamber are summarized in figure 4.4. Because of the manner in which the plasma current is produced, the confinement conditions in the tokamak are pulsed, although the duration of the pulse, that is, the time that the external toroidal field is changing so as to induce the plasma current, can be quite long. Typically, a plasma discharge cycle is initiated by ramping up the current in the central solenoid followed by ramping the current back down. The plasma current serves two important purposes: (1) to generate the poloidal magnetic field that adds to the toroidal magnetic field produced by the poloidal coils to produce the net helical magnetic field lines to stabilize the plasma; and (2) to heat the plasma by ohmic heating due to the resistive properties of the plasma. This is the basic source of plasma heating which is generally supplemented by auxiliary heating methods as described below.

A common coil geometry to produce both the toroidal and poloidal magnetic field that give rise to the net helical field is illustrated in figure 4.3. The exact shape of the helical field lines inside the tokamak depends on the relative strengths of the toroidal and poloidal magnetic fields. If the poloidal field is relatively weak, then the helical field lines will make relatively few twists during one toroidal cycle. If the poloidal field is relatively strong there will be more twist in the helix. The so-called safety factor, q, is defined as the number of helical twists in the magnetic field lines for each path around the torus. Numerically, q is defined in terms of the plasma major and minor radii, R and r, respectively and the toroidal and poloidal magnetic fields, B_t and B_p, respectively, as

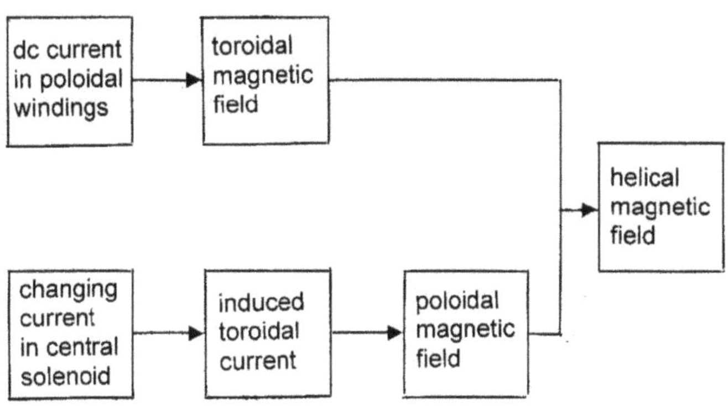

Figure 4.4. Creation of the helical magnetic field in a tokamak.

$$q = \frac{r}{R} \cdot \frac{B_t}{B_p} \tag{4.33}$$

The safety factor is an important parameter in determining the stability of the plasma as discussed below.

A major difficulty in controlling plasma properties in a magnetic confinement fusion reactor arises as a result of instabilities. One common instability is the kink instability as shown in figure 4.5. Fluctuations in the plasma column can grow as a result of differences in magnetic pressure as illustrated in the figure. This type of occurrence can result in energy loss and can make it more difficult to maintain plasma conditions. In the most severe case, the plasma column can impact the vessel wall causing termination of the plasma current and sudden loss of confinement. Such an event is referred to as a disruption. Designing a tokamak with a sufficiently high safety factor can minimize the possibility of a disruption. The kink limit is $q = 2.2$, so maintaining $q > 3$ is sometimes considered appropriate for mitigating instability problems.

In order to satisfy conditions for scientific breakeven and eventually ignition (or close to it) the plasma parameters of ion density, temperature and confinement time must be optimized. Unfortunately, these three parameters cannot be increased indefinitely without introducing instabilities and eventual disruption of the plasma. The importance of certain tokamak design parameters can be seen by rewriting the expression for the safety factor. We begin by relating the poloidal field, $\boldsymbol{B}_p$, to the (toroidal) plasma current density, $\boldsymbol{J}_t$, by Ampère's circuital law;

$$\nabla \times \boldsymbol{B}_p = \mu_0 \boldsymbol{J}_t \tag{4.34}$$

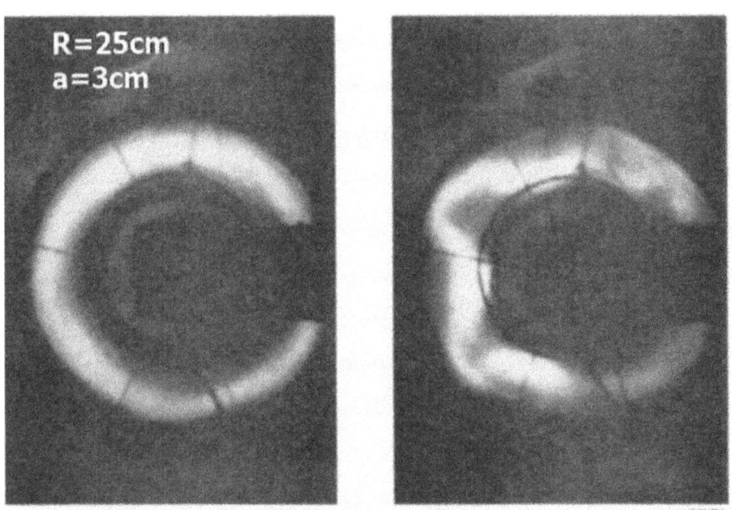

Figure 4.5. Formation of a kink instability in an early z-pinch device. This kink Instability image has been obtained by the author from the Wikimedia website https://commons.wikimedia.org/wiki/File:Kink_instability_at_Aldermaston.jpg, where it is stated to have been released into the public domain. It is included within this book on that basis.

This may be written more simply in terms of the total plasma current, I_t, and the minor plasma radius as

$$2\pi r B_p = \mu_0 I_t \tag{4.35}$$

Substituting equation (4.35) into equation (4.33) gives

$$q = 2\pi \frac{r B_t}{I_t} \cdot \frac{r}{R} \tag{4.36}$$

In this expression it is implied that the plasma column cross section is circular. We can approximate a more general form of equation (4.36) as

$$q \sim 2\pi \frac{r B_t}{I_t} \cdot \frac{r}{R} \cdot \kappa \tag{4.37}$$

where κ is the elongation, that is, the ratio of the height to the width of the plasma column. For a circular plasma column cross section, $\kappa = 1$. It is obvious that plasma geometry is an important factor in optimizing q. From equation (4.37) we see that a 'fat' plasma column, that is, with a large ratio r/R, is advantageous. As well, an elongated plasma, i.e. large value of κ, is also beneficial. For modern tokamaks, a D-shaped plasma is generally utilized (see description of the Joint European Torus, below) with a typical value of $\kappa \approx 1.8$. Equation (4.37) also emphasizes the point that the toroidal field must be large and/or the plasma current must be small. This puts a limit on plasma current that is acceptable in order to avoid instabilities in terms of the available toroidal magnetic field.

In order to achieve fusion conditions, as defined by the plasma triple product, it is important to consider limits that exist on the ion density. While there is no clear theoretical analysis of the maximum plasma density that is consistent with stability, detailed studies of tokamak operation have provided an empirical relationship known as the Greenwald limit, n_G, for density. This is expressed as

$$n_G = \frac{I_t}{\pi r^2} \tag{4.38}$$

Here n_G is expressed in units of 10^{20} m^{-3}, I_t is in MA and r is in m. Using parameters for a large tokamak, i.e. design parameters for ITER as given in the discussion of this project below ($I_t = 15$ MA, $r = 2.0$ m), the Greenwald density limit is found to be 1.2×10^{20} m^{-3}. This limit is generally interpreted as the limit for the line-average electron density (approximately the same as the total electron density). Plasma conditions for breakeven and ignition refer to the fusion fuel ion density. The fuel ion density may be less than the electron density in the plasma because of the presence of impurities. Impurities, when ionized, contribute to the electron density but do not contribute to the fusion fuel. This is one of the important reasons for eliminating plasma impurities. See further discussion below concerning divertors.

A final plasma parameter that we can consider in terms of its relationship to plasma stability is the plasma pressure. Plasma pressure enters into the calculation of the reactor β as given in equation (4.30) and the maximum β that is consistent with plasma stability is referred to as the Troyon limit. This limit may be calculated on the basis of the fundamental magnetohydrodynamic properties of the plasma and is given as

$$\beta_{\max}(\%) = \frac{CI_t}{rB_t} \tag{4.39}$$

where C is the dimensionless Troyon factor, which has a value in the range of 2.8–5.0, and the units for I_t, r and B_t are MA, m and T, respectively. For common tokamak designs, the Troyon limit is around 3.5%.

The above discussion provides guidelines for tokamak design and operational parameters in order to minimize the possibility that instabilities will adversely affect reactor performance or, in the worst case, lead to plasma disruption.

Tokamak technology

Since the earliest tokamaks were constructed in the 1960s, these magnetic confinement fusion reactors have made consistent progress toward the goal of achieving scientific breakeven and to the eventual construction of a commercial power reactor. These advances have been the result of continuous development of the various component systems of the reactor. In the present section, the technical details of some of the most important component systems of an operational tokamak are described. Further details of a viable power reactor are presented in chapter 7.

Magnet technology

Early tokamaks utilized conventional non-superconducting electromagnets to provide the external magnetic field required for operation. While this was a suitable approach for relatively small experimental research tokamaks as it simplified their designs, it has become increasing less appropriate in recent years. As the tokamak design is scaled-up, the magnet system uses an increasingly large fraction of the total power for operation. This becomes a serious concern for the economics of a commercial fusion power plant. It is, therefore, necessary to transition tokamak designs to the utilization of superconducting magnet technology. Superconductors have the advantage over conventional conductors that they can carry much larger currents without resistive loss. This means that they are capable of producing larger magnetic fields and that much less electrical energy is required to operate them. The first tokamak to incorporate superconducting magnets was the Tore Supra in Cadarache, France, which operated from 1988 to 2010. Two major tokamaks that are under construction, the JT-60SA (Japan Torus-60 Super Advanced) and the ITER (International Thermonuclear Experimental Reactor), will also utilize superconducting magnets.

In 1911, superconductivity was discovered by Heike Kamerlingh Onnes (1853–1926) (Kamerlingh Onnes 1911). He observed that the electrical resistivity of elemental mercury dropped to a value that was too low to be measured when the temperature was lowered below 4.1 K. Following this discovery numerous elements, alloys and compounds have been discovered which demonstrate superconductivity below a given temperature, the critical temperature, T_c. Much effort has been made in the past century or more to discover materials with as high a critical temperature as possible. A major advance in superconductivity research came in 1986 when the first high temperature superconductor was discovered by J Georg Bednorz (b. 1950) and K Alex Müller (b. 1927). They found that the perovskite LaBaCuO (known as LBCO) exhibited a critical temperature of 35 K. Since then, other perovskites have shown superconductivity at temperatures as high as around 150 K. In 2015 superconductivity was observed in H_2S under pressure at a temperature of 203 K.

It is well known that superconductivity is closely related to magnetism. A sufficiently large applied magnetic field, the critical field, B_c, is known to destroy the superconducting state. At the critical temperature the critical field is zero and at zero temperature it attains a value $B_c(0)$. The temperature dependence of the critical field can be expressed as

$$B_c(T) = B_c(0)\left[1 - \left(\frac{T}{T_c}\right)^2\right] \quad (4.40)$$

This relationship is illustrated in figure 4.6(a).

Superconducting materials, as shown in figure 4.6(a), where the superconductivity is suddenly destroyed when the magnetic field reaches the critical value, are referred to as 'type-I' superconductors. Figure 4.6(b) illustrates the behavior of a 'type-II' superconductor. In this case, as the magnetic field is increased, the sample becomes partially normal (non-superconducting) at a field B_{c1} and completely normal at a field B_{c2}. In the region between B_{c1} and B_{c2}, referred to as the mixed state, superconducting current will still flow through the superconducting regions. Thus, the material can be used as a superconductor up to a field of B_{c2}. In general, B_{c2} for a type-II superconductor is much greater than B_c for a type-I superconductor. Therefore, superconductors used for magnet construction are virtually always type-II superconductors.

Table 4.2 gives the critical properties of some superconductors that are potentially useful for tokamak magnets. NbTi and Nb_3Sn are traditional superconductors while

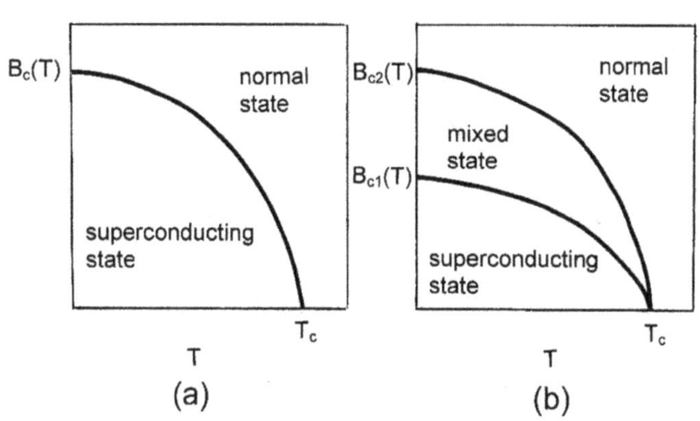

Figure 4.6. Phase diagram of (a) type-I superconductor and (b) type-II superconductor.

Table 4.2. Properties of some useful conventional type-II superconductors.

Material	T_c (K)	B_{c2} (T) at $T = 0$ K
NbTi	9.3	15
Nb_3Sn	18.3	30
$YBa_2Cu_3O_7$	93	168
$Bi_2Sr_2Ca_2Cu_3O_{10}$	110	200

Table 4.3. Critical fields for some useful superconductors at liquid helium (4.2 K) and liquid nitrogen (77 K) temperature. The traditional superconductors are in their normal state at liquid nitrogen temperature.

Material	B_{c2} (T) at $T = 4.2$ K	B_{c2} (T) at $T = 77$ K
NbTi	12	
Nb_3Sn	28	
$YBa_2Cu_3O_7$	268	53
$Bi_2Sr_2Ca_2Cu_3O_{10}$	200	102

$YBa_2Cu_3O_7$ (YBCO) and $Bi_2Sr_2Ca_2Cu_3O_{10}$ (BSCCO) are high temperature perovskite superconductors. In order to utilize the superconducting properties of these materials they must be cooled, typically with an appropriate cryogen, either liquid helium (4.2 K) or liquid nitrogen (77 K). Thus, it is the critical field at the cryogen temperature that is of relevance to the superconductor performance. Table 4.3 gives the relevant properties of the superconducting materials from table 4.2.

Traditional superconductors, e.g. NbTi and Nb_3Sn, must be cooled using liquid helium while high temperature superconductors possess a substantial critical field even at liquid nitrogen temperature. In fact, the high temperature superconductor critical field at 77 K is greater than the traditional superconductor critical field at 4.2 K. While this may appear to be significant incentive to utilize high temperature superconductors for tokamak magnets, there are a number of factors to consider. The technology of fabricating and utilizing traditional superconducting magnets has existed for a significant number of years. On the other hand, technology for the fabrication and use of high temperature superconducting magnets is relatively new. As high temperature superconductors are ceramics, they are brittle and the preparation of a suitable wire from these materials is complex and expensive. Thus, at present, the use of traditional superconductors for experimental tokamak magnets is the preferred choice. Although liquid helium is substantially more expensive than liquid nitrogen, the cost of cryogenic facilities and liquids is a small fraction of the overall cost of the construction and operation of an experimental tokamak.

The choice between NbTi and Nb_3Sn is largely determined by the difference in critical field at the 4.2 K operating temperature. Although NbTi is less expensive and easier to fabricate than Nb_3Sn the greater critical field of Nb_3Sn is a factor for many applications. Often Nb_3Sn is required for toroidal field coils but NbTi can be used for poloidal field coils and correction coils. The actual magnet construction for superconducting tokamaks, such as the ITER, utilizes well established technology. The so-called cable-in-conduit design for the superconducting wires is used in the ITER magnet design. This design as illustrated in figure 4.7, consists of multiple superconducting filaments enclosed in a metal conduit. The superconducting wire is cooled by liquid helium flowing through the hollow center of the cable.

While traditional superconductors are an appropriate choice at the present time for the magnets utilized in experimental tokamaks in terms of technological and financial considerations, this may not be the situation for future commercial fusion power reactors. The world supply of helium is limited (see section 7.4) and the future large scale implementation of power from magnetic confinement fusion reactors may not be feasible if traditional superconducting materials are used. It is likely, therefore, that the development of appropriate high temperature superconducting magnets that are cooled with plentiful and inexpensive liquid nitrogen will be a requirement, both from an economic and a resource utilization standpoint, for widespread implementation of magnetic confinement fusion power.

A final point concerning the use of superconducting magnets in magnetic confinement fusion reactors is the necessity for appropriate shielding. This is a factor for both traditional and high temperature superconductor use in both experimental reactors, as well as future commercial reactors. Deuterium–tritium fusion produces significant quantities of high energy neutrons. The majority of these neutrons exit the plasma vessel and their influence on the superconducting magnet

Figure 4.7. Cross section of the liquid helium cooled Nb_3Sn cable-in-conduit superconductor for use in the toroidal field coils of the ITER. The cable has a 39.7 mm outer diameter and a 9 mm inner diameter. There are 900 strands of 0.82 mm diameter Nb_3Sn superconducting wire. Reproduced from Rolando *et al* (2013).

coils must be considered. There are two potential adverse consequences of neutron interactions with superconducting magnets.

Firstly, neutrons which interact with atoms in the superconductor will lose energy and deposit heat in the superconductor. This will require additional cooling resources to maintain the superconductor temperature. In the case of helium cooled magnets this can be a significant contribution to cooling costs. For liquid nitrogen cooled magnets, this is not such an important consideration.

The second concern about neutron superconductor interactions deals with radiation damage, as noted in section 7.3. Neutrons can introduce structural defects in solids and the microstructure of superconductors can play an important role in their superconducting properties. Generally, total neutron dose, that is, integrated neutron flux, is a factor in evaluating longevity of magnet performance.

For the reasons noted above, it is important to shield superconducting magnets from neutrons produced by fusion reactions in order to reduce neutron flux to an acceptable level.

Fueling system

The introduction of fusion fuel, e.g. a deuterium–deuterium mixture or deuterium–tritium mixture (as would be needed for a commercial power reactor) is not a simple straightforward matter. The most commonly accepted approach to replenishing fuel in an operational tokamak is the injection of fuel pellets. A system to introduce fuel pellets into a tokamak consists of three basic components as follows:

- A pellet source which forms a solid fuel pellet by cooling gas stored in a suitable reservoir. Pellets of the appropriate size may be formed individually, or a continuous ribbon of solid fuel may be produced for subsequent processing.
- An accelerator is used to accelerate the individual pellets to an appropriate velocity for injection into the reactor. In the case of a continuous ribbon of fuel, appropriately sized pieces are cut from the ribbon before acceleration. Mechanical systems such a pneumatic guns or centrifugal force devices have been considered for accelerating fuel pellets.
- A transfer system which transports the fuel pellet to the appropriate launch position at the boundary of the plasma.

Perhaps the most extensive investigation of fuel injection technology to date has been undertaken at the Joint European Torus (JET) reactor. Figure 4.8 shows a diagram of the design of the pellet injection system developed for JET. Further details of the JET project are discussed in the next section.

For a commercial power reactor based on magnetic confinement technology, fuel pellets of an appropriate size must be delivered at a rate compatible with the desired power output of the reactor. In general, an assessment of fueling parameters would include:

- pellet material;
- mass, m, and dimensions of the pellets;
- pellet velocity, v;
- pellet repetition rate, f.

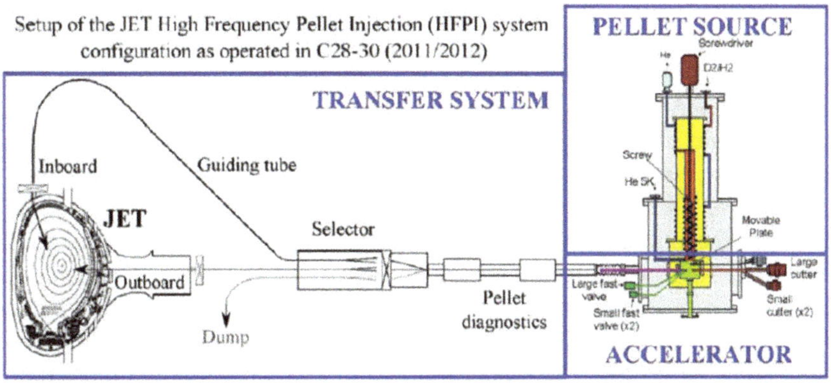

Figure 4.8. Pellet injection system for the Joint European Torus (JET). Reproduced from Lang et al (2019).

The most important parameter for determining energy output is the fuel flux, Γ, given as mass per unit time by the relation $\Gamma = f{\cdot}m$. For the construction of a suitable commercial reactor, millimeter sized fuel pellets with a repetition rate of tens of Hz will be necessary. The theoretical considerations and technology for such a system are similar to that required for a possible future commercial laser fusion reactor as described in chapter 5. The Japan Torus-60 Super Advanced (JT-60SA), which has recently been constructed as detailed below, will be a test bed for investigating fuel injection systems for eventual use on the planned DEMO reactor (see section 4.6).

Current drive

The primary source of heating for the plasma in a tokamak is the induced plasma current that circulates in the toroidal direction, as illustrated in figure 4.3. This current is induced by a changing magnetic field in the axial direction. This axial magnetic field is created primarily by the central solenoid shown in figure 4.3. During operation, the electric current in the central solenoid is ramped up. This changing current produces a time varying axial magnetic field, and this changing magnetic field induces an electric current in the conductive plasma, that is, the toroidal plasma current, which results from the movement of the free electrons through the plasma. The strength of the plasma current depends on a number of parameters. Most importantly, these parameters include the time rate of change of the axial field and the resistive properties of the plasma. The outer coils shown in figure 4.3 are used primarily for adjusting the plasma position and shape. Other contributions to the plasma current from neutral beam injection and electromagnetic waves are described below.

Auxiliary heating

The plasma in a tokamak is subject to ohmic heating by the plasma current. Ohmic heating can raise the temperature of the plasma to temperatures of about 2.5 keV. This is not sufficient for the plasma to reach and maintain suitable conditions for breakeven or ignition. Therefore, additional heating is necessary. Several approaches to auxiliary plasma heating have been taken on various experimental tokamaks. Some of the most useful of these are discussed in the present section.

Adiabatic compression—A gas can be heated by compressing it adiabatically. In the case of plasma heating in a tokamak, the compression results from a sudden increase of the magnetic field that confines the plasma, and the process is sometimes referred to as magnetic compression. This compression is achieved in a tokamak by manipulating the plasma so that it moves into a region of higher magnetic field. The resulting compression has the result of increasing the plasma temperature and also has the advantage of increasing plasma density and thereby moving plasma conditions closer to achieving breakeven. Although magnetic compression was utilized in some early tokamak designs in the 1970s, it has, for the most part, been replaced in more recent designs by the other heating methods described below.

Neutral beam injection—Neutral beam injection refers to a method of plasma heating in a magnetic confinement reactor that uses a high energy beam of neutral atoms or molecules. As there is no convenient method of accelerating neutral particles to high energy, the atoms are first converted into ions. These ions may be accelerated electrostatically, after which they are neutralized and injected into the plasma chamber. The neutral atoms interact with the charges in the plasma and transfer energy to the plasma ions, thereby increasing their temperature. If appropriate neutral beam atoms are utilized (e.g. deuteron ions), then the process also leads to an increase in plasma ion density.

Neutral beam injection can start with either positive or negative ions. In the former case atoms are first ionized by removing electrons to create positive ions and in the latter case atoms pick up electrons to become negative ions. A diagram of the steps involved in positive ion source neutral beam injection are shown in figure 4.9. Heating by neutral beam injection has been a common component in many experimental magnetic confinement reactors and has been employed for both tokamaks and stellarators.

When the beam of neutral particles enters into the plasma chamber, the neutral atoms are quickly ionized to create ions and electrons. Often the neutral beam is

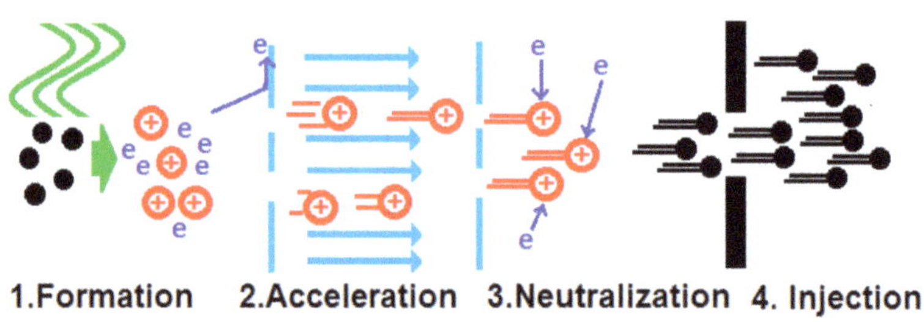

Figure 4.9. Steps to neutral beam injection to heat plasma in a magnetic confinement reactor such as a tokamak. In this case, the process is positive ion source based neutral beam injection. This Neutral Beam Injection image has been obtained by the author from the Wikimedia website https://commons.wikimedia.org/wiki/File:Neutral_Beam_Injection.png where it was made available by WikiHelper2134 under a CC BY-SA 3.0 license. It is included within this book on that basis. It is attributed to WikiHelper2134.

injected more or less tangentially to the toroidal direction of the plasma chamber. Thus, when the neutral atoms are ionized, they not only heat the plasma and increase its density, they also contribute to the plasma current. This non-inductive component of the plasma current can contribute to the steady-state operation of the tokamak.

Ion cyclotron resonance heating—Energy may be imparted to the ions in a plasma through their interaction with a high frequency electromagnetic wave, typically in the microwave regime. These electromagnetic waves can be generated outside of the tokamak using oscillators such as gyrotrons or klystrons and introduced into the plasma chamber through antennae as shown in figure 4.10.

The ion cyclotron resonance frequency is given by

$$\omega_{\mathrm{i}} = \frac{ZeB}{m_{\mathrm{i}}} \tag{4.41}$$

where ω_{i} is the angular electromagnetic wave frequency, Z is the absolute number of elementary charges on the ion, e is the electronic charge, B is the magnetic field and m_{i} is the ion mass. At the resonant frequency, typically in the range of 40 MHz to 70 MHz in typical tokamaks, the ions will spiral around the magnetic field lines due to the Lorentz force, and thereby increase their kinetic energy. As the toroidal magnetic field in a tokamak decreases with major radius, the frequency can be tuned to deposit heat at specific location within the plasma. For both ion cyclotron resonance and

Figure 4.10. Interior of the plasma chamber of the Alcator C-Mod tokamak at the Massachusetts Institute of Technology. Ion cyclotron resonance heater antennae are seen on the walls of the plasma chamber to the right and left of the central column. Divertor channels are seen outside of the central column in the floor of the chamber. This Alcator C-Mod Tokamak image has been obtained by the author from the Wikimedia website https://commons.wikimedia.org/wiki/File:Alcator_C-Mod_Fisheye_from_Fport.jpg where it was made available by Robert Mumgaard under a CC BY-SA 3.0 license. It is included within this book on that basis. It is attributed to Robert Mumgaard.

electron cyclotron resonance as described below, interaction of the electromagnetic waves, with charges in the plasma can contribute to the plasma current.

Electron cyclotron resonance heating—Similar to ion cyclotron resonance, electrons in the plasma can interact with electromagnetic waves giving rise to electron cyclotron resonance heating. The resonant frequency is given by

$$\omega_e = \frac{eB}{m_e} \tag{4.42}$$

where ω_e is the angular electromagnetic wave frequency and m_e is the electron mass. Because of the smaller mass of the electron, compared with the ions, electron cyclotron resonance frequencies are proportionately higher, typically 60 GHz to 120 GHz. Energy is imparted in the plasma as kinetic energy of the electrons and, therefore, increases the electron temperature. However, Coulombic interactions between the electrons and the ions will transfer energy to the ions and also increase their temperature. Electron cyclotron emission is the opposite process to electron cyclotron resonance and is an important diagnostic tool for observing the radial dependence of the electron temperature.

Lower hybrid drive current—Lower hybrid waves have a frequency that is intermediate between those related to ion cyclotron resonance and electron cyclotron resonance as discussed above. The lower hybrid resonance frequency is approximated in terms of the frequencies in equations (4.41) and (4.42) as (Ushigusa 1999),

$$\omega_{LH} \approx \sqrt{\omega_e \omega_i} \tag{4.43}$$

and is typically in the range of a few GHz. Lower hybrid waves are longitudinal oscillations of the ions and electrons in a magnetized plasma where the direction of propagation is perpendicular to the magnetic field. The use of lower hybrid oscillations for the purpose of heating the plasma has been implemented in a number of recent tokamak designs. The intermediate frequency has certain advantages compared to the much lower ion cyclotron frequency (which requires a large antenna) and the much higher electron cyclotron resonance frequency (which requires more complex radio frequency (RF) transmission technology).

Divertor technology

A divertor is a component of a tokamak design that allows for the removal of fusion waste during the reactor operation. Thus, the divertor may be considered as the tokamak exhaust system. The exhaust can consist of fusion by-products or heavy ions that accumulate in the plasma as a result of contamination from the vessel walls. Presence of these by-products and impurities in the plasma vessel will degrade the quality of the plasma and lead to an untimely termination of the plasma discharge. It is important, therefore, to remove such waste in order to maintain the operating conditions of the reactor. The divertor can be seen in the photograph of the interior of the Alcator C-Mod tokamak at the Massachusetts Institute of Technology in figure 4.10. The details of a possible future divertor design are seen in figure 4.11.

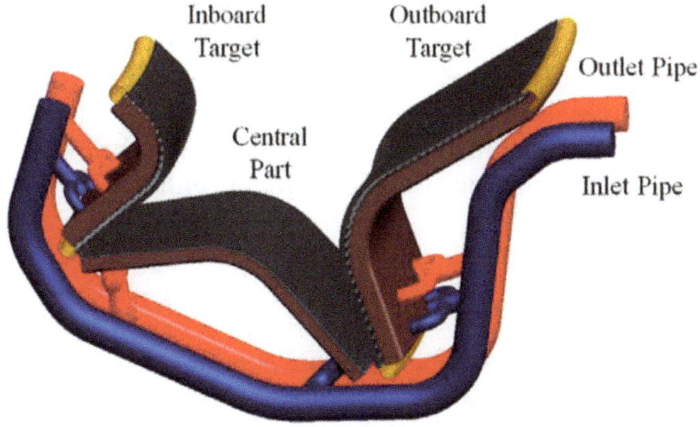

Figure 4.11. Proposed divertor design for the K-DEMO reactor, a possible future demonstration tokamak reactor in Korea. Reproduced from Kim *et al* (2015). Copyright IOP Publishing. All rights reserved.

The design of a suitable divertor for a future commercial power reactor is a significant engineering challenge. The plasma must be cool enough when it exits the plasma vessel through the divertor that its impact on the integrity of the vessel walls is minimized. This is not the case for current experimental reactors. If such reactors were used continuously at power levels expected for economical commercial operation, divertor plates would need to be replaced as a result of degradation more often than would be practical. The MAST spherical tokamak reactor as described below is an important test bed for potential commercial divertor design.

Breeder blankets

For d–t fueled tokamaks, the deuterium is a naturally occurring isotope of hydrogen (about 0.016% naturally abundant) but the tritium must be created artificially. Deuterium–tritium fusion produces neutrons at an energy of about 14 MeV by the reaction

$$\mathrm{d} + \mathrm{t} \rightarrow \alpha + \mathrm{n} \tag{4.44}$$

and these neutrons may be used to breed tritium from naturally occurring lithium. The most viable reaction is

$$\mathrm{n} + {}^{6}\mathrm{Li} \rightarrow \alpha + \mathrm{t} \tag{4.45}$$

Breeding reactions and lithium resources are discussed in more detail in section 7.4.

Experimental fusion reactors (either magnetic confinement or inertial confinement) which utilize tritium and a fuel component, obtain tritium that has been produced off-site. This approach has been appropriate for experimental research facilities but would not be suitable for a commercial power reactor. A modest 1000 MW power reactor would consume about 0.38 kg of tritium per day. This may be compared to current world tritium reserves which are about 20 kg. It is essential, therefore, that commercial fusion reactors utilize the neutrons that they produce to breed the tritium that is needed for future reactions from a blanket of lithium surrounding the reactor. Although relatively little research has been conducted with the goal of designing a practical commercial reactor breeder blanket, as progress continues toward a commercial power reactor the need for breeder blanket design investigations becomes increasingly important. The ITER has incorporated a breeder blanket test bed into its design as detailed further below in the section on that reactor. Blanket designs typically include lithium either in the form of a ceramic or in a Li–Pb eutectic alloy.

Some notable tokamaks

The present section overviews some of the more notable experimental tokamaks that have been constructed. Two tokamaks that are currently under construction are also discussed in this section. Several other tokamaks are in the planning stages and some of these are discussed in section 4.6. Previous devices, for the most part, were intended as proof-of-concept experiments and were utilized to understand the properties of the plasmas that were formed, rather than as attempts to set records for specific plasma conditions. As such, some devices utilized only deuterium as a fuel (i.e. d–d fusion) rather than more commercially viable d–t fuel. Table 4.4 gives basic information about the devices discussed in the present section.

PLT (Princeton Large Torus)

The Princeton Large Torus (PLT) was one of the first significant tokamaks constructed outside the Soviet Union. It is illustrated in figure 4.12. Previous stellarators and tokamaks generally did not incorporate any type of auxiliary heating to heat the plasma. It is acknowledged that such auxiliary heating is a

Table 4.4. Properties of some notable tokamak experimental devices. (* = under construction).

Name	Years operational	Country	Location	Major radius (m)	Minor radius (m)	Magnetic field (T)	Plasma current (MA)
PLT	1975–86	USA	Princeton Plasma Physics Laboratory	1.32	0.4	4.0	0.7
TFTR	1982–97	USA	Princeton Plasma Physics Laboratory	2.4	0.8	6.0	3.0
JET	1983–present	UK	Culham Centre for Fusion Energy	2.96	0.96	4.0	7.0
JT-60	1985–2010	Japan	Japan Atomic Energy Research Institute	3.4	1.0	4.0	3.0
MAST	2000–13	UK	Culham Centre for Fusion Energy	0.85	0.65	0.55	1.35
JT-60SA	*	Japan	Japan Atomic Energy Research Institute	2.96	1.18	2.25	5.5
ITER	*	France	Cadarache	6.2	2.0	5.3	~15

Figure 4.12. Exterior view of the Princeton Large Torus (tokamak) at the Princeton Plasma Physics Laboratory. Image taken in 1975. This Princeton Large Torus image has been obtained by the author from the Wikimedia website https://commons.wikimedia.org/wiki/File:Princeton_Large_Torus_1975.jpg where it is stated to have been released into the public domain. It is included within this book on that basis.

necessary requirement for achieving fusion conditions that are suitable for viable power production. A major goal of the experiments surrounding PLT, was the investigation of the effectiveness of different auxiliary heating approaches. Specifically, PLT utilized neutral beam injection (3 MW), ion cyclotron resonance heating (5 MW) and lower hybrid current drive (1 MW), as described in the section above.

The PLT was very successful during its eleven-year operation. Not only did it demonstrate the usefulness of various plasma heating approaches but achieved several important milestones in plasma conditions. For example, although PLT normally operated at a plasma current of 0.7 MA it was the first tokamak to achieve a plasma current of 1 MA.

In 1978 PLT obtained an ion temperature of 5 keV, the maximum of any device at the time, using neutral beam heating. This temperature falls short of the ideal fusion temperature of around 14 keV but represented a major step in magnetic confinement fusion research. This temperature was achieved using 3 MW of 40 keV neutral deuterium atoms. Four neutral beams were used, with two parallel to the direction of the plasma current (referred to as co-injection) and two antiparallel to the direction of the plasma current (referred to as counter-injection).

In 1981 PLT achieved the first discharge where the plasma current was driven completely by lower hybrid RF waves. This result provided the basis for this approach to plasma heating used in many subsequent tokamaks, although the overall efficiency of the technique fell below expectations.

The final approach to plasma heating investigated by PLT used ion cyclotron resonance heating. RF energy is provided to the plasma by means of an antenna.

An important discovery from this work was that a small concentration of an ion species with a resonant frequency at the fundamental RF frequency could provide a substantial heating effect when imbedded in a non-resonant host plasma. In 1984 an ion temperature of 5 keV along with an electron temperature of 3.5 keV, was obtained using ^{3}He minority ions in a deuterium plasma and an RF frequency of 30 MHz. Heating efficiencies were comparable to those obtained using neutral beam injection.

TFTR (Tokamak Fusion Test Reactor)

The Tokamak Fusion Test Reactor (TFTR) was constructed on the basis of information gained from experiments conducted on the PLT. TFTR became operational in 1982 and was designed and constructed with the goal of achieving breakeven (that is, scientific breakeven, as described above). As well, previous magnetic confinement fusion experiments had used deuterium almost exclusively as a fuel. A major reason for this fact is the radioactive nature of the fuel, as well as the radioactivity of reactor materials caused by fusion neutrons. These factors, as discussed in further detail in chapter 7, complicate operation and maintenance of the reactor and necessitate additional shielding and robotic servicing. TFTR, along with JET, as described below, were the first fusion reactors designed and constructed with the specific plan of using a 50:50 deuterium–tritium mixture for fuel. As d–t fusion is a requirement for a viable commercial magnetic confinement fusion reactor, the designs of TFTR and JET were a step toward this ultimate goal. A photograph of TFTR is shown in figure 4.13.

Figure 4.13. Exterior view of the Tokamak Fusion Test Reactor at the Princeton Plasma Physics Laboratory. Image taken in 1989. This Tokamak Fusion Test Reactor image has been obtained by the author from the Wikimedia website https://commons.wikimedia.org/wiki/File:TFTR_1989.jpg where it was made available by Princeton Plasma Physics Laboratory under a CC BY 3.0 license. It is included within this book on that basis. It is attributed to Princeton Plasma Physics Laboratory.

Initial experiments on TFTR utilized only ohmic heating from the plasma current. It was found, as had been evidenced by PLT results, that this is not sufficient to achieve temperatures compatible with a useful fusion reactor. Additional heating in the amount of 51 MW was incorporated into TFTR. This additional heating was mostly from neutral beam injection with some additional RF heating.

Although TFTR did not actually achieve scientific breakeven, it did provide valuable insight into tokamak design and operation and it also achieved some important milestones in tokamak performance.

In 1993 TFTR was the first tokamak to use d–t fuel and obtained an output power of 3 MW. The following year TFTR produced 10.7 MW output using 51 MW of heating for a gain of $Q = (10.7\ \text{MW})/(51\ \text{MW}) = 0.21$, a record at the time. In 1995 TFTR recorded an ion temperature of 44 keV, well beyond the optimal temperature required for breakeven. TFTR has also created ion densities of about $10^{20}\ \text{m}^{-3}$ and confinement times of 0.21 s. While in principle, these values should easily exceed the requirements for the triple product necessary to achieve scientific breakeven, they could not all be obtained at the same time. Thus, TFTR provided evidence that achieving scientific breakeven was a goal realistically expected to be achieved by future reactors.

Following the decommissioning of TFTR in 1997, Princeton Plasma Physics Laboratory constructed the National Spherical Torus Experiment (NSTX) which became operational in 1999. NSTX was a spherical tokamak of similar design to MAST as described below.

JET (Joint European Torus)

The JET at the Culham Centre for Fusion Energy in the United Kingdom was constructed more or less contemporaneously with TFTR in the United States. A photograph of JET is shown in figure 4.14. Construction of JET began in 1979 as a collaborative project of eleven countries within the European Union. The collaborating countries were Belgium, Denmark, France, Ireland, Italy, Luxembourg, Netherlands, Sweden, Switzerland, United Kingdom and West Germany. JET became operational in 1983 and is still operational today. As such, it has been one of the longest operating tokamaks ever constructed. The long-term future of JET is, at present, somewhat uncertain. As JET was a collaborative project, involving joint funding from a number of European Union countries, the exit of the United Kingdom from the E.U. makes the future funding situation for JET unclear.

JET was the first major tokamak to be designed with a 'D-shaped' rather than circular plasma geometry. The D-shaped geometry allows for a larger plasma volume and plasma current, compared to the circular geometry. It also facilitates the inclusion of divertors in the plasma chamber and the D-shape improved the structural stability of the magnets and vacuum vessel. A major feature in the design of JET was a robust heating system similar to that incorporated in TFTR. The JET heating system consists of neutral beam injection (34 MW), ion cyclotron resonance heating (10 MW) and lower hybrid current drive (7 MW). As part of upgrades completed on JET in 1991, divertors were added to aid in the operation of the machine with d–t fuel.

Figure 4.14. The Joint European Torus (JET) magnetic fusion experiment at the Culham Centre for Fusion Energy. Image taken in 1991. The large orange structures are the armatures for the magnets to produce the toroidal field. This Joint European Torus image has been obtained by the author from the Wikimedia website https://commons.wikimedia.org/wiki/File:The_JET_magnetic_fusion_experiment_in_1991.jpg where it was made available by EFDA JET under a CC BY-SA 3.0 license. It is included within this book on that basis. It is attributed to EFDA JET.

Like its counterpart at Princeton, TFTR, JET was designed with the concept of achieving scientific breakeven. Also, like TFTR, JET failed to reach that goal. However, much was learned from experiments on JET and its operation has made significant progress to the ultimate production of a viable commercial fusion power plant.

In 1991 JET became the first tokamak to operate using a 50:50 mixture of deuterium–tritium as fuel. Later, in 1997, JET produced 16 MW of output power using 24 MW of heating power. This gave a value of $Q = 0.67$, which, at present, remains a record for the closest a tokamak has come to scientific breakeven. Also, in that year, JET produced a sustained output of 4 MW for nearly 5 s, demonstrating a significant increase in plasma confinement time and a major step toward possible tokamak steady state operation.

In 2009 JET was shut down to allow a number of modifications to be incorporated into its design. It resumed operation in 2011 and since that time has served primarily as a test bed for design concepts to be incorporated into ITER (see below). Much of this work has been involved in the study of the materials to be used in the ITER construction. Specifically, the carbon plasma chamber walls of JET were replaced by components made from beryllium and tungsten and tungsten divertors were installed to replicate materials to be used in ITER. In recent years JET has planned additional d–t experiments in order to better understand plasma conditions that will be present in ITER.

JT-60 (Japan Torus-60)

The Japan Torus-60 (JT-60) was located in the Ibaraki Prefecture and was operated by the Japan Atomic Energy Agency (previously named Japan Atomic Energy Research Institute). JT-60, like TFTR and JET, became operational in the mid-1980s and was designed and constructed with the expectation of achieving scientific breakeven. It utilizes a D-shaped plasma geometry and over the years it has been upgraded and referred to as JT-60A and later JT-60U.

Early experiments on JT-60 were conducted using a hydrogen (^1H) plasma. Since 1991 experiments were conducted using a deuterium plasma and a number of important milestones were accomplished. In 1996 JT-60 achieved an ion temperature of 45 keV, slightly exceeding the record set the previous year by TFTR. Also, in 1996, the JT-60 reached a triple product of 1.53×10^{21} keV s m^{-3}, a record which remains valid today. In 1998 JT-60 produced conditions in a d–d plasma that would have achieved breakeven if the reactor had contained d–t fuel. This experiment produced a value of $Q = 1.25$ scaled to a d–t plasma. The operation of the reactor was also shown to be very stable and reproducible as 15 discharges yielded scaled values of $Q > 1.0$. More recent research on JT-60 concentrated on achieving high β steady state operation. In 2006 JT-60 reported that it had maintained a plasma discharge for 28.6 s.

In 2010 JT-60 (at that time, named JT-60U, 'U' for upgraded) was decommissioned to make way for the substantially redesigned Japan Torus-60 Super Advanced (JT-60SA) described below.

MAST (Mega-Ampere Spherical Tokamak)

Parallel to the development of the traditional tokamak, research has been ongoing on the development of the so-called spherical tokamak. The traditional tokamak had a major radius which was significantly greater than the minor radius giving the device a distinct toroidal or doughnut shape. The spherical tokamak has major and minor radii that are similar in magnitude. This gives the device the shape of a sphere with a hole through the center. This is sometimes referred to as a 'cored apple' geometry. In the spherical tokamak design many separate toroidal field coils are replaced by a single central conductor which serves as a common connection to the individual toroidal coils on the outside of the plasma vessel. The first full sized operational spherical tokamak was the Small Tight Aspect Ratio Tokamak (START) at the Culham Centre for Fusion Energy in the United Kingdom. This device was operational from 1990 to 1998.

During the 1990s interest in the spherical tokamak geometry grew and led to the design and construction of two similar devices—the National Spherical Torus Experiment (NSTX) at the Princeton Plasma Physics Laboratory and the Mega-Ampere Spherical Tokamak (MAST) at Culham Centre for Fusion Energy. These devices had very nearly identical dimensions and specifications, although NSTX had a smaller external magnetic field but generated a larger plasma current. Some information about the MAST facility is presented here.

Figure 4.15 shows an image of a typical plasma in MAST and illustrates the geometry of a plasma in a spherical tokamak. The figure shows the features of the

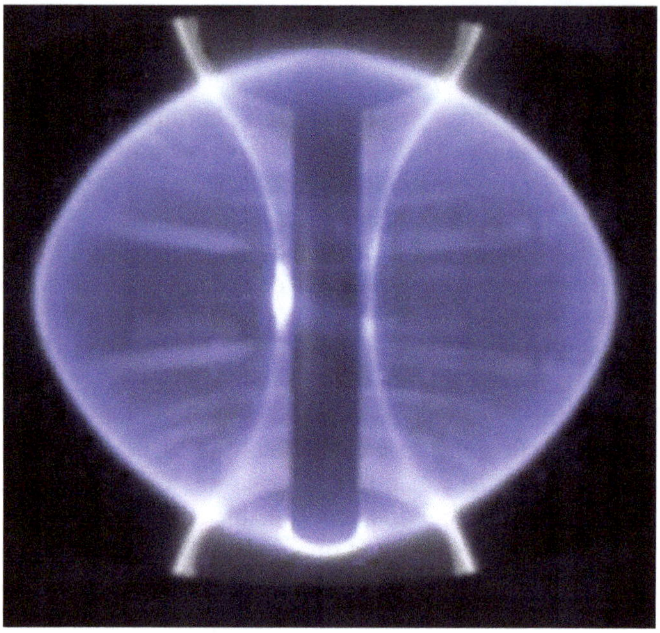

Figure 4.15. Image of a typical plasma in the MAST spherical tokamak at the Culham Centre for Fusion Energy. The roughly spherical main plasma appears in the center of the image. The extensions to the plasma at the top and bottom of the main plasma are areas where the plasma flows into ring-shaped 'diverters'. The diverters help to shape the plasma and clear it of impurities and fusion ash. Some details of the test chamber structure can be seen in image of the main spherical plasma. This MAST Spherical Tokamak image has been obtained by the author from the Wikimedia website https://commons.wikimedia.org/wiki/File:MAST_plasma_image.jpg where it was made available by Culham Centre for Fusion Energy under a CC BY-SA 3.0 license. It is included within this book on that basis. It is attributed to Culham Centre for Fusion Energy.

plasma where it flows into the reactor's divertors. Experiments conducted on MAST extended the results previously obtained using START and confirmed the high efficiency and high β that could be obtained on a spherical tokamak. MAST also contributed to the understanding of instabilities at the edges of the plasma and how they could be controlled.

MAST operated from 2000 to 2013. Since that time, construction has been underway on an upgraded version of MAST with increased heating power, magnetic field, plasma current and discharge duration. In addition to bringing the plasma parameters closer to breakeven, a major focus of research on the upgraded MAST will be to investigate divertor design. The so-called 'Super-X Divertor' is intended to decrease heat loads on the divertor and thereby substantially increase divertor durability and life expectancy.

JT-60SA (Japan Torus-60 Super Advanced)

While previous tokamaks, as described above, have made significant progress in the development of magnetic confinement fusion, they have not achieved scientific breakeven, although the JT-60 achieved a value of $Q = 1.25$ when plasma parameters were scaled from d–d fuel to d–t fuel. While various machines have reported impressive numbers for temperature, density and confinement time, it has not been possible to obtain high values for all of these parameters at the same time. The ITER as described below, is intended to reach and exceed the conditions for d–t fusion scientific breakeven. The design of the Japan Torus-60 Super Advanced (JT-60SA) has been undertaken in collaboration with the design of ITER so that the capabilities and goals of the two machines complement each other to best make use of financial and scientific resources. For the most part, JT-60SA, which, like its predecessor, is being constructed by the Japan Atomic Energy Agency, will be taking a supporting role to ITER in order to best provide the knowledge necessary for the construction of DEMO (DEMOnstration Power Station), as discussed in section 4.7 below.

The principal role of the JT-60SA reactor is for the testing of components that will be needed for DEMO. To achieve this goal JT-60SA has been designed to utilize only d–d fuel. The specific systems that JT-60SA is intended to optimize for future commercial reactor use are described below.

Magnet design—Most previous tokamaks utilized non-superconducting electromagnets to provide the necessary magnetic field. It is generally believed that the implementation of superconducting magnet technology is necessary for the construction of a viable magnetic confinement fusion power reactor. The JT-60SA will incorporate superconducting Nb_3Sn for the 18 toroidal field coils and 3 central solenoid coils and superconducting NbTi for the 6 poloidal field coils and 18 correction coils in order to thoroughly investigate the utilization of superconducting magnets for magnetic confinement fusion reactors. In order to operate these superconducting magnets, as discussed in the section on tokamak technology above, a reliable source of liquid helium as a cryogen is needed. The JT-60SA facility will include such a cryogenic plant and will evaluate the efficacy of such a system.

Remote handling systems—An essential component of an operational d–t fusion reactor is the availability of a robust remote handling system for reactor maintenance and repair. The use of d–t fusion fuel, as noted above and described in detail in section 7.3, presents safety hazards due to the radioactive nature of the tritium and radioactive by-products from neutron irradiation of reactor materials. It is, therefore, essential that reactor work be conducted remotely in order to ensure the safety of personnel.

Fueling systems—JT-60SA will evaluate the suitability of pellet injection for fueling the reactor.

Neutral beam injection and electron cyclotron resonance heating—Heating for JT-60SA will be provided by a combination of neutral beam injection and electron cyclotron resonance heating. There will be 34 MW of neutral ion beam injection, 24 MW from positive ion source based neutral beams and 10 MW from negative ion source based neutral beam. In addition, there will be 7 MW of electron cyclotron resonance heating at dual frequencies of 110 GHz and 138 GHz. This gives the total heating power of 41 MW, which can be introduced to the plasma for up to 100 s.

Figure 4.16. External view of the JT-60SA tokamak under construction at the Japan Atomic Energy Research Institute. Image taken in 2016. This JT-60SA image has been obtained by the author from the Wikimedia website https://commons.wikimedia.org/wiki/File:JT-60SA_Tokamak.jpg where it was made available by Michel Maccagnan under a CC BY-SA 4.0 license. It is included within this book on that basis. It is attributed to Michel Maccagnan.

A photograph of JT-60SA during construction is shown in figure 4.16. In April 2020 the Japan Atomic Energy Agency announced that construction of the JT-60SA reactor had been completed. As of the end of 2020 operational information has not yet been released.

ITER (International Thermonuclear Experimental Reactor)
The ITER is by far the most ambitious magnetic confinement fusion project ever undertaken. The acronym ITER (pronounced 'eater') is also Latin for 'the way'. A detailed description of the ITER project has been published by Claessens (2020).

ITER is a collaborative project funded and operated by seven parties—the European Union, China, India, Japan, Russia, South Korea and the United States. Canada was originally a member of the collaboration but withdrew due to lack of funding. Since Brexit, the status of the United Kingdom in the ITER project is uncertain and is the subject of ongoing negotiations. ITER also has collaboration agreements with Kazakhstan, Switzerland and Australia. ITER is located at the Cadarache facility near the village Saint-Paul-lès-Durance in southern France.

ITER is a very long-term project. The concept for a large-scale international magnetic confinement fusion reactor collaboration originated at the 1985 Geneva Summit Meeting between Soviet leader Mikhail Gorbachev (b. 1931) and United States president Ronald Reagan (1911–2004). The past and projected future timeline for the project is summarized in table 4.5. As with many large-scale scientific projects, the timeline, budget and scientific expectations evolve over time. At the time of the engineering design phase, ITER was expected to cost €5 billion and take 10 years to construct. Table 4.5 gives an estimate of the current estimate for the length of construction. The total cost is still difficult to assess but current estimates are as high as €41 billion.

The changes in ITER's design and scientific expectations are obvious from a comparison of the initial 1998 design and the baseline from 2011 as summarized in

Table 4.5. Milestones for the past and projected timeline for the ITER project.

Year	Milestone
1985	Geneva summit
1988	ITER project initiation
1992	Engineering design begins
1998	Engineering design completed
2008	Site preparation begins
2013	ITER complex construction begins
2020	Tokamak assembly initiated
2025	Completion of construction and first d–d plasma
2035	First d–t plasma

Table 4.6. Comparison of the 1998 engineering design of ITER and the baseline design from 2011. Adapted from data in Parker (1998) and online at http://fusionwiki.ciemat.es/wiki/ITER.

Parameter	1998 design	2011 design
Major radius (m)	8.14	6.2
Minor radius (m)	2.8	2.0
Plasma volume (m^3)	2000	870
Toroidal magnetic field (T)	5.68	5.3
Plasma current (MA)	21	15
Average electron density (m^{-3})	0.98×10^{20}	1.0×10^{20}
Average ion temperature (keV)	13	9.2
Plasma beta (β)	0.022–0.03	0.025
Confinement time (s)	5.9	3.7
Maximum auxiliary heating (MW)	100	73
Fusion power (MW)	1500	500
Gain (Q)	∞ (ignition)	10

table 4.6. It is clear that compromises in the design had to be made to the 1998 plan before construction started about fifteen years later. Some additional minor design changes have occurred since 2011. For example, plasma volume of the present design has been decreased to 840 m^3. Certainly, there was a downsizing of the dimensions of the tokamak, but the most significant difference was the change in the expected reactor gain, Q, as defined in equation (4.9). The original design was intended to reach ignition and, therefore, no external heating would be required to maintain fusion conditions and the gain would be $Q = \infty$. Current expectations are for $Q \approx 10$, certainly well above scientific breakeven, but probably not compatible with a viable power reactor. This change in expectations for ignition are responsible for the change in the expected fusion power output, as shown in the table.

ITER is obviously designed to achieve plasma conditions that exceed those for scientific breakeven and even approach those that are necessary for ignition. As such it will surpass the performance recorded thus far for other magnetic confinement fusion reactors. However, as a step toward a commercial fusion power reactor, ITER will also investigate a number of systems that are not directly related to the creation of plasma conditions appropriate for fusion but will be necessary components of a commercial power reactor. Thus, ITER will be a lead-up to the design and construction of DEMO. Overall, the goals of ITER are as follows:

- To produce fusion with a reactor gain of $Q = 10$;
- To produce steady state fusion output with a minimum gain of $Q = 5$;
- Maintain a plasma discharge for a minimum of 8 min;
- Develop superconducting magnet technology appropriate for a demonstration fusion power plant;
- Develop appropriate remote handling technology;
- Investigate breeder blanket technology;
- Develop and refine neutron shielding and neutron energy recovery technology;
- Investigate the response of different materials to the intense neutron flux from a fusion reactor.

Other systems necessary for the development of a commercial power reactor will be investigated as part of the JT-60SA tokamak project conducted in collaboration with ITER and described above.

Figure 4.17 shows a concept model for ITER at about the time reactor construction began. The ITER incorporates some important design features as described below which are essential to the fulfillment of its specified goals.

Plasma vessel—ITER utilizes a double walled steel toroidal plasma vessel with a D-shaped cross section as illustrated in figure 4.18. The torus will have an outside diameter of 19.4 m and a height of 11.3 m. The overall mass of the plasma chamber with associated port structures and radiation shielding will be over 5200 tons. This is 16 times the mass of any previous tokamak plasma vessel. The space between the two walls of the plasma vessel will include channels for cooling water as described below.

Cooling system—The object of a commercial power reactor is to generate electricity from the energy produced by nuclear fusion. Traditionally this is accomplished in fossil fuel and nuclear fission generating stations by using the heat produced to boil water to drive turbine/generators. ITER will not produce electricity. However, the fusion heat that it produces must still be dissipated. Cooling water circulating through channels in the space between the reactor vessel's double walls will be used for this purpose. Cooling water will be pumped from Canal de Provence through a 5 km long pipeline. Cooling water output will be to the Durance (river). Supplementary cooling will be provided by liquid nitrogen and liquid helium cooling systems.

Divertor—The design of the ITER divertor was a major challenge. Two heat resistant materials were considered as covering materials for the divertor. Firstly, carbon fibers with a sublimation temperature of 3915 K and tungsten with a melting

Figure 4.17. Concept model of the ITER exhibited at the International Fusion Energy Days in Monaco, December 2013. This ITER image has been obtained by the author from the Wikimedia website https://commons.wikimedia.org/wiki/File:ITER_Exhibit_(01810402)_(12219071813)_(cropped).jpg where it was made available by IAEA Imagebank under a CC BY-SA 2.0 license. It is included within this book on that basis. It is attributed to ITER Exhibit (01810402).

temperature of 3695 K. Although carbon is somewhat more heat resistant than tungsten, it has the disadvantage that it readily absorbs substantial quantities of hydrogen (including tritium). This is undesirable for two reasons—it leads to degradation of the carbon and it leads to accumulation of radioactive tritium in the reactor materials. Consequently, tungsten was chosen as the cladding material for the divertor in ITER. The general divertor design is seen in figure 4.18.

Magnets—The magnets used in ITER will be made from the traditional superconducting materials, Nb_3Sn and NbTi. The central solenoid and 18 toroidal field magnets will produce maximum magnetic fields of 13.5 T and 11.8 T, respectively. Since both these field requirements are near or above the critical field of NbTi at liquid helium temperature as noted in table 4.3, the higher critical field material Nb_3Sn will be used. Magnets will be the cable-in-conduit design as illustrated in figure 4.7. Lower field poloidal coils and correction coils will utilize superconducting magnets fabricated from the more economical and more convenient superconductor NbTi.

Figure 4.18. Cross section of the ITER plasma vessel. The divertor, which is sheathed in tungsten, is shown by the region in light grey at the bottom of the vessel. This design may be compared to the divertor shown in figure 4.11. This ITER Plasma Vessel image has been obtained by the author from the Wikimedia website https://commons.wikimedia.org/wiki/File:NIST_Light_Source_Illuminates_Fusion_Power_Diagnostics_(5940503177).jpg where it is stated to have been released into the public domain. It is included within this book on that basis.

Auxiliary heating—Three types of auxiliary heating systems are included in the ITER design. Firstly, there are two negative ion source neutral beam injectors, each providing about 17 MW. The beams consist of 1 MeV neutral deuteron atoms and can provide heating for up to 1 h at a time. There is also a possibility of adding a third neutral beam at a later time. In addition to the neutral beam injectors, ITER also incorporates a 20 MW electron cyclotron resonance heater and a 20 MW ion cyclotron resonance heater. The total auxiliary heating power is 73 MW with the possibility of increasing this to about 100 MW.

Breeder blanket—Breeding of tritium by the reaction of fusion neutrons with lithium will be an essential component of any commercial fusion power reactor. ITER will be the first fusion reactor to include a large-scale testing platform for breeder blanket designs. The reactor will include facilities for readily changing so-called Test Blanket Modules (TBM) which use different blanket designs but the same mounting geometry. Six designs are scheduled for testing to determine the efficacy of the different approaches. Design criteria for breeder blankets include:

- a high efficiency of tritium production;
- ease of tritium extraction;
- mechanical and chemical stability;
- low levels of neutron activation products (see section 7.3).

Potential designs include helium cooled lithium-lead as well as a pebble bed design using lithium titanate (Li_2TiO_3) and lithium orthosilicate (Li_4SiO_4). These ceramics have melting points of 1806 K and 1531 K, respectively, compared to 454 K for elemental lithium.

It is anticipated that the knowledge gained from ITER, along with the collaborative projects undertaken at JT-60SA, will lead to the design and construction of the DEMO discussed in section 4.6.

4.5 Stellarators

Basics of stellarators

The development of the stellarator began in 1951 with the concept proposed by Lyman Spitzer that utilized a figure-eight shaped device to overcome the problems of particle drift in the plasmas of simple toroidal devices. During the early years of stellarator research, alternative approaches to stabilizing particle orbits in toroidal devices using externally applied magnetic fields were considered. A simple, fairly generic, approach is illustrated in figure 4.19. Here the main, poloidal field coils are wrapped around the toroidal chamber, as shown by the green coils in the figure (compare with figure 3.8). An even number of helical windings (i.e. white windings in the figure) are wrapped around the toroid, as shown. The helical windings are arranged in pairs with currents circulating in opposite directions and

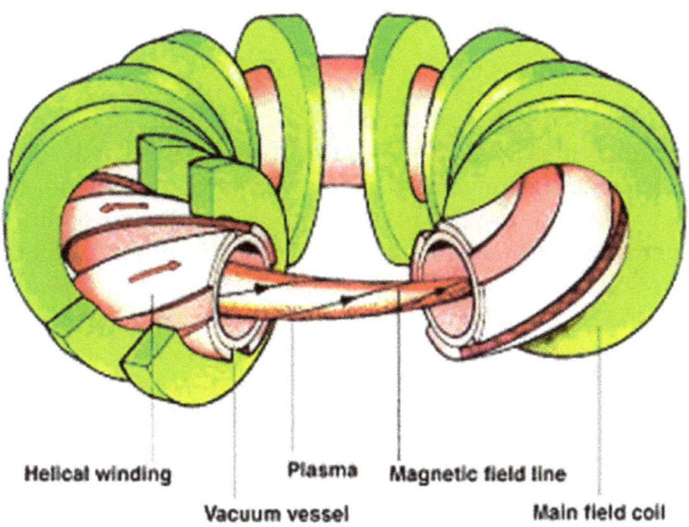

Figure 4.19. Diagram of a simple stellarator utilizing helical field coils (white) and main (poloidal) field coils (green). This configuration is often referred to as a 'classical' stellarator. This Simple Stellarator image has been obtained by the author from the Wikimedia website https://commons.wikimedia.org/wiki/File:Magnetic_field_stellarator.png where it was made available by European Nuclear Society under a CC BY 3.0 license. It is included within this book on that basis. It is attributed to European Nuclear Society.

serve to produce helical trajectories of the charged plasma particles without the introduction of a toroidal plasma current. Modern stellarators often use asymmetric poloidal field coils (rather than ones with a circular cross section) to introduce the helical twist to the magnetic field lines without the helical windings (see description of the Wendelstein 7-X below).

The success achieved by the Soviet T-3 tokamak in the late 1960s shifted research directions in controlled nuclear fusion away from the stellarator and toward this geometrically simpler magnetic confinement device. However, in recent years there has been renewed interest in the stellarator. This interest stems from several factors including the following.

- The inherent non-steady state nature of the conditions in the tokamak make this device less appealing than the stellarator. The pulsed nature of the plasma conditions in the tokamak arise because a toroidal current is generated inductively. The stellarator, however, can operate in steady state mode, as the plasma is kept in equilibrium by the field produced by the external coils and no toroidal current is necessary to ensure plasma stability.
- As tokamaks have become more powerful and have come close to fulfilling the Lawson criterion, additional plasma instabilities have developed, particularly as a result of the large toroidal plasma currents necessary to confine the plasma. Stellarators are much less susceptible to instabilities that can destroy the plasma equilibrium as large toroidal plasma currents are not necessary.
- Finally, the availability of vastly improved computer modeling capabilities in recent years has opened up the possibility of more complex stellarator geometries (compared to the 'classical' stellarator design in figure 4.19). Recent stellarator experiments, as described below, make use of this approach.

While tokamak research has been ongoing for more than half a century, modern stellarator research is in its much earlier stages. The research conducted thus far has concentrated on understanding the basic properties of plasmas in a device with helical symmetry and no toroidal plasma current and the possibility that such a device could ultimately overcome some of the drawbacks of the tokamak design that are difficult to resolve. Experiments to date have often used a hydrogen (^{1}H) plasma, or in some cases a deuterium plasma, to study reactor performance. In addition to the possibility of steady state operation, the stellarator approach to magnetic confinement provides additional potential advantages over the tokamak. In principle, there is no plasma pressure limit, i.e. the Greenwald limit, as discussed for the tokamak. The possibility of higher plasma pressures also leads to the possibility of higher values of β, and this been observed, as noted below, in some stellarator experiments.

Recent stellarator experiments

Beginning in the 1990s several experimental facilities utilizing the stellarator concept have been established. These experimental devices are small compared to major

Table 4.7. Principal design characteristics of some recent experimental stellarators.

Stellarator	Major radius (m)	Minor radius (m)	Maximum magnetic field (T)
Wendelstein 7-AS (W7-AS)	2.0	0.2	3.5
Large Helical Device (LHD)	3.7	0.64	3.0
Helically Symmetric Experiment (HSX)	1.2	0.15	1.25
Wendelstein 7-X (W7-X)	5.5	0.53	3.0

tokamak installations and are mainly aimed at understanding the potential advantages of stellarators over tokamaks by investigating plasma characteristics and studying the potential of these devices to be scaled-up to commercial power reactor size facilities. The present section overviews the four most prominent of these devices. These are covered chronologically in the order in which they became operational. The major characteristics of the stellarators discussed in this section are summarized in table 4.7.

W7-AS (Wendelstein 7-AS)

The Wendelstein 7-AS, located at the Max Planck Institute for Plasma Physics (Max-Planck-Institut für Plasmaphysik) in Garching, Germany, was one of the earliest modern stellarators. It operated from 1988 to 2002, when it was decommissioned to make way for the Wendelstein 7-X. The Wendelstein project was named after The Wendelstein, a 1838 m high mountain in the Bavarian Alps. This naming approach followed from 'Project Matterhorn', an early fusion research program in the United States. The name for Project Matterhorn was suggested by Lyman Spitzer, who was an enthusiastic mountaineer. In 1965 Spitzer along with Donald Morton made the first ascent of Mount Thor (1675 m) which is located in Auyuittuq National Park in Nunavut, Canada. Project Matterhorn was later renamed the Princeton Plasma Physics Laboratory.

The Wendelstein 7-AS project was instrumental in demonstrating the potential of the stellarator design to overcome some of the problems associated with further development of the tokamak design. Specifically, some of the major accomplishments of the project were as follows:

- investigation of properties of a current-free plasma;
- observation of improved plasma stability as a result of careful control of magnetic field geometry;
- investigation of various approaches to plasma heating;
- demonstration of the effectiveness of divertor technology in stellarator design;
- investigation of optimization criteria for plasma stability.

Based on the results obtained on Wendelstein 7-AS, the construction of an improved and larger stellarator, the Wendelstein 7-X was undertaken. Details of this larger machine are presented below.

LHD (Large Helical Device)

The Large Helical Device or LHD (Japanese: Ōgata Herikaru Sōchi) is a superconducting stellarator located in Toki, Gifu, Japan and operated by the National Institute for Fusion Science (Shizenkagaku kenkyuukikou). It became operational in 1998. The design of the LHD uses a single set of helical coils to generate both the toroidal component and poloidal component of the external magnetic field. This design is known as a heliotron in Japan where a number of smaller devices of this geometry have been constructed. Outside of Japan it is commonly referred to as a torsatron. Figure 4.20 shows the design of the vacuum chamber and the important design characteristics are given in table 4.7. The LHD uses neutral beam injection, ion cyclotron RF and electron cyclotron heating to heat the plasma.

Initial experiments were conducted using a hydrogen (^1H) plasma. Since 2017 experiments utilizing a deuterium plasma have also been conducted. Experiments have concentrated on the measurement of plasma properties rather than achieving conditions for nuclear fusion. Ion temperatures of 13.6 keV and electron temperatures of 10 keV have been achieved in the LHD, although not at the same time. Under normal operating conditions, ion temperatures are higher than electron temperatures. This is a desirable situation, since it is the ion temperature that is important for fusion reactions. Perhaps the most impressive achievement of the LHD is a plasma density of 10^{21} m^{-3}. This is substantially greater than the Greenwald limit, which, as noted above, has been typically assumed to be the maximum plasma density achievable in a tokamak. It is also significant that the high plasma pressures that have been obtained, also lead to a high value of $\beta \approx 0.05$ (or 5%). The steady

Figure 4.20. Vacuum chamber of the Large Helical Device at The National Institute for Fusion Science in Japan. This Large Helical Device image has been obtained by the author from the Wikimedia website https://commons.wikimedia.org/wiki/File:Large_Helical_Device_mirrored_(2725741289).jpg where it was made available by Justin Ruckman under a CC BY 2.0 license. It is included within this book on that basis. It is attributed to Infinite Machine.

state operation of the stellarator design has been demonstrated by the LHD, which has maintained plasma conditions for as long as 3900 s.

HSX (Helically Symmetric Experiment)

The Helically Symmetric Experiment (HSX), shown in figure 4.21, is an experimental stellarator in the Department of Electrical and Computer Engineering at the University of Wisconsin—Madison that became operational in 1999. The design features of the HSX are summarized in table 4.7. Unlike the LHD (as noted above) and Wendelstein 7-X (as discussed below), the HSX does not use superconducting magnets but rather 48 non-planar non-superconducting coils. The device utilizes 100 kW of electron cyclotron resonance heating at 28 GHz to achieve plasma conditions.

The HSX has a magnetic field structure that is known as quasi-helically symmetric and which is topologically equivalent to a tokamak without the plasma current. The HSX is the only device worldwide with this geometry. As is the case for other stellarators, the HSX operates in steady state mode and there is not expected to be the plasma density limitations that are characteristic of tokamaks.

Experiments on the HSX have achieved plasma at a temperature of 3 keV and a density of 8×10^{18} m^{-3}. Research has focused on the fundamental physics of plasmas with quasi-helically symmetry.

W7-X (Wendelstein 7-X)

Wendelstein 7-X is the most recent of the major stellarator experiments and, at present, it is the largest. It was constructed in Greifswald, Germany by the Max

Figure 4.21. The Helically Symmetric Experiment (HSX) at the University of Wisconsin—Madison. This Helically Symmetric Experiment image has been obtained by the author from the Wikimedia website https://commons.wikimedia.org/wiki/File:HSX_picture.jpg where it is stated to have been released into the public domain. It is included within this book on that basis.

Planck Institute for Plasma Physics and became operational in October 2015. It followed from the earlier stellarator, Wendelstein 7-AS, located at the Max Planck Institute for Plasma Physics at Garching, Germany. Physical parameters for the Wendelstein 7-X are summarized in table 4.7.

Figure 4.22 shows an illustration of the Wendelstein 7-X design. There are 50 non-planar (blue) and 20 planar (brown) liquid helium cooled NbTi superconducting coils. Heating is accomplished using 10 MW of electron cyclotron resonance heating, with additional heating by ion cyclotron resonance available. Figure 4.23 shows the helical shape of the confined plasma as defined by constant magnetic field surfaces.

Phase 1 of Wendelstein 7-X's experimental investigations occurred between 2015 and 2018. In June 2018 experimental results gave a maximum observed triple product of $nT\tau = 5 \times 10^{19}$ m^{-3} keV s. This may be compared to equation (4.12) for the triple product specified by the Lawson criterion for ignition, $nT\tau = 3 \times 10^{21}$ m^{-3} keV s. Thus, the conditions in Wendelstein 7-X are about two orders of magnitude short of reaching ignition. Toward the end of 2018, experiments on the Wendelstein 7-X reached plasma densities around 2×10^{20} m^{-3}, temperatures of around 2 keV and confinement times of 100 s, although not all at the same time.

During Phase 1, much was learned about the behavior of plasmas in stellarators, and a number of factors that would benefit from improvement had been identified. At the end of 2018, Wendelstein 7-X was shut down to implement upgrades. Initial estimates of the time scale for the completion of upgrades put the

Figure 4.22. Poloidal magnet coils of the Wendelstein 7-X stellarator. Non-planar coils are shown in blue and planar coils are shown in brown. The five-period helical geometry can be seen in the illustration. This Wendelstein 7-X Stellarator image has been obtained by the author from the Wikimedia website https://commons.wikimedia.org/wiki/File:Wendelstein_7-X_schematic_view_of_magnets_system.jpg where it was made available by M Nagel, C P Dhard, H Bau, H-S Bosch, U Meyer, S Raatz, K Risse and T Rummel under a CC BY 3.0 license. It is included within this book on that basis. It is attributed to Nagel *et al* (2017).

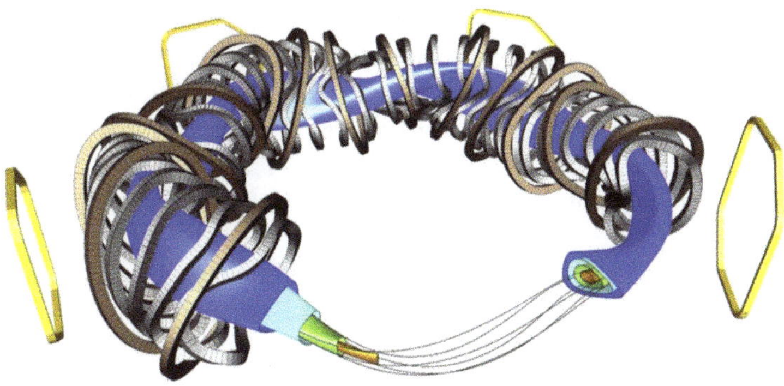

Figure 4.23. Diagram of the magnetic coil configuration and plasma geometry in the Wendelstein 7-X. Planar coils are shown in brown and non-planar coils are shown in grey. Coils in the front portion are not shown in the illustration to allow for clarity of the plasma features. Four of the five trim coils are shown in yellow (the fifth trim coil would be in the foreground of the illustration). Representative constant magnetic field surfaces are shown in blue, teal, green and orange. Magnetic field lines are shown on the green magnetic field surface by the black lines. Reprinted from Sunn Pedersen T *et al* (2016).

beginning of Phase 2 operation at some time in 2021. Phase 2 upgrades are expected to include the following:

- addition of 8 MW of neutral beam injection for up to 10 s;
- installation of water-cooled inner cladding on large sections of the plasma chamber;
- installation of water cooled divertor plates to replace the carbon fiber reinforced divertor plates used in the Phase 1 experiments.

These upgrades should allow for improved plasma heating, increased plasma density and steady-state operation for discharge times up to 30 min.

4.6 Future directions for magnetic confinement fusion

Summary of progress in magnetic confinement fusion

Much progress has been made in the field of nuclear fusion energy. However, experimental reactors are not yet at a point where commercial utilization of fusion power is viable. Speculation of when (or even if) fusion energy would become practical has existed since the early days of research in this field in the 1950s. It has often been anticipated that fusion power reactors would be available in 20 years, although the twenty-year estimate keeps moving forward. It is clear that such continued optimism is not likely realistic. Much of this speculation is related to the history of nuclear fission. Following the development of our knowledge of the nucleus in the early part of the 20th century, neutron induced fission of uranium was discovered in 1938 by Lise Meitner (1878–1968), Fritz Strassmann (1902–80) and Otto Hahn (1879–1968). Motivated by the desire to construct nuclear weapons, the United States undertook the development of a nuclear reactor for the purpose of producing fissile material. In 1942, only four years after the first observation of neutron induced fission, the first nuclear reactor was constructed under the direction of Enrico Fermi (1901–54). After World War II there was rapid development of nuclear energy for the purpose of generating electricity. The first experimental nuclear power reactor became operational in Idaho, United States in 1951. The first commercial nuclear power reactor that provided electricity for the grid became operational in Sellafield, United Kingdom in 1956. Thus, from the first experimental observation of neutron induced fission to the first commercial reactor was a mere 18 years.

The history of the development of nuclear fission energy does not give us much insight into the time scale for the development of nuclear fusion energy. The first laboratory observation of nuclear fusion occurred in 1934, and actually predates the observation of induced fission. However, after more than 80 years, the goal of commercial fusion power is still many years, or more likely many decades, in the future. We can, however, gain insight into the possible future of fusion energy by looking in some detail at developments in the past. Magnetic confinement fusion has the longest history and serves as a good example of progress in this area.

At present, there are two major directions to research on magnetic confinement nuclear fusion reactors—tokamak type devices and stellarator type devices. These two approaches differ in the presence (tokamak) or absence (stellarator) of a toroidal plasma current. While tokamak research has continued quite rigorously since the late 1960s, stellarator research has only fairly recently been approached on a large scale. Thus, stellarator research is, relatively speaking, at the fairly early stages of development. Nevertheless, both devices have made major accomplishments toward understanding the requirements for a successful commercial magnetic confinement nuclear fusion reactor.

The progress that has been made in magnetic confinement fusion research can be summarized by plotting the triple product that has been obtained by various experimental reactors as a function of ion temperature. This is shown in figure 4.24. The time scale roughly follows a diagonal line beginning in the lower left for the T-3 reactor that was operational in the late 1960s to the predicted performance for ITER, in the upper right, that will likely be operational for d–t fusion by the mid-2030s.

The plot in figure 4.24 clearly shows the relationship of performance of recent traditional tokamaks with that of recent spherical tokamaks and stellarators. The future development of these latter two reactor designs is uncertain. Spherical tokamaks have several advantages over traditional tokamaks. The spherical geometry can potentially lead to higher β values and a corresponding reduction in magnet size. As well, the more compact geometry will reduce overall infrastructure costs. However, although the central conductor design benefits overall reactor size, it has the major disadvantage that it makes it very difficult to include suitable neutron shielding inside the central core and to utilize superconducting toroidal field coils. These two factors make it questionable whether the spherical tokamak design could be scaled-up to a size that would be appropriate for a commercial power reactor. The stellarator design has several attractive features. Firstly, the intrinsic lack of a toroidal plasma current means that steady state operation is possible. As well, the fact that stellarator plasma density is not limited by the Greenwald limit means that

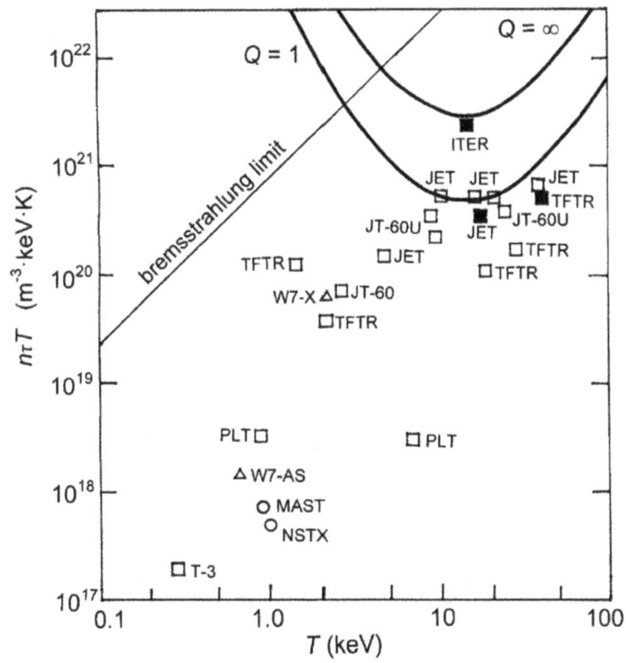

Figure 4.24. Triple product as a function of ion temperature showing results for some magnetic confinement fusion reactors including predicted values for Wendelstein 7-X and ITER. Tokamaks are shown as squares, spherical tokamaks as circles and stellarators as triangles. Open symbols represent hydrogen or deuterium experiments and closed symbols represent d–t experiments. Acronyms are from the text or table 4.4 for tokamaks and from table 4.7 for stellarators. Curves are shown for scientific breakeven ($Q = 1$) and ignition ($Q = \infty$). The bremsstrahlung limit is defined by the conditions for which bremsstrahlung losses exceed fusion power.

higher operating pressures are possible than in the traditional tokamak design. This higher pressure leads to a higher β value and, therefore, less stringent field requirements and a more compact and cost-effective design. Future experimental results from the most recent major stellarator device, Wendelstein 7-X, will be an important factor in directing research efforts on these devices.

For the present, the major effort of magnetic confinement fusion research is directed toward the development of the traditional tokamak. The current construction of ITER is the most significant project in this area to be undertaken thus far. It is also an important step in establishing the approach of relying on international collaboration as a means of moving forward with such massive endeavors. The supporting role of other projects such as JT-60SA is a major component of this collaborative effort. Previous tokamak experiments, i.e. TFTR and JET, were expected to achieve scientific breakeven with a d–t plasma but did not quite reach that goal. As well, initial (i.e. 1998) expectations for the design of ITER were downgraded from expectations of reaching ignition ($Q = \infty$) to achieving a gain of $Q = 10$. However, as indicated in figure 4.25, ITER is expected to significantly improve on previously achieved plasma parameters and lead the way to a

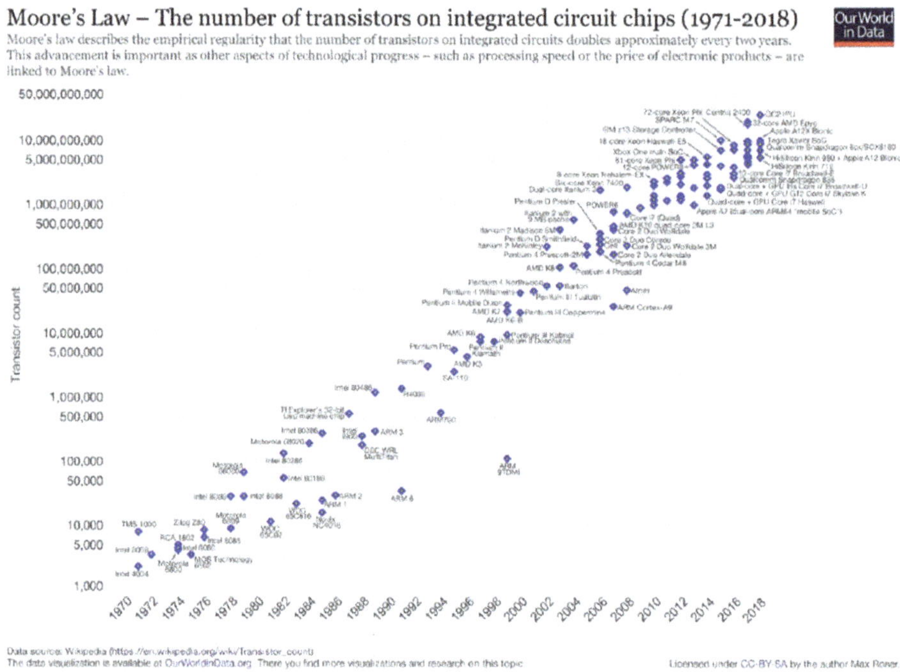

Figure 4.25. Application of Moore's law to the number of transistors on an integrated circuit chip. The number of transistors is plotted on a semi-log plot as a function of year from 1970 to 2018. This Moore's Law image has been obtained by the author from the Wikimedia website https://commons.wikimedia.org/wiki/File:Moore%27s_Law_Transistor_Count_1971-2018.png where it was made available by Max Roser under a CC BY-SA 4.0 license. It is included within this book on that basis. It is attributed to Max Roser.

demonstration commercial fusion power plant. Contributions from ITER, along with JT-60SA, will be an increased understanding of plasma properties and the refinement of designs of systems such as superconducting magnets, breeder blankets, fueling systems, auxiliary heating and divertors.

Moore's law

The development of technology is sometimes described in the context of Moore's law. This law is an empirical relationship based on the observation of past events and is not founded on any basic principle. The law was first proposed by and named after American engineer Gordon Moore (b. 1929) in 1965 and deals with the development of solid state electronic devices. Moore's law states that the number of transistors on a densely packed integrated circuit will double every two years. This relationship is manifested as a straight line on a semi-log plot of transistor number as a function of year. Figure 4.25 shows that developments in electronic technology have followed this law for nearly fifty years. This trend, however, cannot continue indefinitely. As electronic components get progressively smaller and smaller, they will eventually reach a size that is comparable to that of an atom. It is generally believed that a breakdown of Moore's law will probably begin in around 2025.

Moore's law applies (in an empirical sense) to other technological developments, including fusion energy. The relationship between the triple product and time on a semi-log plot is shown in figure 4.26. Moore's law for transistor density in micro-circuits shows a doubling time of 2.0 years. The triple product for magnetic confinement fusion reactor performance shows a doubling time of 1.8 years, indicating that progress in fusion has been somewhat more rapid than in semi-conductor device technology.

While data for the triple product of magnetic confinement nuclear fusion rectors are well described by Moore's law up to the 1990s with the experimental results from reactors such as JT-60 and JET, it is unclear how future fusion reactors will fit in with this trend. As no major tokamak facilities have become operational since JET, data beyond the late 1990s are lacking. Expectations for ITER are included on the graph in figure 4.26. It is seen in the figure that extrapolating the historical line for Moore's law for magnetic confinement fusion reactors would reach the expected triple product value for ITER in around 2002. However, this expectation has clearly not been fulfilled. The earliest experimental results from ITER (for d–d fusion) are not expected before the mid-2020s and d–t fusion experiments are not expected to begin before the mid-2030s. Certainly, a data point for the expected ITER triple product on the graph in figure 4.26, would fall substantially below the extrapolated line for Moore's law. There are several reasons for the breakdown of Moore's law.

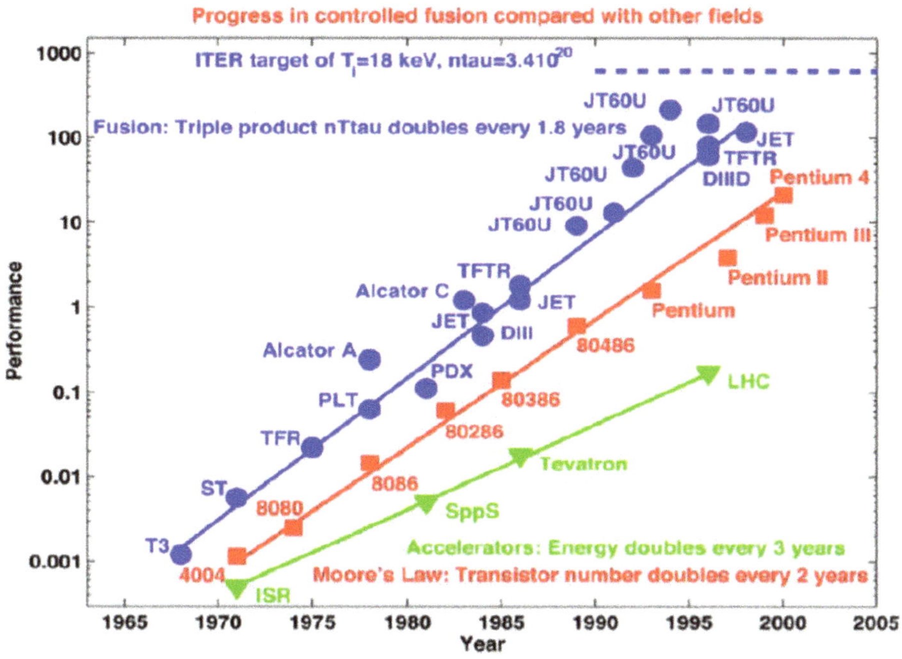

Figure 4.26. Semi-log plot of the scaled triple product for experimental magnetic confinement fusion reactors (tokamaks) as a function of time. Reproduced from Ikeda (2010). Copyright IOP Publishing. All rights reserved.

As noted above for transistors, the factor limiting circuit density will ultimately be the size of the atom. For magnetic confinement fusion reactors there are several possible reasons for the breakdown of Moore's law. These factors include:

1. The initial goal has been reached and there is no further need for addition progress;
2. The scientific or technical limits that are possible have been reached;
3. The scientific or technical challenges are becoming more difficult and progress has slowed;
4. Resources necessary for continued progress are not available;
5. Interest in continuing to progress toward the goal no longer exists.

Clearly, factor (1) is not valid, and there is no evidence that the limits noted in (2) have been reached. Although certain approaches, such as the tokamak, may have scientific limitations (e.g. the Greenwald limit or the Troyon limit) which may ultimately limit the achievable triple product, there are other approaches, such as the stellarator, which will not share the same limitations.

The final three possibilities deserve careful consideration. In fact, they are all, in some ways, related. The scientific and technical challenges are certainly becoming more difficult to deal with and this also makes them more expensive to deal with. Financial resources are an important consideration for the development of any technology, but the development of nuclear fusion energy, at least from a conventional perspective of magnetic confinement or inertial confinement, as discussed in the next chapter, may be more financially demanding than any other previous human endeavor. The financial aspects of commercializing nuclear fusion are discussed in further detail in section 7.5.

The plans for magnetic confinement fusion development past ITER are not well defined and, to a large extent, hinge on the results of this most recent experimental reactor. The remainder of this chapter describes some of the most prominent ideas on the way forward for magnetic confinement fusion research past ITER.

DEMO and PROTO

The DEMO is a proposed tokamak based nuclear fission reactor that will approximate the plasma conditions necessary for a commercial power reactor and will be the first fusion reactor to actually utilize thermal output to generate electricity. It is part of the long-term fusion energy plan promoted by the European Commission. The design of DEMO will be developed on the basis of accumulated experience from previous experimental tokamak reactors. ITER is generally considered as a final step toward the design and construction of DEMO. Although ITER will be the largest magnetic confinement fusion reactor ever constructed, it will produce plasma conditions and fusion energy that will fall short of those predicted for DEMO. Thus, results of experiments conducted on ITER will provide guidance for the design of DEMO but will not be able to predict in detail how DEMO will function. One of the important factors in the design of DEMO is the effect of the intense neutron flux produced by d–t fusion reactions on the properties of the reactor materials. This is particularly the case for the first wall (i.e. the inside wall of the plasma vessel) and the divertors, which will receive the greatest neutron flux. Although much is known about neutron radiation damage on the basis of many years of experimental investigations, studies utilizing fluxes as expected from d–t fusion in DEMO have not been undertaken. Parallel to ITER, the construction of an additional facility, the International Fusion Materials Irradiation Facility (IFMIF) is planned. IFMIF will be an accelerator-based neutron source that will produce a flux of neutrons that is similar to that expected from DEMO and with a similar energy spectrum. A 40 MeV deuteron beam will be incident on a lithium target and will produce neutrons with an energy of about 14 MeV (similar to the d–t fusion reaction) by means of the deuteron stripping reaction Li(d.xn). This facility will be essential for an understanding of neutron damage and activation that will be incurred by materials in DEMO.

A comparison of the design characteristics of DEMO compared with those of ITER is shown in table 4.8. The most obvious factor that illustrates the major step that DEMO represents toward a viable power reactor is the increase in the gain, Q. As noted above, $Q = 25$ is well within the expected range of gain for a commercial fusion reactor.

The proposed overall design of DEMO is shown by the block diagram in figure 4.27. The details of DEMO depend on the acquisition of sufficient technical knowledge from ITER and associated projects such as JT-60SA and IFMIF to ensure that the design and construction materials are consistent with the operational requirements. There are several significant challenges in the design of DEMO and, as a precursor to a functional power reactor, dependability, safety and economics must be design consideration in addition to technical and scientific feasibility.

Specific design criteria that must be determined for DEMO are as follows.

First wall design—The first wall, that is, the inner wall of the plasma vessel, is subject to the most extreme plasma conditions and, therefore, must be designed to

Table 4.8. Comparison of major design characteristics for ITER and DEMO. ITER information is as given in table 4.6. DEMO information adapted from Biel *et al* (2017).

Parameter	Unit	ITER	DEMO
Major radius	m	6.2	9.1
Minor radius	m	2.0	2.9
Plasma volume	m^3	870	2500
Toroidal magnetic field	T	5.3	5.7
Plasma current	MA	15	19.6
Pulse duration	min	8	120
Heating power	MW	50	80
Fusion power	MW	500	2000
Gain (Q)	—	10	25
Electrical power output	MW	0	500

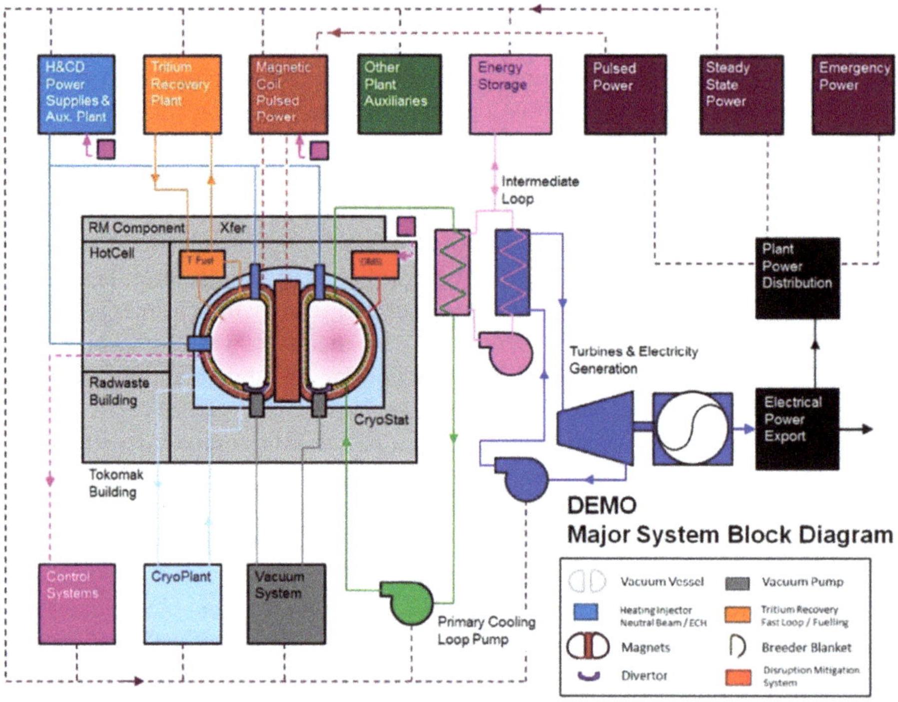

Figure 4.27. Block diagram of proposed DEMO fusion power station. Reproduced from Federici *et al* (2017).

minimize adverse effects caused by this extreme environment. The first wall must also be effectively integrated with the breeder blanket.

Divertor design and materials—It is essential that the plasma that is directed through the divertor is sufficiently cool and the divertor design and materials are sufficiently robust to minimize degradation of the divertor during power production. The frequency of divertor maintenance and replacement is an important consideration in the reactor operating cost.

Pulse duration—The minimum pulse duration that is required to achieve efficient operation, including a suitable power gain, must be determined.

Heating system—A number of heating systems will be tested on the ITER project. These include neutral ion beam heating, electron cyclotron resonance heating and ion cyclotron resonance heating. The effectiveness and efficiency of these approaches will be evaluated to determine which will be most appropriate for DEMO.

Breeder blanket design and materials—The breeder blanket must efficiently produce tritium fuel from fast fusion neutrons, effectively shield outer reactor components from fast neutrons in order to minimize degradation of components as a result of radiation damage and extract energy from neutrons to contribute to electricity production.

Superconducting magnet design—As noted above, high temperature superconducting magnets will be a requirement for the wide-scale application of magnetic confinement fusion energy. The utilization of this technology on DEMO would provide a test bed for refining the design necessary for a commercial power reactor.

Access for remote maintenance—A major safety consideration for reactor maintenance and repair is the design of a suitable remote access system to allow work to be done using robots. This was a minor consideration for experimental reactors that utilized d–d fusion fuel. Recent experiments conducted at TFTR and JET have utilized d–t fusion fuel with the resulting radiation hazard. This will be the case, as well, for ITER and DEMO, as well as a commercial magnetic confinement fusion reactor in the foreseeable future.

The timescale for construction of DEMO has been a subject of discussion since the early 2000s. A DEMO timeline established in 2004 suggested the milestones for development given in table 4.9. It should be noted that appropriate results from ITER and IFMIF are a requirement for completing the engineering design of DEMO. ITER construction has been delayed and operation will not begin before 2025, with d–t experiments not expected before 2035. The plans for IFMIF have not yet been finalized. The 2004 timeline for DEMO is, therefore, not feasible. The DEMO timeline was re-evaluated and updated in 2012 and again in 2018 in view of progress on ITER and IFMIF. The most recent time estimate places the completion of engineering design and the start of construction in 2040 and the likely start of operation around 2054. Development of commercial fusion reactors would follow some time after that. Because of the interconnectivity of the various stages of fusion reactor development, there is considerable uncertainty in the timeline very far in the future. The considerable costs associated with the development of these facilities means that economical and political (as well as social) roadblocks, in addition to potential scientific and technical ones, will likely need to be overcome. This is discussed further in chapter 7.

If ITER and DEMO achieve their specified goals and the way ahead for commercial magnetic confinement fusion power is clear, it is expected that the next stage of fusion energy development would be the construction and testing of a prototype commercial reactor designated as PROTO. PROTO is promoted by EUROfusion, the European Consortium for Development of Fusion Energy, that supports and funds fusion research on behalf of the European Commission. At present, the timeline for PROTO, is uncertain because of the uncertain timelines for

Table 4.9. Milestones for the completion of DEMO as proposed in 2004. Note that phase 2 includes upgraded blanket design.

Year	Milestone
2017	Completion of conceptual design
2024	Completion of engineering design and start of construction
2033	Completion of phase 1 of construction and start of phase 1 operation
2038	Completion of phase 1 of operation
2040	Completion of upgrades and start of phase 2 operation

ITER and DEMO. However, it would certainly be past the 2050s before construction on PROTO would begin. The future of fusion energy is, at present, uncertain. Much depends on the success of ITER and subsequently on the success of DEMO.

Parallel to ITER and DEMO, at least two other major magnetic confinement fusion initiatives are planned. The United Kingdom, whose relationship with EUROfusion is uncertain following Brexit, has planned the construction of a major spherical tokamak reactor. Referred to as STEP (Spherical Tokamak for Energy Production), the reactor is expected to produce net electricity by the 2040s.

Another major planned magnetic confinement fusion project is the China Fusion Engineering Test Reactor (CFETR). Phase 1 of this project is planned to begin in the 2020s and would consist of the construction of a reactor that demonstrates steady state operation and tritium breeding capabilities. Phase 2 of the project is aimed at producing a gain of $Q > 12$ and a demonstration of tritium self-sufficiency from the breeder blanket.

While planned magnetic fusion projects are aimed at progressing toward a commercial fusion power reactor, the time scale for this development is still uncertain. There are many factors that will ultimately affect the possibility of commercial nuclear fusion energy and these are summarized in chapter 7.

References and suggestions for further reading

Biel W, Beckers M, Kemp R, Wenningerd R and Zohme H 2017 Systems code studies on the optimization of design parameters for a pulsed DEMO tokamak reactor *Fusion Eng. Des.* **123** 206–11

Bobin J L 2014 *Controlled Thermonuclear Fusion* (Singapore: World Scientific)

Burns G 1991 *High-Temperature Superconductivity: An Introduction* 1st edn (New York: Academic)

Chen F F 2011 *An Indispensable Truth: How Fusion Power Can Save the Planet* (New York: Springer)

Chen F F 2016 *Introduction to Plasma Physics and Controlled Fusion* 3rd edn (Cham: Springer)

Claessens M 2020 *ITER: The Giant Fusion Reactor—Bringing a Sun to Earth* (Cham: Springer)

Donné T and Morris W 2018 *European Research Roadmap to the Realisation of Fusion Energy—Long Version* (EUROfusion, Munich) available for download at www.euro-fusion.org/eurofusion/roadmap

Dunlap R A 2004 *An Introduction to the Physics of Nuclei and Particles* (Belmont, CA: Brooks/Cole)

Dunlap R A 2019 *Electrons in Solids—Contemporary Topics* (San Rafael, CA: Morgan & Claypool)

Federici G, Biel W, Gilbert M R, Kemp R, Taylor N and Wenninger R 2017 European DEMO design strategy and consequences for materials *Nucl. Fusion* **57** 092002

Fujiwara M *et al* 1999 Plasma confinement studies in LHD *Nucl. Fusion* **39** 1659–66

Geraud A *et al* 2007 The JET high frequency pellet injector project *Fusion Eng. Des.* **82** 2183–88

Hagan J W, Opperman E K and Trego A L 1984 The fusion materials irradiation test (FMIT) facility *J. Nucl. Mater.* **123** 958–64

Harms A A, Schoepf K F, Miley G H and Kingdon D R 2000 *Principles of Fusion Energy—An Introduction to Fusion Energy for Students of Science and Engineering* (Singapore: World Scientific)

Hosea J, Goldston R and Colestock P 1985 The Princeton Large Torus (PLT) *Nucl. Fusion* **25** 1155–60

Huba J D 2007 *NRL Plasma Formulary* (Washington, DC: Office of Naval Research)

Ikeda K 2009 ITER on the road to fusion energy *Nucl. Fusion* **50** 014002

Kim K *et al* 2015 Design concept of K-DEMO for near-term implementation *Nucl. Fusion* **55** 053027

Kleiner R, Buckel W and Huebener R 2016 *Superconductivity: An Introduction* 3rd edn (Berlin: Wiley)

Koizumi N, Nunoya Y, Yoshida K and Barabaschi P 2014 Progress of ITER and JT-60SA magnet development in Japan *Phys. Procedia* **58** 232–35

Lang P T, Nakano T, Garzotti L, Pégourié B, Ploeckl B and Sakurai S 2019 A flexible pellet injection system for the tokamak JT-60SA: the final conceptual design *Fusion Sci. Technol.* **75** 178–96

Li S, Jiang H, Ren Z and Xu C 2014 Optimal Tracking for a Divergent-Type Parabolic PDE System in Current Profile Control *Abstr. Appl. Anal.* **2014** 940965

McCracken G and Stott P 2013 *Fusion: Energy of the Universe* 2nd edn (Waltham, MA: Academic)

Miyimoto K 2016 *Plasma Physics for Controlled Fusion* (Cham: Springer)

Morse E 2018 *Nuclear Fusion* (Cham: Springer)

Nagel M, Dhard C P, Bau H, Bosch H-S, Meyer U, Raatz S, Risse K and Rummel T 2017 Cryogenic commissioning, cool down and first magnet operation of Wendelstein 7-X *IOP Conf. Ser.* **171** 012050

Onnes H K 1911 The superconductivity of mercury *Comm. Phys. Lab. Univ. Leiden* Nos. 122 and 124

Parisi J and Ball J 2019 *The Future of Fusion Energy* (London: World Scientific)

Parker R 1998 ITER a major step towards a fusion reactor *Europhys. News* November/December 1998 228–29

Rolando G, Foussat A, Knaster J, Ilin Y and Nijhuis A 2013 Performance assessment and optimization of the ITER toroidal field coil joints *Supercond. Sci. Technol.* **26** 085004

Sunn Pedersen T *et al* 2016 Confirmation of the topology of the Wendelstein 7-X magnetic field to better than 1:100,000 *Nat. Commun.* **7** 13493

Schmidt G 1979 *Physics of High Temperature Plasmas* 2nd edn (Waltham, MA: Academic)

Takahashi H *et al* 2018 Realization of high T_i plasmas and confinement characteristics of ITB plasmas in the LHD deuterium experiments *Nucl. Fusion* **58** 106028

Tinkham M 2004 *Introduction to Superconductivity* 2nd edn (New York: Dover Books)

Uchida S 2015 *High Temperature Superconductivity: The Road to Higher Critical Temperature* (New York: Springer)

Uhran M 2017 Creating a star—the global ITER partnership *Energy from the Nucleus* ed G M Crawley (Singapore: World Scientific) pp 189–228

Ushigusa K 1999 *Lower Hybrid Current Drive in Tokamak Plasmas* Japan Atomic Energy Research Institute Report JAERI 1339 available for download at https://inis.iaea.org/collection/NCLCollectionStore/_Public/30/045/30045352.pdf?r=1&r=1

Zarnstorff M C and Goldston R J 2017 Magnetic fusion energy *Energy from the Nucleus* ed G M Crawley (Singapore: World Scientific) pp 165–88

Zohuri B 2016 *Plasma Physics and Controlled Thermonuclear Reactions Driven Fusion Energy* (Cham: Springer)

Zohuri B 2017 *Magnetic Confinement Fusion Driven Thermonuclear Energy* (Cham: Springer)

IOP Publishing

Energy from Nuclear Fusion

Richard A Dunlap

Chapter 5

Inertial confinement fusion

5.1 Introduction

Inertial confinement fusion utilizes inertial forces, rather than magnetic forces, as described in the previous chapter, to contain a plasma at sufficient temperature and density for a sufficient period of time to induce nuclear fusion reactions. While several possible approaches to creating a controlled fusion reactor fall into the category of inertial confinement fusion, this terminology is generally used to refer to fusion produced by laser irradiation and it is this approach that forms the subject of the present chapter. Other approaches to inertial confinement fusion include electrostatic inertial confinement fusion, heavy ion drivers and z-pinch machines, all of which are discussed in chapter 6.

As discussed in chapter 3, considerable research was conducted during the 1950s concerning the possibility of controlled nuclear fusion using magnetic confinement. After the development of the laser in the early 1960s, interest in its use as a means of introducing nuclear fusion began. John Nuckols (b. 1930) and collaborators at the Lawrence Livermore National Laboratory undertook extensive theoretical and modeling studies of inertial confinement fusion using laser irradiation. It was in the early to mid-1970s that experimental work in this direction began. Results of these investigations have continued since that time and have provided a potential option to magnetic confinement fusion as a means of providing energy for the future.

5.2 Lasers

History of the laser

The acronym 'laser' refers to **L**ight **A**mplification by **S**timulated **E**mission of **R**adiation. The laser is the optical analog of the 'maser' (**M**icrowave **A**mplification by **S**timulated **E**mission of **R**adiation). The operating principles of the maser were first described in 1953 by American physicist Joseph Weber (1919–2000), best known for his work on gravity wave detection. In the same year Soviet physicist

doi:10.1088/978-0-7503-3307-8ch5

Nikolay Basov (1922–2001) and Australian-born physicist Alexander Prokhorov (1916–2002) also provided a detailed description of maser operation. This first maser was constructed in 1953 by Charles Hard Townes (1915–2015), James P Gordon (1928–2013) and Herbert J Zeiger (1925–2011) at Columbia University. Basov, Prokhorov and Townes shared the 1964 Nobel Prize in Physics for their work on the development of the maser.

For more than 60 years there has been controversary concerning the origin of the concept for the laser. In 1958 Townes, along with Arthur Leonard Schawlow (1921–99) published their theory of optical stimulated emission. This work formed the basis for the construction of an operational laser two years later by Theodore Maiman (1927–2007). However, in 1957, a year prior to the Townes and Schawlow publication, Gordon Gould (1920–2005) had documented his ideas concerning the construction of an optical analog of the maser. The credit for the invention of the laser has never been determined unequivocally, either from an historic or from a legal standpoint.

Since the early 1960s lasers have been constructed from a wide variety of different materials. Traditional lasers have been constructed from solids, liquids and gases, and have involved transitions between atomic states or various molecular states. In recent years lasers have also been constructed from semiconducting junctions involving transitions between energy bands. The next section provides an overview of the basic physics of the operation of a laser. This is followed by some specific examples of lasers that have been, or might be, important for laser fusion reactors.

The physics of lasers

Laser light is monochromatic, that is, of a single frequency, and it is intense. These features are the consequences of the way in which the light is produced. Specifically, it is an understanding of how the light which is produced is stimulated and amplified, as the name implies, that is of utmost importance for its application, not only to laser fusion, but to an almost unlimited number of devices that have influenced our daily lives. It is the stimulated and amplified nature of the radiation that gives rise to the unique properties of the light from a laser, that is, coherence, monochromaticity and intensity.

Simulated emission

We begin with a consideration of a quantum mechanical two-level system (e.g. an atom), consisting of a stable ground state, g, and an unstable excited state, e. If the atom has been excited from the ground state to the excited state, for example by an electrical discharge or an incident photon, it will subsequently spontaneously decay back to the ground state thereby emitting a photon of quantized energy. If the photon encounters another identical atom in the excited state, as shown in figure 5.1, it can cause the atom to undergo stimulated emission to the ground state, thereby creating a second photon of the same energy. An essential feature of the transition in the figure is that the photon that was produced during the stimulated decay is travelling in the same direction as the photon that stimulated the decay and is in

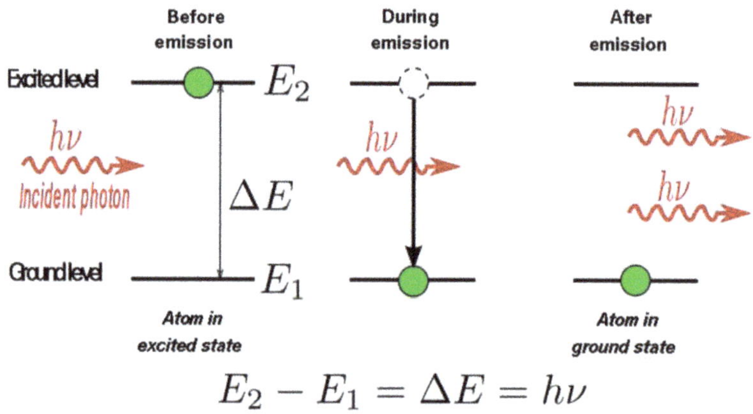

Figure 5.1. Stimulated emission of a photon in the two-level system. This Stimulated Emission image has been obtained by the author from the Wikimedia website https://commons.wikimedia.org/wiki/File:Stimulated_Emission.svg where it was made available by V1adis1av under a CC BY-SA 4.0 license. It is included within this book on that basis. It is attributed to V1adis1av.

phase with it. In a collection of many identical atoms all in the excited state, one spontaneous decay will trigger a chain reaction of stimulated decays and create a large number of photons all of the same energy (i.e. monochromatic) and all traveling in the same direction and in phase. The radiation which is produced in this way will be monochromatic (as all the atomic transitions will be of the same energy, to the extent described below) and coherent (as all of the components of the beam will be in phase and will interfere constructively).

The three-level laser

In order to produce continuous radiation as described above, or even to produce an intense pulse of radiation, it is necessary to create a population inversion, that is, a situation where there are more atoms in the excited state than in the ground state. At a temperature T, atomic levels will be populated according to the Maxwell–Boltzmann distribution given by

$$\frac{N_i}{N} = \frac{\exp\left[-\frac{E_i}{k_B T}\right]}{\sum_j \exp\left[-\frac{E_j}{k_B T}\right]} \tag{5.1}$$

where N is the total number of atoms in the system, N_i is the number of atoms in the ith energy level, E_i is the energy of the ith level and k_B is the Boltzmann constant. Normally, at room temperature, only a very small fraction of atoms exists in levels above the ground state. Atoms can be excited using incident photons of the appropriate energy, but they will continuously be decaying spontaneously back to the ground state. In order to create a population inversion, that is, a situation where a higher energy state has a greater population than the lower energy state below it,

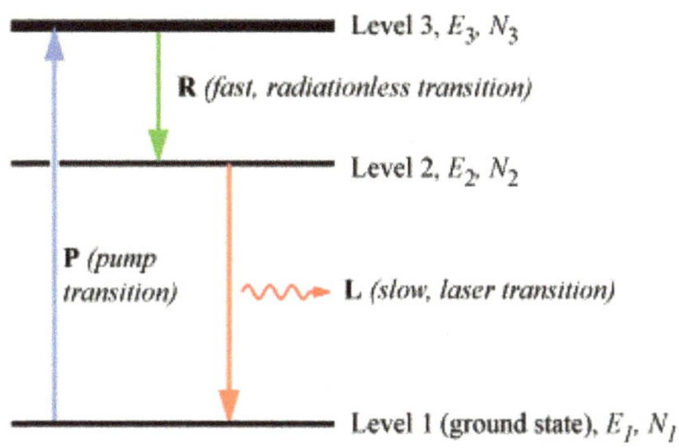

Figure 5.2. Creation of a population inversion in a three-level system. This Three-level Population Inversion image has been obtained by the author from the Wikimedia website https://commons.wikimedia.org/wiki/File:Population-inversion-3level.png where it was made available by Bob Mellish under a CC BY-SA 3.0 license. It is included within this book on that basis. It is attributed to Bob Mellish.

so that stimulated radiation can be produced, it is necessary to have a system with at least three energy levels with appropriate lifetimes. This is illustrated in figure 5.2.

In the example shown in figure 5.2, the excited state (level 3) is a short-lived state with a lifetime of (e.g.) 10^{-8} s. The metastable state (level 2) has a longer lifetime of (e.g.) 10^{-5} s and the ground state (level 1) is stable. A population inversion between the metastable state and the ground state can be created in the following manner. Photons of the appropriate energy, E_0, can be used to pump atoms from the ground state to the excited state. The atoms in the excited state decay very quickly to the metastable state where they collect, because of the metastable state's longer lifetime, thereby creating a population inversion between the metastable state and the ground state. Now a spontaneous decay from an atom in the metastable state to the ground state can trigger stimulated emission of radiation with energy E_γ from all the other atoms in the metastable state. Obviously, $E_0 > E_\gamma$ and this is why it is common to use pumping radiation that might be (for example) green to produce a lasing transition that emits red light. In principle, it should be possible to continuously pump atoms that return to the ground state back into the excited state and thus, maintain continuous emission of stimulated laser radiation. In practice, this is generally not possible in the system as described and three-level lasers typically operate in the pulsed mode, where a pulse of pumping radiation creates the population inversion and this is followed by the emission of a pulse of laser radiation.

The four-level laser

Most continuous lasers are based on a four-level system (although a four-level laser can also be used in the pulsed mode). A typical four-level laser energy diagram is shown in figure 5.3. The excited, metastable and ground states have lifetimes along the lines of those described above for the three-level laser. In addition, an intermediate state, *i*, with a very short lifetime exists between the ground state and the metastable state. The transition between the metastable state and the

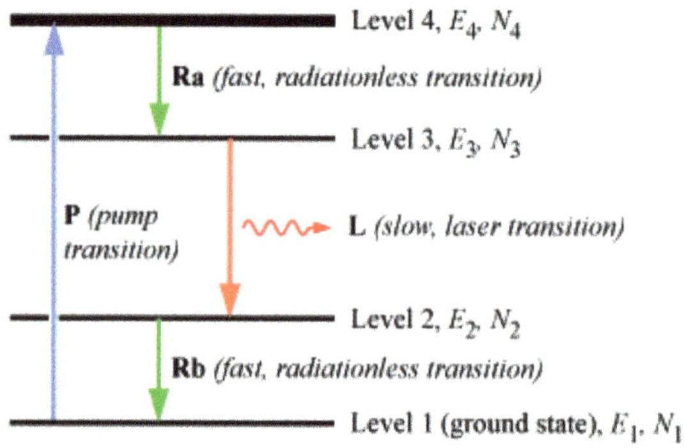

Figure 5.3. Creation of a population inversion in a four-level system. This Four-level Population Inversion image has been obtained by the author from the Wikimedia website https://en.wikipedia.org/wiki/File:Population-inversion-4level.png where it was made available by Bob Mellish under a CC BY-SA 3.0 license. It is included within this book on that basis. It is attributed to Bob Mellish.

intermediate state, as shown in the figure, is the lasing transition. When stimulated emission occurs, leaving atoms in the short-lived intermediate state, they very quickly decay spontaneously to the ground state where they can be pumped back up into the excited state. This allows a population inversion to be maintained between the metastable and intermediate states and for stimulated radiation to be emitted continuously.

Resonant cavities and laser modes

There are two final points concerning laser radiation that we need to address. Firstly, how monochromatic is the radiation, and secondly, how do we optimize the intensity of the beam? To answer the first question, we can begin by applying Heisenberg's principle to the transition between the two atomic states. We write the uncertainty in energy, ΔE, and time, Δt, as

$$\Delta E \Delta t \geqslant \frac{\hbar}{2} \tag{5.2}$$

where $\hbar$ is the reduced Planck constant. If we consider the transition for the three-level laser, as shown in figure 5.2, the uncertainty in time for the transition can be taken to be the metastable state lifetime, τ. Using the relationship between energy and wavelength, λ;

$$E = \frac{hc}{\lambda} \tag{5.3}$$

where c is the speed of light, we find the uncertainty in wavelength as

$$\Delta\lambda \geqslant -\frac{\lambda^2}{4\pi c\tau} \tag{5.4}$$

Using a wavelength for an optical photon of 500 nm and a lifetime for the metastable state, as given above, $\tau = 10^{-5}$ s, gives an uncertainty in wavelength, that is, the width of the wavelength distribution, as 6.6×10^{-9} nm. This amounts to about one part in 10^{11} of the photon wavelength. The Heisenberg width of the energy distribution is a significant underestimate of the actual energy width of a typical laser transition. The excess transition width comes from the properties of the laser medium and is an essential feature of the transitions that actually allows for the construction of an operational laser. In gas lasers, Doppler broadening of the transition as a result of the thermal motion of atoms or molecules in the gas gives rise to the broadening. In the case of solid state lasers, phonon interactions give rise to a broadening of the atomic energy levels. Typical wavelength distributions are in the range of $\Delta\lambda \sim 2 \times 10^{-3}$ nm.

In order to optimize the amplitude of the light produced by a laser it is necessary create a resonant cavity, as shown in figure 5.4. The resonant cavity consists of the lasing medium, which may be a solid, liquid or gas, with mirrors placed on both ends. One mirror is fully reflecting while the other mirror is partially reflecting. The photons produced by stimulated emission will reflect back and forth in the cavity, creating more stimulated photons as they go. Since all of the stimulated photons will be in phase, they will interfere constructively, leading to a beam that continues to grow in intensity. The laser output will result from some fraction of this intense beam leaking out through the partially reflecting mirror.

This method sounds very simple, except for one important point. In order for the beam to remain in phase with itself as it reflects back and forth in the cavity, the length of the cavity must be exactly an integer multiple of half the wavelength of the light. How do we ensure that this is the case? Actually, we do not need to do anything to ensure that this is the case. We can see this by looking more closely at the

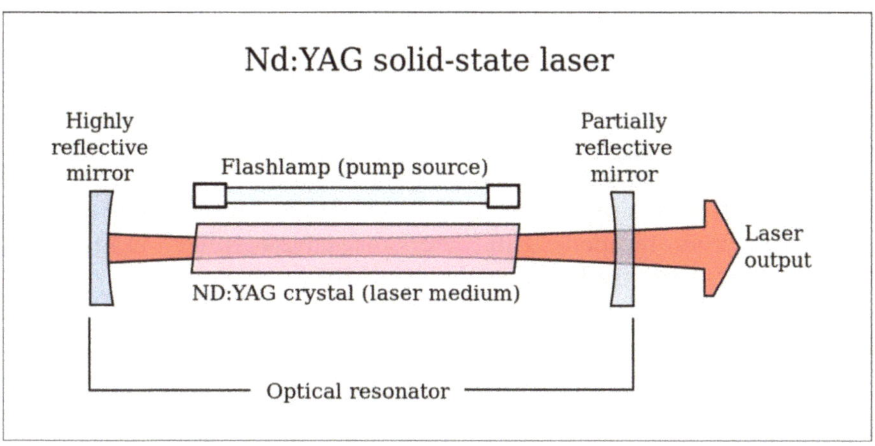

Figure 5.4. Resonant cavity used to construct a laser. This Resonant Cavity image has been obtained by the author from the Wikimedia website https://commons.wikimedia.org/wiki/File:Lasercons.svg where it was made available by Lakkasuo under a CC BY-SA 3.0 license. It is included within this book on that basis. It is attributed to Lakkasuo.

number of half wavelengths of light that can fit into the cavity. If the length of the cavity is L then the number of half wavelengths will be

$$n = \frac{2L}{\lambda} \tag{5.5}$$

Solving for wavelength gives

$$\lambda = \frac{2L}{n} \tag{5.6}$$

Taking the derivative of this expression and substituting for n from equation (5.5) gives

$$\Delta\lambda = -\frac{\lambda^2}{2L}\Delta n \tag{5.7}$$

Thus, in order to fit one more half wavelength of light inside the resonant cavity ($\Delta n = 1$), then the wavelength will have to be reduced by

$$\Delta\lambda = \frac{\lambda^2}{2L} \tag{5.8}$$

Using a typical length for a laboratory laser of $L = 0.5$ m, equation (5.8) gives the spacing between resonant modes in frequency as 2.5×10^{-4} nm. This can be compared to the typical wavelength distribution for an atomic transition as given above, $\Delta\lambda \sim 2 \times 10^{-3}$ nm, and we can see that there are about eight resonant modes within the wavelength distribution. A graphical representation of this analysis is illustrated in figure 5.5. So, we see that no matter what length we make the resonant cavity the lasing transition will produce a number of wavelengths for which the resonant condition will be satisfied. This simple calculation shows why the laser would not be possible if the Heisenberg broadening along determined the monochromaticity of the light.

There are a large number of different materials that exhibit atomic transitions that are suitable for the construction of a laser. These materials may be solids, liquids or gases and the laser light which is produced from these various lasers depends on the detailed properties of the laser medium. Laser materials which allow for the construction of lasers with very intense output, can potentially be used to induce nuclear fusion reactions. Three lasers that have been used in fusion experiments are the CO_2 laser, the KrF laser and the Nd-glass leaser, and the properties of these three lasers are discussed below.

CO_2 lasers

The carbon dioxide (CO_2) laser is an example of a molecular gas laser and is one of the earliest very intense gas lasers to be developed. It is currently the most intense laser available that can be used in the continuous mode. The CO_2 laser contains a mixture of gas with typically 10%–20% CO_2, 10%–20% N_2 and up to a few percent H_2 and/or Xe. The remainder (up to 80%) is He. It is the energy levels of the CO_2

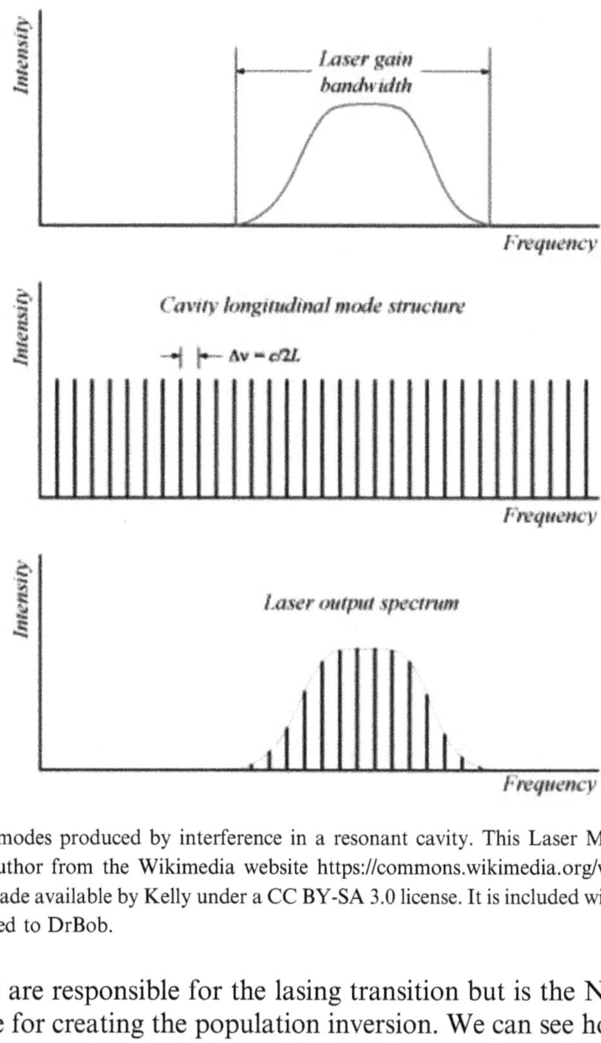

Figure 5.5. Laser modes produced by interference in a resonant cavity. This Laser Modes image has been obtained by the author from the Wikimedia website https://commons.wikimedia.org/wiki/File:Modelock-1.png where it was made available by Kelly under a CC BY-SA 3.0 license. It is included within this book on that basis. It is attributed to DrBob.

molecules that are responsible for the lasing transition but is the N_2 molecules that are responsible for creating the population inversion. We can see how this works by looking at the energy level diagrams of CO_2 and N_2 as shown in figure 5.6.

An electric discharge is used to excite N_2 molecules into an excited state. This state is a metastable molecular vibrational state. This N_2 vibrational state is at very nearly the same energy as a metastable (asymmetric stretch) vibrational state of the CO_2 molecule. The energies of these two states are sufficiently close that the small energy difference is more than compensated for by the thermal energy. Thus, collisions between excited N_2 molecules and CO_2 will transfer energy to the CO_2 molecule, thereby populating the CO_2 metastable state. This asymmetric stretch CO_2 state decays by stimulated emission to one of two intermediate lower energy vibrational states; a symmetric stretch vibrational state by emission of a 10 600 nm photon or a bending vibrational state by emission of a 9600 nm photon. These two intermediate states transition to the CO_2 ground state by transferring energy through collisions to helium atoms. This process allows a population inversion to

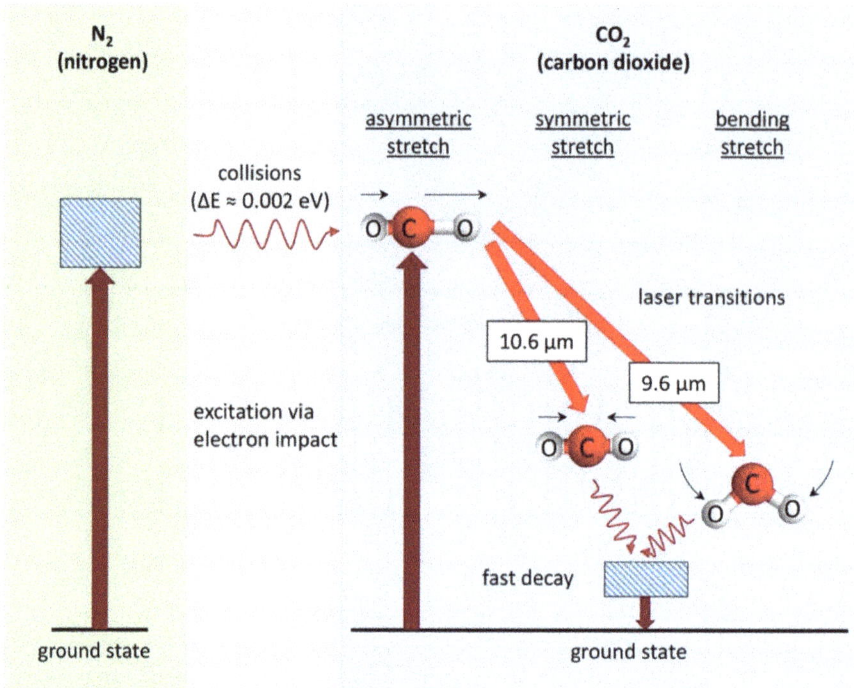

Figure 5.6. Energy levels of N_2 and CO_2 in a CO_2 laser illustrating the creation of a population inversion in the CO_2 excited state by transfer of energy from the N_2 excited state. The lasing transitions at 10 600 nm and 9600 nm are shown. Reproduced from Wieneke and Gerhard (2018). Reproduced with permission.

be maintained between the CO_2 metastable state and the two intermediate states and therefore allows the CO_2 laser to be operated continuously. The high intensity laser radiation produced by CO_2 lasers makes them of interest as a driver for inertial confinement fusion.

Nd–glass lasers

Atomic transitions between states in a Nd^{3+} ion are the basis for the construction of some of the highest intensity solid state lasers. The Nd^{3+} ions are incorporated into a suitable transparent host material. Common hosts for Nd lasers include yttrium aluminum garnet, $Y_3Al_5O_{12}$ (commonly known as YAG), and various glasses. The energy level diagram for the Nd^{3+} ion is shown in figure 5.7. Photons with wavelengths in the range of 800 nm are produced by a flash lamp and are used to pump the Nd^{3+} ions from the ground state to one of two short-lived excited states, as shown in the figure. These short-lived excited states decay to a metastable state from which lasing transitions occur to short lived intermediate states. The principal lasing transition occurs at 1064 nm for Nd^{3+} ions in a YAG host. Energy levels are affected by electronic interactions between the Nd^{3+} ions and the host atoms, thereby influencing the exact energy of the lasing transitions. Nd^{3+} laser output for Nd in

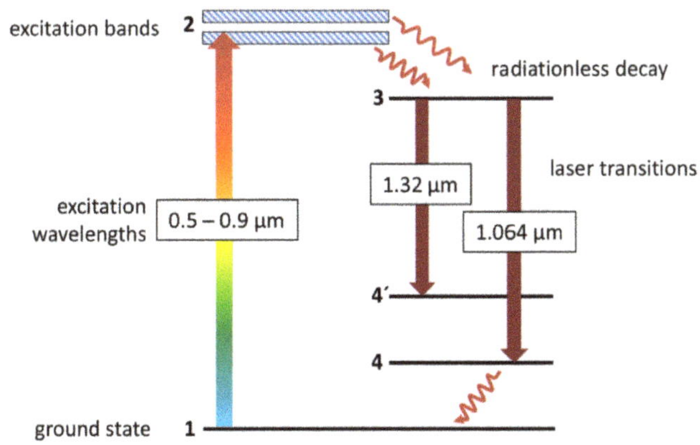

Figure 5.7. Energy levels of the Nd^{3+} ion in YAG showing the lasing transitions. Reproduced from Wieneke and Gerhard (2018). Reproduced with permission.

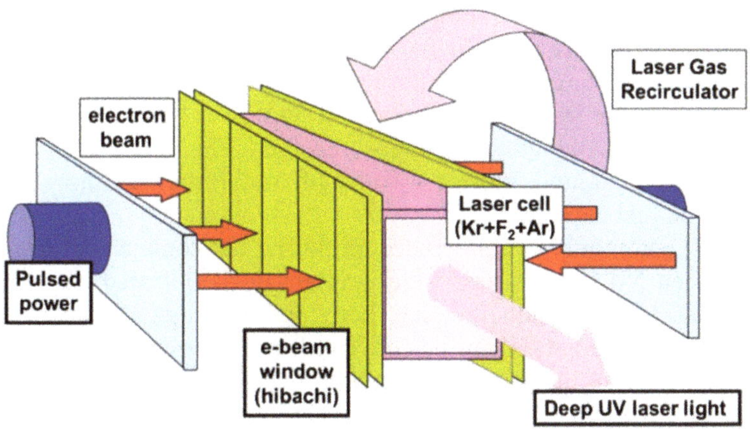

Figure 5.8. Diagram of an electron beam pumped KrF laser. Reproduced from Obenschain et al (2015).

silicate glasses is at about 1062 nm and in phosphate glasses it is at about 1054 nm. Nd lasers can be operated in both pulsed and continuous modes.

KrF lasers

The KrF laser is a type of ultraviolet laser that is referred to as an exciplex laser (which is short for 'excited complex laser'). An exciplex laser contains a mixture of a noble gas and a halogen gas. Because exciplex lasers emit very intense short wavelength laser radiation, they have found common use in the production of high-density semiconducting devices by laser lithography. Their high intensity and short wavelength also make them of potential interest for laser fusion.

A KrF laser is typically pumped using an electron beam using an arrangement, as shown in figure 5.8. The electrons interact with the Kr atoms, which transition to an excited state. Normally, noble gas atoms are inert and do not bond to other atoms. However, when they are in an excited state they can interact with some other atoms, e.g. halogens, and form a metastable stable state of a pseudo-molecule (the exciplex). In the present case the reaction is

$$2Kr + F_2 \rightarrow 2KrF \tag{5.9}$$

This metastable state can undergo stimulated emission to an unstable ground state emitting laser radiation at 248 nm. The ground state is very short lived and undergoes dissociation

$$2KrF \rightarrow 2Kr + F_2 \tag{5.10}$$

Thus, a population inversion can be formed between the metastable exciplex and the unstable ground state.

Harmonic generation

The inertial confinement fusion experiments conducted at the Shiva laser facility in the late 1970s showed that radiation in the infrared portion of the electromagnetic spectrum was not effective at heating a capsule (sometimes called a pellet) of fusion fuel. This is because the laser radiation coupled very strongly with electrons that were formed when the laser initially created a plasma in the fuel and this led to a substantially reduced ability to compress the fuel. This effect was found to be wavelength dependent and shorter wavelength radiation was more effective at compressing the fuel capsule.

As the most commonly used lasers for recent inertial confinement fusion experiments have been Nd based lasers, which radiate in the infrared, the efficiency of energy transfer to the fusion fuel has been an important consideration. Long wavelength infrared radiation can be converted into shorter wavelength visible or ultraviolet radiation using harmonic generation. Two techniques are of particular relevance in this respect, second harmonic generation (sometimes called frequency doubling) and sum frequency generation (in this case, frequency tripling), both of which make use of the non-linear optical properties of some crystalline materials.

Second harmonic generation occurs when two photons of a given energy are combined to create one photon of twice the energy. This process can be viewed from a classical standpoint by looking at the way in which atoms in a dielectric material respond to irradiation by electromagnetic waves. The positive and negative charge distributions of the atoms that arise from the nuclear and electronic charges, respectively, are not spatially coincident in a dielectric material. This gives rise to an electric dipole moment associated with the atom. When electromagnetic radiation passes through such a dielectric material the electric field associated with the radiation induces a polarization, P, that is proportional to the magnitude of the electric field, E. For a linear optical material this relationship is

$$P = \varepsilon_0 \chi E \tag{5.11}$$

where ε_0 is the permittivity of free space and χ is the relative electric susceptibility of the medium. For a nonlinear optical material, the relationship between the induced polarization and the applied electric field is not linear and can be expanded as a Taylor series of the form

$$P = \varepsilon_0 \chi_1 E + \varepsilon_0 \chi_2 E^2 + \varepsilon_0 \chi_3 E^3 + \cdots \quad (5.12)$$

Second harmonic generation corresponds to the second term on the right-hand side of equation (5.12). While the magnitude of this term relative to the linear response (first term on the right-hand side of the equation) depends on the properties of the material, it is easy to see that in materials where this second harmonic term exists, it becomes more significant for larger values of the electric field, i.e. more intense radiation. Figure 5.9 shows the production of second harmonic radiation at twice the frequency, along with the residual fundamental frequency radiation. Common materials used for second harmonic generation include monopotassium phosphate (KH_2PO_4, commonly known as KDP), lithium triborate (LiB_3O_5), barium borate (β-BaB_2O_4), lithium niobate ($LiNbO_3$), lithium tantalate ($LiTaO_3$) and potassium titanyl phosphate ($KTiOPO_4$, commonly known as KTP). Practical second harmonic generation devices can have efficiencies as high as about 80% for high input intensities. This approach may be used for a Nd laser, which radiates in the infrared at about 1060 nm, to produce visible light at about 530 nm, which is in the green portion of the electromagnetic spectrum.

The wavelength of light from (e.g.) a Nd laser can be reduced further by combining three photons to create a photon of one third the wavelength. This process is referred to as frequency tripling and corresponds to summing the frequency of the original wave and the second harmonic wave. In terms of photon energy

$$\hbar\omega = \hbar\omega_1 + \hbar\omega_2 \quad (5.13)$$

where $\omega_2 = 2\omega_1$. This process is most efficient when the phase matching condition is satisfied. This condition is expressed as

$$\hbar k = \hbar k_1 + \hbar k_2 \quad (5.14)$$

where the k_i are the angular wavenumbers of the respective waves.

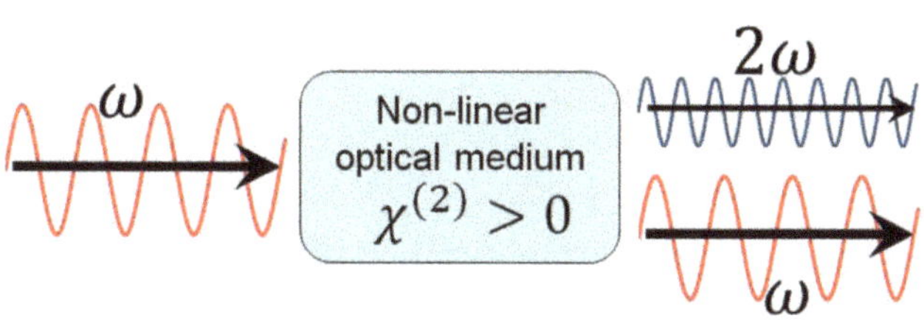

Figure 5.9. Representation of second harmonic generation. This Second Harmonic Generation image has been obtained by the author from the Wikimedia website https://commons.wikimedia.org/wiki/File:Schematic_of_the_SHG_conversion_of_an_excited_wave_in_a_non-linear_medium.png where it was made available by BP-Aegirsson under a CC BY-SA 4.0 license. It is included within this book on that basis. It is attributed to BP-Aegirsson.

A diagram of a frequency tripler is shown in figure 5.10. Monopotassium phosphate (KDP) is the most common material used for this application. The application of the frequency tripling method to the output from a Nd laser will yield radiation at 353 nm which is in the near ultraviolet. Practical efficiencies of around 50% are typical for frequency tripling.

5.3 The physics of laser fusion

Basic principles of laser fusion

While laser fusion, like magnetic confinement fusion, must achieve appropriate plasma condition in order to produce viable fusion energy, it does so in quite a different manner. Typical confinement times for laser fusion are 9–10 orders of magnitude shorter than for magnetic confinement fusion and, subsequently, the required densities are 9–10 orders of magnitude greater.

The basic process associated with laser fusion is illustrated in figure 5.11. As the figure shows, the process begins with the irradiation of the fuel capsule (1). This may

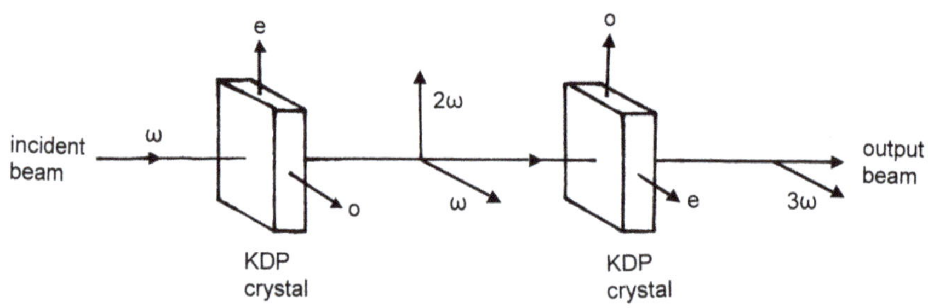

Figure 5.10. Method of frequency tripling. The relative ordinary (o) and extraordinary (e) directions for the KDP crystals are shown.

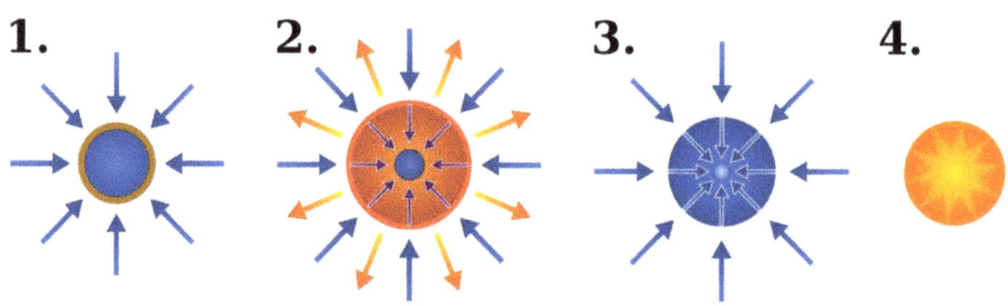

Figure 5.11. Basic sequence of events leading to fusion by laser irradiation. (1) Laser radiation or x-rays produced from the hohlraum (see below), heat the fuel capsule from the outside creating a surrounding plasma envelope. (2) Fuel is compressed by ablation of the hot surface material. (3) Compression and heating of the inner core of the fusion fuel. (4) Thermonuclear fusion reaction spreads through the compressed fuel. Blue arrows represent incident radiation, orange arrows represent ablated material and purple arrows represent thermal energy transport. This Fusion by Laser Irradiation image has been obtained by the author from the Wikimedia website https://commons.wikimedia.org/wiki/File:Inertial_confinement_fusion.svg where it is stated to have been released into the public domain. It is included within this book on that basis.

be direct irradiation with the laser beam (referred to as *direct drive*) or it may be irradiation with x-rays produced by the laser beam (referred to as *indirect drive*) as described in more detail below. The absorption of energy from the beam heats the outer portion of the capsule. This outer portion ablates (2) and the resulting inertial force compresses the core of the fuel capsule. Continued irradiation of the core (3) causes the remaining capsule material to heat, thereby reaching density and temperature conditions over a sufficient period time such that nuclear fusion will occur (4).

In order to achieve the necessary conditions for nuclear fusion in accordance with the series of events as described above, it is necessary to consider some important factors, as follows:

- It is important to ensure that the surface of the fuel capsule is uniformly radiated. Since it is the heat transfer to the capsule that causes the ablation and subsequently the compression of the fuel, it is necessary that this compression be evenly distributed over the surface. If there are nonuniformities then it is rather like trying to squeeze a piece of plasticine with your hand. As you squeeze it, the material tends to ooze out between your fingers. In the case of the compression of fusion fuel, this is analogous to the creation of undesirable instabilities.
- It is essential to optimize the coupling of the incident energy to the fuel capsule in order to maximize compression and heating.
- It is important that the inertial forces from the ablation compress the fusion fuel core to sufficiently high density before excessive heating of the fuel occurs. The generation of heat within the core produces outward pressure which causes the core to expand and, therefore, opposes the production of sufficiently high density within the fuel. Typically, a plastic (or similar material) outer layer on the fuel capsule acts as an ablation layer which is ablated at the beginning of the energy pulse and serves to compress the +core consisting of fusion fuel.
- The amount of fusion, and therefore energy production, within the core must be maximized before the core expands and the fuel dissipates. This dissipation of the fuel occurs on a time scale of a sound wave traversing the diameter of the remaining fuel core. This typically amounts to a time on the order of 10^{-8} s.

The ability to control these features of the capsule irradiation depends on a number of experimental characteristics. Specifically, important design considerations include:

- the energy associated with the incident pulse and the distribution of power over the duration of the pulse;
- the number and geometry of the incident beams;
- the wavelength of the incident radiation (for direct drive) or energy of the incident x-rays (for indirect drive);
- the size, composition and structure of the fuel capsule.

The importance of these different experimental parameters will be discussed in more detail below.

Instabilities

The ability to optimize the compression of the fuel capsule depends on maintaining spherical symmetry as the capsule is compressed. This, unfortunately, is a difficult task. As noted above, various experimental parameters can help to mitigate the effects that lead to the development of a non-spherical shape of the imploding capsule. Here we consider some of the fundamental reasons why the instabilities that give rise to non-spherical compression can occur.

Instabilities arise at the interface between the inner surface of the expanding ablation material and the outer surface of the imploding fusion fuel as a result of their different densities. There are three major instabilities that can occur, Rayleigh–Taylor instabilities, Kelvin–Helmholtz instabilities and Richtmyer–Meshkov instabilities. Figure 5.12 shows how these instabilities arise in a conventional fluid system. These instabilities and how they apply to inertial confinement laser fusion are discussed in some detail below.

Rayleigh–Taylor instabilities

Rayleigh–Taylor instabilities, as seen in figure 5.12(a), arise in conventional fluid systems when a denser fluid (density ρ_2) is above a less dense fluid (density ρ_1) in the presence of a gravitational field (with gravitational acceleration, g) that is perpendicular to the fluid interface. The growth rate, R, for instabilities defined by a wavenumber, k, can be shown to be

$$R = (gk)^{1/2} \tag{5.15}$$

indicating that instabilities with greater wavenumber (i.e. shorter wavelengths) develop most quickly.

In the case of a fusion fuel capsule, the density of the imploding fuel is greater than that of the expanding ablation layer. The acceleration of the fuel and ablatant are analogous to a gravitational force in the radial direction and, therefore, give rise

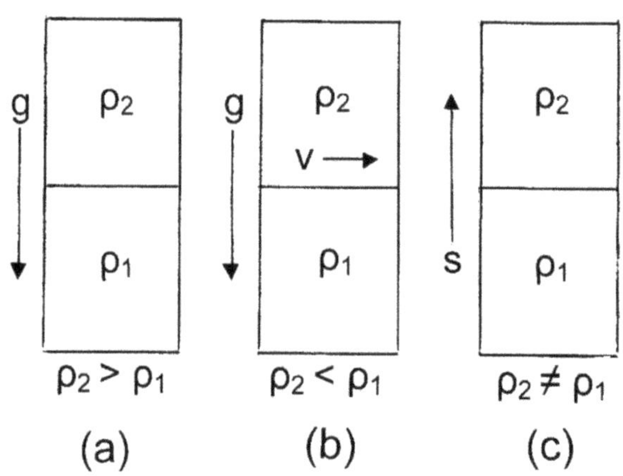

Figure 5.12. Illustration of (a) Rayleigh–Taylor instability, (b) Kelvin–Helmholtz instability and (c) Richtmyer–Meshkov instability.

to the formation of Rayleigh–Taylor instabilities at the interface. These instabilities have two detrimental effects on the fusion yield. Firstly, the instabilities limit the maximum compression of the fuel that is achievable by reducing the sphericity of the imploding core and, secondly, mixing of the fuel with the ablatant material reduces total potential energy production.

Kelvin–Helmholtz instabilities

In the case as shown in figure 5.12(b) where a less dense layer of fluid sits on top of a denser layer of fluid in a gravitational field, the conditions for the development of Rayleigh–Taylor instabilities do not exist. However, in the case where there is a shear velocity of one layer with respect to the other, as seen in the figure, then Kelvin–Helmholtz instabilities can develop. It can be shown that instabilities will have a threshold wavenumber given by

$$k_{\min} = g\frac{(\rho_1 - \rho_2)(\rho_1 + \rho_2)}{\rho_1\rho_2 v} \tag{5.16}$$

Thus, short wavelength instabilities develop and lead to reduction in fusion yield for the reasons described above.

Richtmyer–Meshkov instabilities

Richtmyer–Meshkov instabilities are analogous to Rayleigh–Taylor instabilities in a fluid system with a density difference across an interface that is subject to shock acceleration. Instabilities develop for either positive or negative density differences relative to the direction of the shock, as shown in figure 5.12(c). Short wavelength, that is high wavenumber, instabilities develop most rapidly.

Direct drive

There are two, fundamentally different, approaches to irradiating fusion fuel; direct drive, where the fuel is irradiated directly with the laser, and indirect drive, where the fuel is irradiated with an x-ray produced by the laser when it is incident on a target. Direct drive is the simplest method of imparting energy to the fuel capsule and was the first experimental approach to inertial confinement laser fusion. An example of direct drive geometry is illustrated in figure 5.13. The major advantage of direct drive (compared to indirect drive, as described below) is a more efficient laser-fuel coupling. This means that a greater portion of the laser energy is acquired by the fuel capsule than in the indirect case. Typically, a fuel capsule irradiated by direct drive with a 2 MJ ultraviolet laser pulse would typically acquire 80–100 kJ of energy (that is, 4%–5% of the laser energy). A principal consequence of this greater energy coupling is the ability to utilize larger fuel capsules.

Some disadvantages of direct drive compared with indirect drive are a lower mass ablation rate, greater nonuniformity of the energy distribution on the surface of the fuel and cross-beam energy transfer. The means by which the adverse effects of these factors may be mitigated are discussed below.

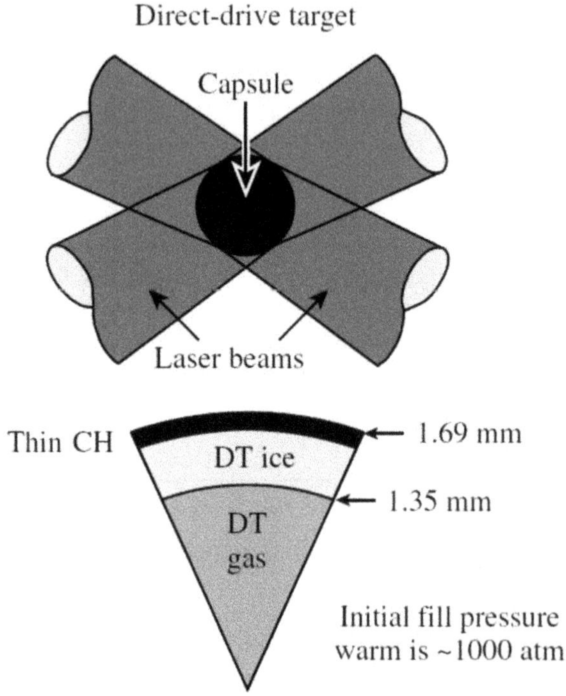

Figure 5.13. Irradiation of fuel in the direct drive scheme (top) and typical internal structure of fuel capsule (bottom). CH = carbon–hydrogen plastic ablator, DT = deuterium–tritium fuel. Reproduced from Sangster *et al* (2007). Copyright IOP Publishing. All rights reserved.

Ablation rate

A typical fuel capsule design for a direct drive system is illustrated in figure 5.13. This illustration shows an ablation layer made of a carbon–hydrogen based plastic. When the laser beam is incident on the fuel capsule, the ablation layer is heated and is blown-off as shown in part (2) of figure 5.11. In a direct drive system, the ablation rate is less than for the indirect drive system described below. As a result, the ablation layer is thinner (relative to the overall capsule size). This thinner ablation layer provides less protection of the inner fuel from preheating by hot electrons. This works against the requirement, as stated above, that maximum compression of the fusion fuel should occur before there is significant heating.

Nonuniformity

Nonuniformity of radiation on the capsule surface leads to hot spots in the fuel and corresponding instabilities. Irradiation of the target with a large number of beams from different directions will minimize this problem but it is difficult to eliminate problems associated with such instabilities. Approaches which deal with other issues, as indicated below, may be counterproductive in dealing with nonuniformities. Nonuniformities can be mitigated to some extent by adding a high atomic number impurity in the plastic ablation layer.

Cross-beam energy transfer

Cross-beam energy transfer occurs because the laser beam coming from one direction is refracted by the sample and this refracted beam interferes destructively (to some extent) with the incident beam from the opposite direction. This interference reduces the incident intensity by typically about 20%. Some possible approaches to mitigating this problem are to detune beams incident from opposite directions or reducing the laser spot size on the target. This latter approach has the advantage that it improves energy coupling between the beam and the fuel, but it is somewhat counterproductive as it tends to increase beam nonuniformity.

Indirect drive

In the indirect drive system, the fuel capsule is contained within a centimeter-sized hollow metal cylinder referred to as a 'hohlraum' (German for 'hollow space'). This arrangement is illustrated in figure 5.14. The structure of a typical indirect drive fuel capsule is also shown in figure 5.14. The laser beams enter the hohlraum through plastic windows (typically 1 μm thick polyimide) at either end of the device and are incident on the inner walls of the chamber. Hohlraums are made of a metal with a high atomic number (Z-value) and high density. The requirement for these properties is the need to minimize energy loss outward through the hohlraum. For this same reason it is important that the dimensions of the plastic windows are the minimum necessary to allow the laser beams to effectively enter the hohlraum. Gold is commonly used to construct the hohlraum, although depleted uranium (often with a gold coating) is also sometimes used. As the figure also shows, the interior of the

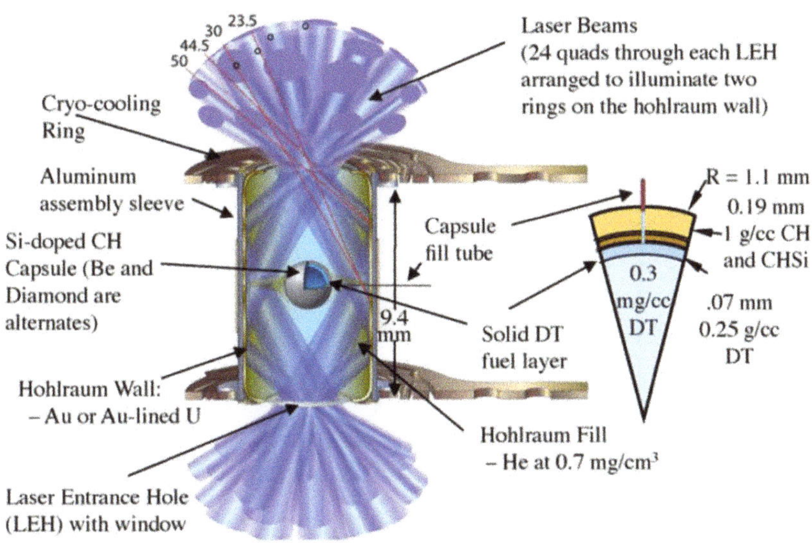

Figure 5.14. Irradiation of fuel in the indirect drive scheme (left) and typical internal structure of fuel capsule (right). Design from the National Ignition Facility. CH = carbon–hydrogen plastic ablator, DT = deuterium–tritium fuel. Reproduced from Landen *et al* (2012). Copyright IOP Publishing. All rights reserved.

hohlraum is filled with helium gas. This is to slow the transport of plasma produced by the vaporized hohlraum to the fuel capsule.

The laser beams are incident on the inside of the hohlraum where they generate x-rays. The x-rays are produced by two processes when the hohlraum metal is rapidly heated. Firstly, blackbody radiation in the x-ray region of the electromagnetic spectrum is produced as a result of the high temperature of the hohlraum material. Secondly, characteristic x-rays associated with the hohlraum material are produced as a result of induced inner shell electronic transitions. Figure 5.15 shows a computer simulated energy spectrum of a gold hohlraum and a uranium hohlraum. This figure shows the predominant blackbody energy spectrum with additional characteristic elemental x-ray lines.

The properties of indirect drive inertial confinement fusion may be compared with the properties discussed above for direct drive. Specifically, indirect drive has the following advantages:

- Improved uniformity of radiation on the surface of the fuel capsule. One important design parameter which greatly influences radiation uniformity is the ratio of the hohlraum inner diameter to the capsule outer diameter (sometimes referred to as the 'case-to-capsule ratio'). Recent work by Jing *et al* (2018) on hohlraums with six (rather than two) openings has shown improved drive symmetry and improved stability.
- Ability to utilize a relatively thicker ablation layer in order to mitigate adverse effects due to preheating of the fuel by hot electrons.
- No reduction of beam intensity due to cross-beam energy transfer.

Unfortunately, there is one very substantial disadvantage of indirect drive—poor energy transfer between the laser and the target. For example, it is estimated that a

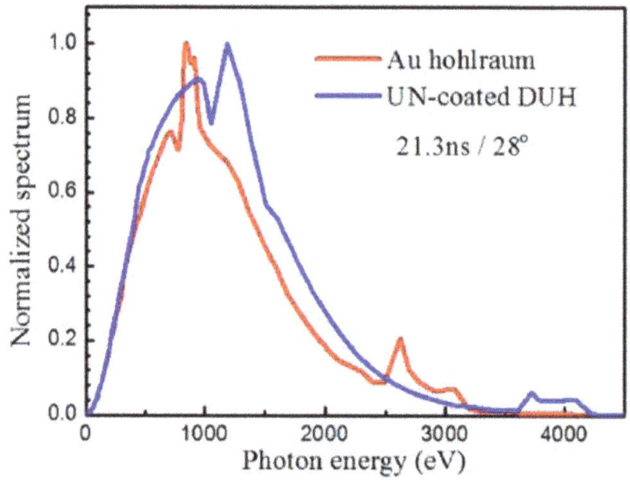

Figure 5.15. Computer simulated spectra of Au hohlraum and uranium-nitride coated depleted uranium hohlraum. UN = uranium nitride, DUH = depleted uranium hohlraum. Reproduced from Guo *et al* (2015).

2 MJ ultraviolet laser pulse gives up 12–16 kJ or 0.12%–0.16% of its energy to the fuel capsule. However, the ability to control the implosion of the fuel capsule in the indirect drive method has been beneficial to our understanding of inertial confinement fusion by optimizing fuel compression. Consequently, much effort has been directed toward indirect drive laser fusion experiments in recent years, as described below.

In order to understand the mechanism by which the fuel is compressed, it is important to consider the ablation pressure, P_{ab}. The ablation pressure is related to the radiation temperature, T_r, as

$$P_{ab} = \left[\frac{A_{ave}}{Z_{ave} + 1}\right]^{1/2} T_r^{7/2} \tag{5.17}$$

where A_{ave} and Z_{ave} are the average mass number and atomic number, respectively, of the ablatant material. Figure 5.16 shows a typical time development of the radiation temperature over the duration of the irradiation pulse used for indirect drive. Each step in the radiation temperature corresponds to a step in the ablation pressure, as defined by equation (5.17), which sends a compression shock wave into the fuel.

It is actually the laser power, P, that is controlled directly during the irradiation and this can be related to the desired radiation temperature by means of the Stefan–Boltzmann law as

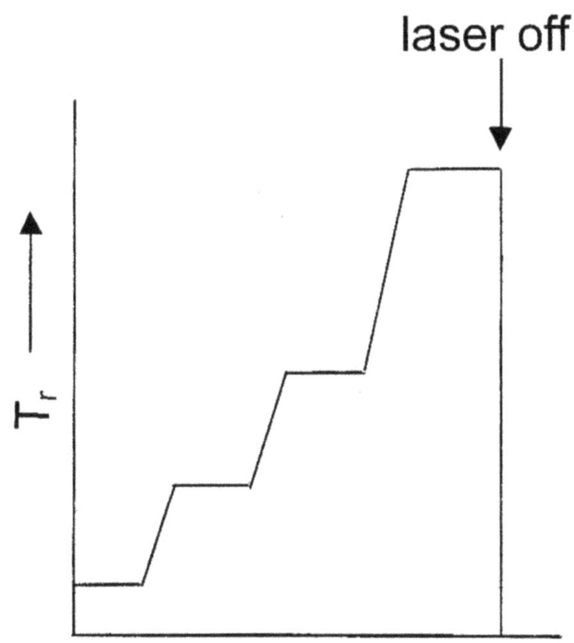

Figure 5.16. The radiation temperature as a function of time for indirect drive. The total pulse duration is of the order of a nanosecond.

$$P = \sigma A T_r^4 \tag{5.18}$$

where σ is the Stefan–Boltzmann constant and A is the inside area of the hohlraum. The magnitude and duration of the steps in T_r will determine the implosion velocity of the adiabat and will establish the level of stability characterizing the compression. Through the use of appropriate time development of the irradiation pulse, instabilities, specifically, Rayleigh–Taylor instabilities, can be minimized during fuel compression. Experimental results, specifically from the National Ignition Facility (NIF), that are related to progress in our understanding in this area are discussed later in this chapter.

Fusion criteria

Ultimately, a viable inertial confinement fusion power reactor must produce more energy than it consumes. In fact, as discussed in more detail in chapter 7, it is important that the reactor produces much more energy than it consumes in order to make the energy as economical as possible. Several steps can be defined along the path to a successful laser fusion system. Firstly, of course, is the requirement of introducing fusion reactions in the fusion fuel. The creation of fusion, at least in the case of d–d and d–t reactions (which have been used in laser fusion experiments) is evidenced by the emission of neutrons at the appropriate energy. The second step in producing a viable fusion reactor is achieving ignition. Ignition, as described in the previous chapter, is a condition where fusion in a portion of the fuel produces sufficient energy to cause a chain reaction in a significant portion of the remaining fuel, leading to substantial energy release. The final goal is to produce more energy than is expended, that is, to achieve and exceed breakeven (that is commercial breakeven). Achieving this goal is a particular challenge for inertial confinement laser fusion because of the energy conversion efficiencies. This is described in detail below.

As in the case of magnetic confinement fusion, appropriate conditions must be met in order for inertial confinement fusion to be viable. There are, however, fundamental differences in the approach to satisfying fusion conditions for magnetic and inertial confinement fusion. Inertial confinement fusion is intrinsically a pulsed method where the fuel achieves a high density for a short period of time. This is in comparison to typical magnetic confinement fusion conditions where densities are much lower and confinement times are much longer. As a result, the approach to the analysis of fusion conditions in inertial confinement devices is somewhat different than for magnetic confinement devices. However, the determination of a condition analogous to the Lawson criterion, as described below, is necessary to understand experimental progress in this field.

We begin by looking at the details of how the fuel in a capsule undergoes fusion. When the capsule is compressed and heated, the heating does not occur uniformly throughout the compressed fuel. Rather, a very small 'hot spot' occurs near the center of the fuel. It is in this hot spot that fusion begins. We use d–t fusion,

$$\mathrm{d} + \mathrm{t} \rightarrow \alpha + \mathrm{n} \tag{5.19}$$

as an example. The energy released in this reaction is 17.6 MeV, of which 3.5 MeV is carried away as kinetic energy of the alpha particle and 14.1 MeV is carried away as kinetic energy of the neutron. The neutron energy may be recovered by subsequent reactions in a reactor blanket, but has no further role in the fusion process, as such. The alpha particle, which travels outward from the central hot spot, reacts with the remaining cooler compressed fuel, depositing some fraction of its energy and causing the fuel to heat and undergo additional fusion. It is the amplification of the fusion rate from this self-heating (or alpha particle heating) of the fuel that increases the energy output from the fuel and leads (in the ideal case) to satisfying the ignition conditions. It should be noted, however, that the potential fusion energy of the entire fuel capsule can never be realized, as some fuel will always remain unignited. It is generally assumed that for an efficient commercial reactor one might expect a d–t fuel burn of 20%–30%.

In order to relate conditions for breakeven for inertial confinement (laser) fusion compared to magnetic confinement fusion, it is necessary to understand the differences in the way in which fusion energy is utilized to maintain fusion reactions in a reactor and to provide net electrical output. As discussed in some detail in chapter 4 for a magnetic confinement reactor, heat produced by fusion reactions is converted into electricity with an efficiency of up to 40%. Although there may be some losses in the system, most of this electricity becomes available to maintain the plasma conditions (through ohmic heating) and to provide end user electricity. The conversion of fusion energy into electricity is discussed in further detail in chapter 7. The situation for an inertial confinement fusion reactor is quite different. The heat produced from fusion reactions can be converted into electricity in the same manner as for a magnetic confinement reactor. However, the fraction of electricity that must be returned to the reactor to maintain the fusion reaction is typically much greater. Rather than using electricity with near 100% efficiency for ohmic heating of the plasma, electricity is used to drive the laser which in turn deposits energy (either directly or indirectly) to subsequent fuel capsules. The conversion of electrical energy to laser energy and the coupling of the laser to the fuel is far less than 100%. This is a major disadvantage of inertial confinement fusion as it means that a much higher ratio of fusion energy output to electrical energy input is needed to achieve breakeven. In order to assess operating conditions for inertial confinement fusion reactors as discussed below and in section 5.7, it is essential that this factor is included in the analysis.

The derivation of the Lawson criterion for magnetic confinement fusion has been discussed in detail in chapter 4. The Lawson criterion is derived from the requirements on the triple product, $n\tau T$, where n is particle density, τ is confinement time and T is temperature, by the fact that the minimum value of $n\tau$ for breakeven shows a minimum in temperature at around 30 keV (for d–t fusion) as illustrated in figure 4.1. Similarly, for inertial confinement fusion, a condition on the minimum value of the product $n\tau$ for breakeven exists for an optimal temperature. This may be shown by the requirement for d–t fusion that

$$P\tau > \frac{11f(T)}{\theta_\alpha} \tag{5.20}$$

where P is the plasma pressure, τ is the plasma confinement time and θ_α is the fraction of alpha particle energy that is deposited in the fuel. The dimensionless function $f(T)$ is related to the temperature of the fuel and typically has a minimum at around 20 keV. This provides an estimate of the ideal fusion temperature for inertial confinement fusion.

In order to quantify the fusion conditions for inertial confinement fusion we begin with the Lawson criterion for the product of particle number density and confinement time, $n\tau$, as given by equation (4.10) in chapter 4 for d–t magnetic confinement fusion,

$$n\tau > 7 \times 10^{-4}\frac{T}{\eta\langle\sigma v\rangle} \tag{5.21}$$

where the fusion rate, $\langle\sigma v\rangle$, is the cross section averaged over particle velocities and η is the efficiency (as a fraction) of returning fusion energy to the fuel for subsequent fusion reactions. Here temperature is expressed in keV and $n\tau$ is in m^{-3} s. See the discussion in section 4.2 for more details on the use of units. In the case of inertial confinement laser fusion, the fuel will begin to undergo fusion at the central hot spot, and this reaction will expand outward, as described above, as the cooler fuel is heated to ignition by energetic alpha particles emitted by the reaction. As this occurs, the fuel will expand as it is heated, and its density will decrease. The average particle density of the plasma may be expressed in terms of the fuel capsule radius, r, as

$$n \sim r^{-3} \tag{5.22}$$

From chapter 4, it was shown that the alpha particle heating (equation (4.4)) is related to the particle density as

$$P_\alpha \sim n^2 \tag{5.23}$$

meaning that

$$P_\alpha \sim r^{-6} \tag{5.24}$$

So, as the fuel expands the power output decreases rapidly. It is customary to assume that the fusion reaction will stop, i.e. $P_\alpha \approx 0$, when the fuel capsule has expanded by 25% of its compressed radius. If the expansion velocity is v, then the burn time is easily determined to be

$$\tau = \frac{r}{4v} \tag{5.25}$$

The expansion velocity is the velocity of the ions and this may be obtained by equating the ion kinetic energy, $mv^2/2$ where m is the average ion mass, and the thermal energy, kT, to give

$$v = \left[\frac{2kT}{m}\right]^{1/2} \tag{5.26}$$

Combining equations (5.21), (5.25) and (5.26) for a 50:50 d–t fuel mixture gives

$$nr > 545\frac{T^{3/2}}{\eta\langle\sigma v\rangle} \tag{5.27}$$

where the units for nr are m^{-2}. Following from the discussion above, the minimum value for nr for which breakeven is achieved occurs at a temperature of around 20 keV for which

$$\frac{T^{3/2}}{\langle\sigma v\rangle} \approx 2.1 \times 10^{23}\ \mathrm{keV^{3/2}\ m^{-3}\ s} \tag{5.28}$$

Substituting equation (5.28) into (5.27) and using the fact that the mass density is given by the number density as $\rho = mn$ yields the usual breakeven condition for inertial confinement fusion

$$\rho r > \frac{0.48}{\eta} \tag{5.29}$$

in units of kg m^{-2}. The quantity ρr is referred to as the areal density (i.e. mass per unit area) and is defined as an integral over the density distribution of the compressed fuel capsule as

$$\rho r \equiv \int_0^r \rho(r')\mathrm{d}r' \tag{5.30}$$

Another approach to determining the requirements for a viable inertial confinement fusion reactor is to consider the necessary compression of the fuel prior to ignition. If a fuel capsule has a radius r_0 prior to compression and a radius r after compression, we can define the compression factor as

$$c = \frac{r_0}{r} \tag{5.31}$$

The mass density of the fuel after compression, ρ, is related to the mass density of the fuel before compression, ρ_0, as

$$\rho = \rho_0 c^3 \tag{5.32}$$

combining equations (5.31) and (5.32) gives

$$\rho r = \rho_0 r_0 c^2 \tag{5.33}$$

Finally, substituting into equation (5.29) gives the compression requirement for breakeven (in SI units, as above) as

$$c > \left[\frac{0.48}{\eta\rho_0 r_0}\right]^{1/2} \tag{5.34}$$

Using typical values for initial fuel radius of 0.001 m, an initial density of 270 kg m^{-3} (typical of a solid 50:50 d–t mixture) and $\eta = 0.01$ yields a minimum compression factor for breakeven of about 13. The parameters used in this calculation, particularly the low value of η, and the result of this analysis will be discussed in further detail in section 5.7.

Advanced ignition schemes

In addition to conventional direct drive and indirect drive approaches as described above, a number of other methods of producing nuclear fusion using lasers have been investigated. These are sometimes generically referred to as 'advanced ignition schemes'. These schemes can involve a laser pulse followed by a second pulse of high energy electrons or ions (referred to as *fast ignition*), significantly altered laser pulse shapes (referred to as *shock ignition*), or laser irradiation combined with an externally applied magnetic field (such as *magnetized liner inertial fusion*). The present section describes shock ignition and the next section describes fast ignition. Some recent work on magnetized liner inertial fusion (or *MagLIF*) is discussed in chapter 6.

As indicated previously in this chapter, a major challenge to successfully achieving energy gain from a laser fusion reactor is to sufficiently compress the fusion fuel prior to ignition and to do so without excessive preheating. Fast ignition and shock ignition (described below) are both approaches to dealing with this challenge by separating the energy that is supplied to the target to compress the fuel from the energy that is provided to the target to ignite the fuel.

Fast ignition

Fast ignition is a scheme that uses two separate drivers, one to compress the fuel and the second to ignite the fuel. The concept of fast ignition followed from the development of high-intensity short-pulse lasers that could be used to produce the ignition pulse of electrons or ions as described below. The first driver is a conventional laser that is used to compress the fuel by heating the ablation layer. Typically, a high-intensity short-pulse laser is then used to provide the ignition pulse. This laser pulse is incident on a target that releases a coronal plasma consisting of high energy ions and electrons. The electrons, which are in the MeV range, are relativistic and have been used as the ignition irradiation in recent experiments. A major problem of this two-step irradiation process is the fact that the second beam must penetrate the plasma cloud formed by the expanding ablatant and this cloud may be less than fully transparent to the ignition beam. Figure 5.17 shows a solution to this problem that has been used with the OMEGA laser (see description of the OMEGA project below). A gold cone is used to allow the laser to penetrate the coronal plasma where it produces the high energy electrons at the tip of the cone, very close to the imploding fuel capsule. The fuel is ignited when the electron beam is incident on the hot spot in the compressed fuel. In principle, this approach may be adaptable to both direct drive and indirect drive systems.

Experimental work on fast ignition was first undertaken at Gekko XII and more recently has taken place at the OMEGA facility. These are discussed in further detail in the relevant experimental sections below.

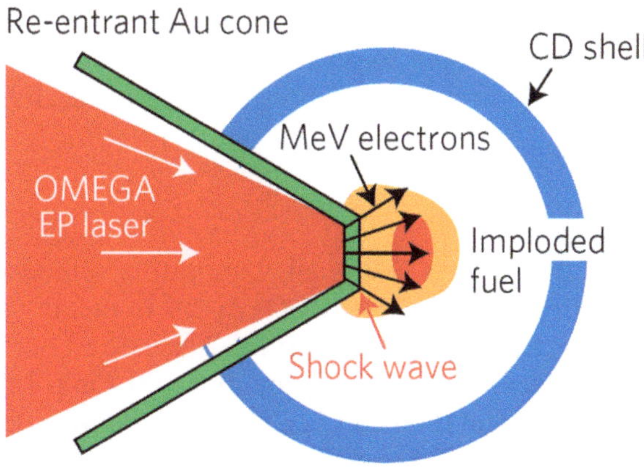

Figure 5.17. Fast ignition arrangement used at the OMEGA laser facility. (CD = deuterated plastic). Reprinted with permission from Macmillan Publishers Ltd. Nature Communications. Betti and Hurricane (2016). Copyright (2016).

Shock ignition

Shock ignition is a more recent development that follows along the lines of fast ignition but uses only a single laser. In the shock ignition scheme, a single laser provides both the compression pulse and the ignition pulse. The laser produces an initial laser pulse containing about 80% of the total irradiation energy and this is used to compress the fuel by heating the ablation layer. The remaining 20% or so of the laser energy is contained in a spherically symmetric short laser pulse at the end of the compression phase and is used to ignite the hot spot. For symmetry reasons, it is most likely that this approach would be applicable only to systems using direct drive. The ignition pulse is timed to arrive at the end of the compression pulse so that the ignitor shock converges with the return shock from the compression pulse at the target surface. These colliding shock waves produce an inward travelling shock that further compresses the fuel and causes ignition at the hot spot. The most significant experimental investigations of the shock ignition concept have been carried out using the OMEGA system as described below.

5.4 Overview of some CO_2 laser fusion experiments

In this and the next two sections, we review some experimental investigations of inertial fusion utilizing lasers. While the examples given here are not intended to be comprehensive, they summarize the history of the field and represent a cross section of the different approaches. These experiments are categorized in terms of the type of laser which was utilized. The present section deals with experiments using CO_2 lasers. Sections 5.5 and 5.6 deal with experiments using Nd lasers and KrF lasers, respectively. The ultimate goal of inertial confinement fusion research is to achieve ignition and to produce viable carbon-free energy. It is important to note, however, that the goal of some experiments is not to actually achieve ignition. Some experiments are meant to investigate the details of specific problems facing internal confinement fusion, while others are designed to provide insight into the basic physics of problems related to astrophysics, plasma physics or nuclear weapons. Future inertial confinement fusion experiments that are in the planning or construction stages, as well as directions for possible future research in this field, are discussed in section 5.7.

Helios

In the early 1970s Los Alamos National Laboratory initiated a program to investigate the possibility of using pulsed CO_2 lasers to create controlled thermonuclear fusion in deuterium–tritium fuel capsules. The earliest operational system in 1973 was a single-beam facility capable of providing 100 J in a 1 ns pulse. This was followed by a two-beam system (named Gemini) which became operational in 1976 and was capable of providing 0.5 kJ in a 1 ns pulse. These two systems were intended as a means of developing the technology of high-power CO_2 lasers. Gemini was the first CO_2 laser based inertial fusion experiment to actually produce nuclear fusion in a deuterium–tritium fuel capsule as evidenced by the observation of 14 MeV fusion neutrons.

Based on the technology developed during the single-beam and Gemini experiments, Los Alamos National Laboratory constructed the Helios laser fusion experiment. This was an eight-beam system that became operational in 1978. Each of the eight beams can provide 1250 J, for a total of 10 kJ in a 0.5–1 ns pulse. Maximum power was in excess of 20 TW. A similar 10 kJ eight-beam CO_2 laser system constructed at about the same time at Osaka University was also used to conduct experiments on inertial confinement fusion.

The eight beams utilized by Helios allowed for much improved spatial distribution of laser energy on the target compared to the earlier Gemini system. The target is located at the center of a virtual cube and the eight mirror-directed beams enter the cube through the eight cube corners.

The Helios system was used to irradiate a sample of deuterium–tritium fusion fuel. The fuel consisted of deuterium–tritium gas enclosed in a Parylene (a poly (p-xylylene) polymer) capsule. Results of these experiments showed that imploding fuel capsules could achieve densities of as much as 20 times the density of liquid hydrogen. This density is about 40% higher than the density of water. A common measure of fusion yield for inertial confinement laser fusion experiments is the measured number of fusion neutrons per shot. Helios produced about 10^9 neutrons per shot. While not sufficient to achieve breakeven, the fuel density obtained in the studies conducted with Helios was viewed optimistically and these experimental results along with modeling studies led the way to a larger scale CO_2 laser based inertial confinement fusion system, Antares, as described below.

Antares

The CO_2 laser based system Antares was constructed at Los Alamos National Laboratory on the basis of experience gained from the Helios system. It became operational in 1983 and was designed with the expectations of achieving scientific breakeven. The Antares system consists of 24 laser beams that can provide a 100 kJ pulse with a duration of up to 1 ns. Maximum power of 200 TW was achieved for 0.3 ns pulses. The beams irradiate a 400 μm deuterium–tritium target located at the center of a 7 m diameter test chamber. While breakeven was anticipated on the basis of scaling up the performance of the Gemini and Helios laser fusion experiments conducted at Los Alamos, the Antares system did not perform up to expectations. It was found that because of the long wavelength of the incident radiation the target was preheated by electrons emitted during plasma formation in the early part of the irradiation. This preheating was detrimental to the subsequent compression of the fuel capsule, as described in section 5.3, and confirmed results of earlier experiments conducted using the Nd-based laser systems Argus and Shiva, as described below. Consequently, the resulting fusion energy was much less than had been anticipated and the Antares system fell far short of expectation to achieve breakeven. Following these unfortunate discoveries, the Antares program was disbanded in 1985 and subsequent laser fusion experiments have concentrated almost exclusively on drivers that provide short wavelength radiation, either frequency doubled or tripled Nd-laser radiation of KrF laser radiation as described below.

5.5 Overview of some Nd laser fusion experiments

Nd-based laser experiments at Lawrence Livermore National Laboratory

Nuclear fusion induced by laser irradiation was first achieved in 1974 at KMS Fusion using a Nd:YAG laser at 1064 nm as the driver. KMS Fusion was a private sector company founded by American physicist Kip Siegel (1923–75) and has been the only independent company to pursue inertial confinement fusion research. The initial observation of fusion, as evidenced by fusion neutrons, occurred for a pure deuterium fuel capsule. Shortly thereafter similar observations were made using deuterium–tritium as a fuel. Later in 1974 the Janus laser (a Nd-silica glass laser) at Lawrence Livermore National Laboratory successfully achieved nuclear fusion in a deuterium fuel capsule. Since that time, a number of notable laser fusion experiments have been undertaken, many of them in United States National Laboratories. The present section reviews some of these experiments, as well as some of those conducted in other countries that have utilized Nd laser drivers.

Argus

Argus was a two beam Nd-silica glass laser constructed at Lawrence Livermore National Laboratory in 1976. It was capable of producing about 2 kJ in a 1 ns pulse. Maximum power was about 4 TW for a 100 ps pulse. While the Argus laser was quickly surpassed in total energy by CO_2 lasers, as described above, it provided very important information concerning laser target interactions that led to the development of future Nd-based laser fusion experiments. Argus was the first laser system to undertake indirect drive experiments by using a hohlraum to generate x-rays. It was also the first system to discover that the long wavelength infrared radiation produced by Nd-based lasers (and CO_2 lasers, as well) was disadvantageous to fuel compression because of target preheating by hot electrons. Argus observed the reduction in x-ray production by hot electrons when using frequency doubled and tripled laser radiation. As noted above, this was confirmed by CO_2 laser experiments and led to the utilization of harmonic generation methods for subsequent Nd-laser fusion systems. Maximum d–t fusion neutron yield from Argus was about 10^9 neutrons per shot. Argus was dismantled in 1981, but the knowledge that it provided about laser-target interactions was instrumental in the design of future laser fusion experiments.

Shiva

The Shiva laser was constructed at Lawrence Livermore National Laboratory and became operational in 1978. This 20-beam Nd-silica glass laser, which produced up to 10.2 kJ in a 0.5–1.0 ns pulse, was based on early experimental information from the Argus laser, and was intended to provide improved results because of its increased power and improved distribution of irradiation on the fuel capsule. While the goal of Shiva was to achieve ignition, it was unsuccessful because of the poor coupling of infrared radiation to the fuel. The necessity of utilizing frequency doubling or tripling techniques, as also demonstrated using the Argus laser at Lawrence Livermore and CO_2 lasers at Los Alamos, was confirmed by experiments with Shiva. Shiva was also used for early experiments using indirect drive methods. The maximum d–t fusion neutron yield from Shiva was between 10^{10} and 10^{11} neutrons per shot. Shiva was dismantled in 1981 to make way for Lawrence Livermore's subsequent laser experiments, Novette and Nova, which utilized components recovered from Shiva.

Novette/Nova

The Novette laser was constructed at Lawrence Livermore National Laboratory and became operational in 1983. It followed from experiments conducted on the Argus and Shiva laser systems and utilized components recovered from these previous experiments. Novette was a two-beam frequency tripled Nd-phosphate glass laser. It was used briefly as a test bed for the Nova laser in order to validate the utilization of frequency tripled radiation.

The Nova laser became operational at Lawrence Livermore in 1984 and reused parts from the Novette laser. Nova utilized ten beams and was the first facility to effectively make use of beamline folding to reduce the overall length of the device. Each of Nova's ten 180 m long beamlines was folded once to yield a 90 m long device. Figure 5.18 shows an image of one of Nova's laser amplifiers. The unit contains two Nd-glass plates set at Brewster's angle in opposite directions, as seen in the figure. Brewster's angle, θ_B, is the angle of incidence of a light beam on an interface between two materials of different indices of refraction at which no reflection occurs. For a material with an index of refraction, n_2, in a surrounding medium with an index of refraction, n_1, Brewster's angle is given by

$$\theta_B = \tan^{-1}\left[\frac{n_2}{n_1}\right] \tag{5.35}$$

Population inversions in the Nd^{3+} energy levels are created using xenon flash tubes. Nova's lasers were capable of producing up to 100 kJ in 2–4 ns pulses at 1054 nm.

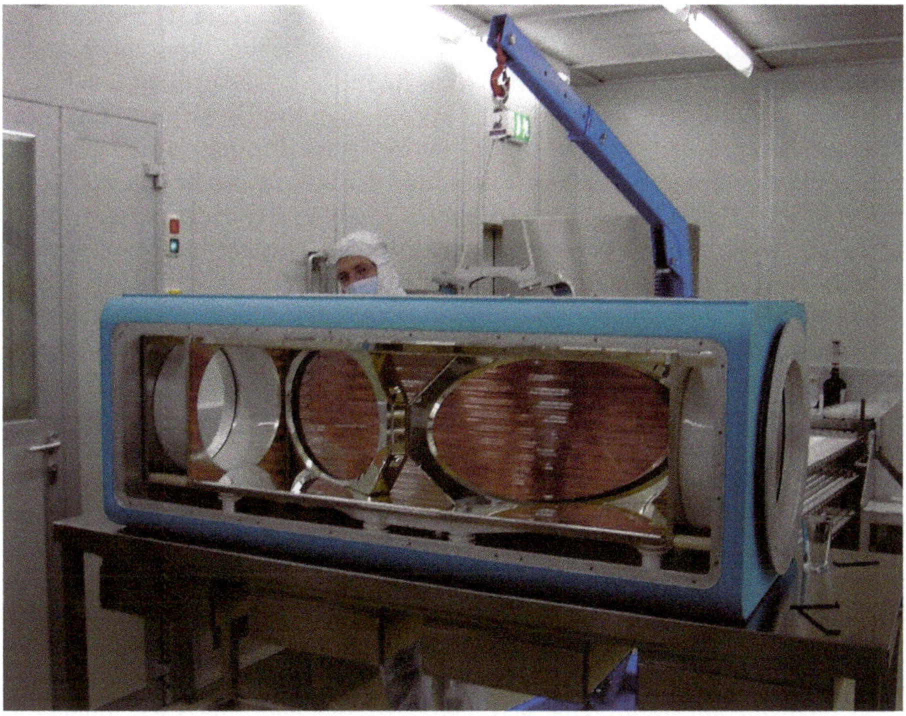

Figure 5.18. An opened A315 amplifier from the Nova laser system, loaned in 2003 to the PHELIX (Petawatt High Energy Laser for heavy Ion eXperiments) laser facility at the GSI Helmholtz Centre for Heavy Ion Research in Germany. This A315 Amplifier image has been obtained by the author from the Wikimedia website https://commons.wikimedia.org/wiki/File:A_315_nova_amplifier_opened_2009-03-27.jpg where it was made available by Commander-pirx under a CC BY-SA 3.0 license. It is included within this book on that basis. It is attributed to Commander-pirx.

After frequency tripling, lasers could provide pulses of 40–45 kJ at 351 nm with a maximum ultraviolet power of 16 TW.

The research associated with the Nova project concentrated largely on the indirect drive method. Although Nova was intended to achieve ignition, it did not reach this goal. Experimental results showed that this failure was primarily the result of the existence of Rayleigh–Taylor instabilities. These instabilities were the result of the inability to match the intensity of the multiple beamlines sufficiently well. This in turn led to temperature inhomogeneities and non-uniform compression of the fuel capsule. The maximum d–t fusion neutron yield from Nova was about 10^{13} neutrons per shot.

Two concerns about laser fusion that are specifically relevant to Nd-laser systems are efficiency and cycle rate. In section 5.3 it was noted that the efficiency of returning energy from fusion reactions to drive subsequent reactions in inertial confinement laser fusion reactors is very low. In the discussion above, 1% was used as a typical value. For Nd-glass lasers, efficiencies of 1%–1.5% for converting electrical input to laser energy are fairly typically (see further discussion on this topic in section 5.7).

The second important concern for Nd-laser fusion is cycle rate. The low efficiency for the conversion of electrical energy to light by xenon flash tubes and the subsequent emission of stimulated radiation by the laser, means that a substantial amount of energy that is input into the system is generated as heat. This heat is distributed between the flash lamps and the Nd-glass. In order to avoid damage, which could be caused by excessive heating, the laser must be allowed to cool between firings. The Nova system was limited to about six firings per day to avoid overheating. In section 5.7 we will show that this is an unacceptably low cycle rate for the purpose of constructing a useful power reactor.

Nova was dismantled in 1999 but results of research conducted using this system led to the subsequent development of Lawrence Livermore's National Ignition Facility, as described below.

National Ignition Facility (NIF)

The NIF at Lawrence Livermore National Laboratory continues the research on inertial confinement fusion that was conducted on the Nova laser system. The initial concept for a new facility to replace Nova began around 1994 when it was realized that any viable upgrades to Nova would not be sufficient to achieve ignition. Construction on the new facility, the NIF, began in 1997, before Nova was officially decommissioned. At the time construction of the facility began, the cost was estimated to be $1.1 billion (USD) with an additional $1 billion for related research. The estimated completion date was 2002. In the end, the total budget was $4.2 billion, and the facility was finally commissioned, with the first large-scale laser experiments, in 2009.

The NIF has 192 beams provided by frequency tripled Nd-doped phosphate glass lasers with a total energy of 1.8 MJ per pulse and is capable of delivering up to 500 TW of laser power. It is interesting to note that this is about 200 times the average world electric power use, although only for a nanosecond or so. To uniformly irradiate the target, it is important that the beams arrive within a few picoseconds of each other. In order to ensure that this is the case, a single source, known as the master oscillator, provides the initial laser pulse for all of the beams. This master oscillator is a low power (nanojoule) ytterbium-doped optical fiber laser operating at 1053 nm. The output from the master oscillator is split into 48 beams which are amplified to a total of around 6 J. Each of the 48 beams is split into four beams giving a total of 192 beams. These 192 beams are amplified by the main amplifier, which consists of xenon pumped Nd-phosphate glass amplifiers. Each amplifier incorporates 40 xenon flash tubes for a total of 7680 flash tubes in the main amplifier. The amplifiers are located in two bays, one of which is shown in the photograph in figure 5.19. The main amplifiers increase the total infrared laser energy to around 4 MJ per pulse. The total path length of the laser beam is about 1500 m. Most of this total path length is occupied by tubes which are spatial filters. Each beam is focused at the center of the tube and an aperture masks out any portion of the beam that extends outside a well-defined focal point. This filtering process ensures that the beam is very uniform in intensity when it reaches the target.

Figure 5.19. Laser bay number 2 (one of two bays) at the National Ignition Facility at Lawrence Livermore National Laboratory. This Laser Bay Number 2 image has been obtained by the author from the Wikimedia website https://commons.wikimedia.org/wiki/File:NIF_Laser_Bay.jpg where it is stated to have been released into the public domain. It is included within this book on that basis.

A system of mirrors organizes the 192 beams into 48 groups of 4 beams each. Before entering the target chamber, the beams pass through sets of two KDP crystals. The first KDP crystal frequency doubles a large portion of the 1053 nm infrared beam into green 527 nm light. The second KDP crystal combines the 527 nm radiation with the remaining 1053 nm radiation to produce the frequency tripled 351 nm beam. Just before entering the target chamber the beams pass through a series of mirrors that provides the proper directional distribution to uniformly impinge on the target.

The target chamber is a 10 m diameter steel sphere with a mass of about 1.3×10^5 kg and the hohlraum containing the fuel capsule mounted at the center. The details of the hohlraum and fuel capsule have been described previously in section 5.3 and are illustrated in figure 5.14. In the case of the NIF, the fuel capsule is cooled to about 18 K. This creates a thin layer of solid d–t mixture inside the ablation layer with a gaseous mixture of deuterium and tritium inside of that.

Upon its completion in 2010, the NIF began the National Ignition Campaign with the goal of achieving ignition. During the period of 2010–12, the operation of the system was optimized with regard to laser performance, fuel capsule design and irradiation approach. By 2012 the conditions in the imploding fuel were estimated to be about 10% of that needed to reach ignition. As it appeared unlikely that there would be any significant progress toward ignition in the foreseeable future, the National Ignition Campaign was terminated in September 2012. Since that time, the focus of research at the NIF has been in the areas of materials research and weapons studies.

To understand the reasons for the failure of the National Ignition Campaign to achieve ignition, we can begin by looking at the flow of energy during a laser shot. In a typical experiment, approximately 3 MJ of the laser's 4 MJ infrared capacity per pulse might be realized. Of this, about half, or 1.5 MJ, might be converted to ultraviolet radiation. While about 85% of the incident ultraviolet radiation is converted to x-ray radiation by the hohlraum, it is the coupling of this radiation to the fuel capsule which is an inherently inefficient process. In the end only 10–14 kJ of the energy is actually deposited in the fuel capsule. That is, typically less than about half of one percent of the original laser energy goes toward producing fusion. In addition to the very low efficiency of coupling the laser output to the fuel, it was believed that the mixing of ablator material with the fuel as a result of instabilities was also detrimental to achieving ignition.

In parallel with the experimental investigations conducted at the NIF, Lawrence Livermore undertook the Laser Inertial Fusion Energy (LIFE) program. This program was initiated in 2008 as the construction of the NIF was nearing completion and continued throughout its operation. The program provided a detailed evaluation on the basis of the results obtained during the National Ignition Campaign of the path forward for the construction of a commercially viable inertial confinement fusion reactor. The results of this program have been summarized by Meier *et al* (2014). Based on their investigations, LIFE suggested that a commercial laser fusion power plant could be realized using diode pumped Nd-glass lasers (see section 5.7) which delivered 2.2 MJ pulses of ultraviolet radiation with a repetition rate of 16.7 Hz to an indirect drive d–t target. Fusion energy would be converted to electric energy using thermal conversion (see chapter 7) with an efficiency of 47%. LIFE ended in April 2014 when it became apparent that the National Ignition Campaign failed to achieve its desired goals and that a major rethinking of the approach to indirect drive inertial confinement fusion using lasers was needed.

Although most of the NIF's activities have since focused on other research areas, some work on improving conditions leading to ignition have continued. In September 2103 it was announced that breakeven had been achieved. This evaluation, however, was based on a different definition of breakeven where the fusion energy output exceeds the total energy input into the system. In the announcement, breakeven was defined as the ratio of the fusion energy output to the laser energy actually absorbed by the fuel. As noted above, the laser energy absorbed by the fuel is in the range of 10 kJ, so the fusion energy output observed was only a very small fraction of the energy required to achieve scientific breakeven and an even smaller fraction of the energy required for commercial viability. It is important to note, however, that results have shown that measurable alpha particle heating occurred, and this is an important step toward ignition. The most recent results from 2018, have shown the production of 1.9×10^{16} neutrons per shot and a fusion energy output of 54 kJ for a 1.5 MJ ultraviolet laser pulse.

Nd-based laser experiments at the Laboratory for Laser Energetics (LLE)

The Laboratory for Laser Energetics (LLE) is a research facility founded in 1970 that is part of the University of Rochester campus in Brighton, New York. The laboratory has the mandate to:

- conduct experiments in support of the National Inertial Confinement Fusion program;
- develop new laser technologies;
- provide educational opportunities to university students in the fields of high-power lasers, electro-optics, plasma physics and nuclear fusion;
- conduct research related to high-energy-density physics;
- operate the National Laser Users' Facility.

This laboratory is unique, as it is the only non-government facility in the United States to undertake large-scale research on inertial confinement laser fusion.

One of the most significant developments to come out of LLE has been the development of *chirping* or *chirped pulse amplification*. Canadian physicist Donna Strickland (b. 1959) and French physicist Gérard Mourou (b. 1944) were awarded the 2018 Nobel Prize in Physics for the invention of chirping at LLE during the mid-1980s. Chirping is a technique which is now in common use at laser fusion facilities to produce ultrashort-pulse, very high-intensity laser pulses. Prior to the development of chirping, non-linear optical effects were used to increase the power density in a laser beam. Non-linear effects can increase power density by a mechanism known as self-focusing. Self-focusing occurs in non-linear media where the index of refraction increases with electromagnetic field strength, leading to a convergence of rays within the material. Chirping is a much more effective method of increasing laser power density. In this method, a laser beam is stretched out temporally and spectrally, followed by amplified and then compression. A simple way of accomplishing this process is by means of diffraction gratings. Figure 5.20 shows a diffraction grating based pulse stretcher, where it can be seen how the different spectral components of the beam can be separated as a result of the different paths that they follow. Figure 5.21 shows a diffraction grating based pulse compressor where the temporally stretched pulse is compressed because of the different paths followed by different spectral components.

The LLE has constructed several laser systems with the goal of investigating inertial confinement fusion. The most notable of these, as described below, are the OMEGA system and the OMEGA EP system.

OMEGA/OMEGA EP

The OMEGA laser system at LLE became operational in 1995. It is a 60-beam frequency tripled Nd-glass laser capable of producing about 40 kJ pulses in the ultraviolet. Its maximum power is about 60 TW in sub-ns pulses. OMEGA's energy

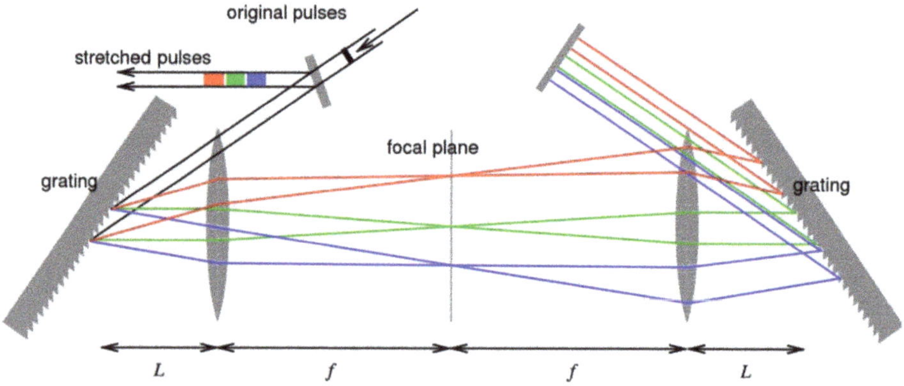

Figure 5.20. Diagram of a diffraction grating based pulse stretcher where the red spectral component is first to exit the device. This Pulse Stretcher image has been obtained by the author from the Wikimedia website https://commons.wikimedia.org/wiki/File:Cpa_stretcher.svg where it was made available by Han-Kwang Nienhuys under a CC BY-SA 3.0 license. It is included within this book on that basis. It is attributed to Han-Kwang Nienhuys.

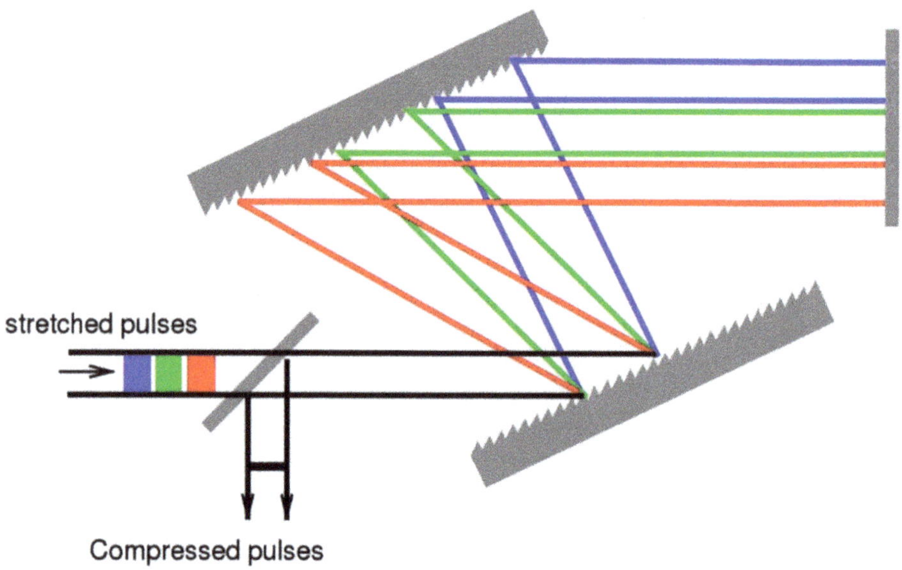

Figure 5.21. Diagram of a diffraction grating based pulse compressor where the red spectral component is first to enter the device. This Pulse Compressor image has been obtained by the author from the Wikimedia website https://commons.wikimedia.org/wiki/File:CPA_compressor.svg where it was made available by Han-Kwang Nienhuys under a CC BY-SA 3.0 license. It is included within this book on that basis. It is attributed to Han-Kwang Nienhuys.

output is slightly less than that of the previously operational Nova system at Lawrence Livermore. However, when Nova was decommissioned in 1999, OMEGA became the highest energy laser in the world. It was surpassed in 2005 when portions of the NIF laser became operational.

OMEGA research has concentrated on the use of direct drive and, therefore complements the research activities that have been carried out at the NIF. As a direct drive system, OMEGA has been instrumental in investigating the advantages associated with fast ignition. Specifically, in combination with computer modeling studies, OMEGA has made significant progress in increasing fusion yield. To date, the maximum neutron yield from d–t fusion that has been achieved by OMEGA is 10^{14} neutrons per shot.

A final point concerning the OMEGA laser system that is unique to its placement in the research community and its mandate is its availability to outside researchers. The National Laser Users' Facility has provided an important research tool to the broader research community that has been used for a wide range of studies related to both inertial confinement fusion and other fields including high-energy-density physics, plasma physics and electro-optics.

In 2008 the OMEGA EP (extended performance) laser system was dedicated. This is a four-beam laser system that extends the capabilities of the OMEGA system. OMEGA EP incorporates large-aperture pulse compression gratings (see figure 5.21) that allow for very short pulse experiments. Although the total energy available per pulse is relatively low, the short pulse length means that the maximum power can be quite high. The availability of this high power short pulse in addition to the main laser pulse from OMEGA is ideal for fast ignition experiments.

Laser Mégajoule

Laser Mégajoule Nd-glass laser is a laser system constructed by French Commissariat à l'Energie Atomique et aux Energies Alternatives (CEA) at Centre d'Etudes Scientifiques et Techniques d'Aquitaine (CESTA) near Bordeaux, France. It began with the construction of a smaller prototype device, Ligne d'Intégration Laser (Laser Integration Line), which became operational in 2002. laser Mégajoule which followed from preliminary studies at Ligne d'Intégration Laser was commissioned in 2014. The first phase of the project will consist of 176 beams of which 80 beams have been assembled as of 2020. Each beam provides about 15 kJ per pulse in the infrared or about 7.5 kJ in frequency tripled ultraviolet. The pulse duration is 0.325 ns with a maximum ultraviolet power of about 400 TW. In the long term, it is expected that the number of beams will be increased to 240 with a total ultraviolet output of 1.8 MJ per pulse. The PETAL laser (PETawatt Aquitaine Laser) is being constructed at the same site as Laser Mégajoule. PETAL is a single short pulse (500 fs to 10 ps) ultra-high-power laser. Although it only produces a few kJ of energy per pulse, the short pulse duration means that the peak power is very high, in the PW range. The high power is obtained using chirped pulse amplification as described above. The addition of the PETAL laser to the Laser Mégajoule facility will greatly increase the ability to understand phenomena related to high energy density physics including inertial confinement fusion reactions and, as noted above for the OMEGA EP laser, this makes it an important platform for fast ignition experiments.

The primary focus of research at Laser Mégajoule is to provide information on nuclear fusion that is relevant to the French nuclear weapons program. However, additional research programs are directed toward astrophysics, planetary science, medical physics and inertial confinement fusion power. The fusion power experiments concentrate on indirect drive methods using a hohlraum constructed from a uranium–gold mixture with a gold surface layer.

Inertial confinement fusion research at Laser Mégajoule has utilized experimental results to verify modeling studies. These studies have suggested that ignition for a d–t target using indirect drive will require 0.9 MJ of ultraviolet radiation per pulse with a maximum power of 260 TW. The energy yield from this ignition is expected to be about 20 MJ per shot.

5.6 Overview of some KrF laser fusion experiments

The above discussion of experiments involving CO_2 and Nd-glass lasers have made clear the particular concerns with using these devices for inertial confinement fusion. Specifically, the low efficiency, the long wavelength without harmonic generation, existence of Rayleigh–Taylor instabilities and the low repetition rate. Electron beam pumped KrF excimer lasers have properties that deal with these potential drawbacks. Specifically, KrF lasers have the following properties:

- efficiencies of 10% or more for conversion of electrical energy to laser light (compared to 1%–1.5% for typical Nd-glass lasers);
- wavelength of 248 nm (in the ultraviolet);
- high beam uniformity (which particularly benefits direct drive configurations)
- repetition rate of ~5 Hz as a result of minimal heating due to the continuous flow of the laser medium and the use of electron beam excitation.

The U.S. Naval Research Laboratory in Washington, DC has undertaken a series of experiments involving the application of KrF excimer lasers for inertial confinement fusion in order to take advantage of these beneficial properties. In addition, these experiments have been able to make use of direct drive because of high beam uniformity, and this has the advantage of:

- simplicity of design;
- simplicity of sample preparation;
- lack of hohlraum debris;
- improved laser-to-target coupling, up to ~85% energy transfer.

The Naval Research Laboratory has developed two KrF laser inertial confinement fusion systems, *Nike* and *Electra*, as described below. These are both relatively low power laser systems (by inertial confinement fusion standards) which have been designed to investigate two specific problems that have been concerns for Nd–glass laser fusion experiments, spatial homogeneity and repetition rate. A long-term goal of the KrF laser fusion research program is the production of a laser capable of providing a few MJ of ultraviolet radiation per pulse with a repetition rate of 10 Hz, as would be suitable for a viable commercial fusion power plant.

Nike

After a number of preliminary experiments in the 1970s, the Naval Research Laboratory began development of the Nike laser system in the early 1980s. This system was operation by the late 1980s and has been used for inertial confinement fusion and other experiments. Nike is a 56 beam electron beam pumped excimer KrF laser that provides 3 kJ in a 4 ns pulse.

A main objective of the Nike laser system was to produce a beam which is spatially highly uniform and, therefore, minimized Rayleigh–Taylor instabilities. Spatial nonuniformity exists in all large laser systems. Optical elements in the beam line are not perfect and will introduce phase distortion. When the beam is focused on the target, these phase distortions lead to spatial variations in intensity. The Nike system deals with nonuniformities using induced spatial incoherence. Induced spatial incoherence is a technique which introduces spatial and temporal incoherence in the beam, and this tends to average out spatial variations in intensity, at least on a time scale relevant to fusion target irradiation. This approach is effective in dealing with nonuniformities at all spatial wavelengths. Experimental results from Nike have shown a spatial variation in laser intensity over the target dimensions of less than 0.3% when utilizing 44 of the systems 56 beams. This nonuniformity is a factor of 10 less than that which has been achieved in Nd-glass laser systems such as the NIF.

A major focus of the Nike research program has been inertial confinement fusion. While the low power of the Nike laser means that achieving ignition is not a realistic possibility, Nike, and its direct drive approach, has been important in establishing directions for future inertial confinement fusion research, as discussed in the following sections. Specifically, Nike has shown that KrF lasers in combination with induced spatial incoherence beam smoothing techniques can provide irradiation of the target which is sufficiently uniform to reduce the occurrence of Rayleigh–Taylor instabilities to a point where direct drive methods of laser fusion will likely be possible.

In addition to its contributions to laser fusion research, Nike has other possible applications, largely as a result of its high beam uniformity. These applications include:

- investigation of material properties under uniform high compression;
- evaluation of aging effects on materials used in nuclear weapon construction;
- simulation of nuclear weapon effects.

Electra

Electra is a 700 J KrF laser that has been constructed at the U.S. Naval Research Laboratory. Its major research goal is to investigate the repetition rate of KrF lasers. As discussed further in the next section, repetition rates of 5–10 Hz will be necessary for commercial scale power reactors based on inertial confinement laser fusion. As noted above in the sections on Nova and the NIF, repetition rates for Nd–glass lasers of sufficient power are in the range of a few cycles per day. Experiments conducted with the Electra laser have shown that the laser can be run continuously at 2.5 Hz for ten hours (for a total of 90 000 shots) or for 100 min at 5 Hz (for a total of 30 000 shots). At present, the limit to the duration of operation is due to the performance of the spark gap switches used to trigger the laser. The Naval Research Laboratory has developed a solid state triggering system that has operated continuously at 10 Hz for 1.1×10^7 shots or about 13 days, which can be used to increase the repetition rate of lasers such as Electra.

5.7 Future directions in inertial confinement fusion research

Progress in laser fusion

The experiments on inertial confinement fusion using a laser driver that have been conducted over the past nearly half century have made major steps toward achieving a viable fusion reactor. Despite substantial understanding of the problems involved in reaching this goal, it has also become apparent that there are still many technical roadblocks that must be overcome and experimental results have, in general, fallen short of expectations. A summary of the major factors that need to be dealt with in the design of a viable laser fusion reactor and the knowledge that has been gained since the 1970s in these areas is considered below.

Wavelength of radiation

While the earliest laser fusion experiments were largely concerned with providing sufficient energy in the laser pulse to compress and heat the fuel, it became obvious from the results of CO_2 laser systems and early Nd-glass laser systems that infrared laser radiation was not appropriate for producing a significant fusion reaction. This is because the long wavelength of the incident radiation caused preheating of the target by electrons emitted during the initial phases of the irradiation. Subsequent work on laser fusion has concentrated on either frequency doubled or frequency tripled radiation from Nd–glass lasers or short wavelength radiation from ultraviolet KrF excimer lasers.

Laser energy

Through experiments such as those conducted at the NIF which have measured fusion energy output as a function of laser energy input, the requirements for a system capable of commercial energy production can be estimated. Several goals can be established along this path. Scientific breakeven is defined as the condition where fusion energy output is equal to (or exceeds) laser energy input. Ignition is defined as the condition where a fusion chain reaction occurs. This results from the absorption of fusion energy by remaining fuel and its heating to fusion conditions. As described in the section above on the NIF it has been estimated that the 1.5 MJ of ultraviolet laser radiation per pulse that is provided to the fuel target is about one third of that needed to achieve ignition. Breakeven, as defined for a commercially viable fusion power reactor is the condition where electrical output exceeds the total energy input into the facility, as discussed in section 5.3. This is a much more stringent condition than scientific breakeven or ignition, partly because of the efficiency of converting fusion energy into electrical energy (which might be around 40%), but mainly because of the efficiency of converting electrical energy into laser energy and coupling this to the fusion fuel (which may be only a few percent). The design of a successful inertial confinement fusion power reactor must deal with these efficiencies.

Laser efficiency

As noted previously, low laser efficiency is a significant detriment to the development of a viable inertial confinement fusion reactor and Nd–glass lasers have intrinsically low efficiency. Combined with efficiencies of around 50% for frequency tripling, overall efficiencies around 1% are typical. A major factor limiting the efficiency of Nd–glass lasers is the inherent low efficiency of the flash tubes used to create the population inversion. Currently available KrF lasers can be an order of magnitude more efficient than flash tube pumped Nd–glass lasers and are certainly advantageous from the standpoint of achieving breakeven.

Laser-target coupling efficiency

Laser energy may be coupled to the target either directly or indirectly. While indirect drive has certain advantages, such as improved uniformity of radiation on the target surface, ability to mitigate effects of preheating by hot electrons and elimination of cross beam energy transfer, it has much reduced efficiency of laser-to-target energy transfer. Recent work using KrF lasers with induced spatial incoherence beam smoothing techniques has made much progress in reducing effects of the adverse features of direct drive while taking advantage of its greater efficiency.

Laser repetition rate

As noted above, repetition rates in the order of 5–10 Hz are necessary for the construction of a commercial fusion reactor. It is relatively straightforward to understand the reasons these rates are necessary for a commercial reactor. The electrical power, P_e, produced by a laser fusion reactor may be expressed as

$$P_e = E_f f \eta_e \tag{5.36}$$

where E_f is the fusion energy produced per shot, η_e is the efficiency of converting fusion energy to electrical energy and f is the repetition rate. The fusion energy produced per shot is expressed as

$$E_f = E_L Q \tag{5.37}$$

where E_L is the laser energy per shot and Q is the energy gain, as for magnetic confinement fusion. Increasing the fuel capsule size and increasing the laser energy per shot accordingly will increase the fusion energy per shot and, in principle, will reduce the necessary repetition rate in equation (5.36). There is, however, a maximum amount of energy per shot that is compatible with maintaining the integrity of the facility. The energy that is released from the fusion event occurs over a short period of time. Many smaller pulses of energy over a period of time are much less damaging than one huge pulse, even though the total energy is the same. The LIFE program, as described above in the section of the NIF, considered such factors in detail. They determined that an indirect drive system using 2.2 MJ laser pulses would yield 132 MJ of fusion energy per shot for a gain of about 60. For a repetition rate of 10 Hz and a fusion energy to electrical energy conversion efficiency of 40%, equation (5.36) gives an electrical power output of

$$P_e = (132\ \text{MJ}) \times (10\ \text{s}^{-1}) \times (0.4) = 528\ \text{MW} \tag{5.38}$$

This represents a modest size commercial generating facility. Typical large nuclear fission or coal-fired generating stations produce about 1000 MW output.

Future approaches for laser fusion

Future laser fusion research that follows from existing technologies may utilize either direct drive or indirect drive and may use either Nd–glass or KrF lasers. The above discussion shows that no path to achieving a viable reactor is perfect, at least not at present. In moving ahead with future work, it is important to fully understand the concerns and limitations for each approach.

Drive concerns

Direct drive is the simplest approach to laser fusion and has the best efficiency but suffers from instabilities during fuel compression that result from beam nonuniformity. Power loss as a result of cross beam energy transfer is also a concern. While indirect drive mitigates these concerns to some extent, its low efficiency is a serious problem. As well, indirect drive requires complex target preparation, careful target orientation and removal of hohlraum debris from the reactor chamber. For these reasons, indirect drive is likely not viable from an economic standpoint for commercial power production.

Laser concerns

Nd–glass lasers have the highest energy output of the lasers used for inertial confinement fusion experiments and have come the closest to reaching ignition, as described above in the section on the NIF. Three major concerns need to be considered with regard to the use of Nd–glass lasers for fusion reactors—efficiency, beam uniformity and repetition rate.

Traditional Nd–glass lasers, which are pumped using a xenon flash lamp, are highly inefficient. Much of this inefficiency comes from the flash lamp. The need for frequency tripling further reduces the efficiency for conversion of electricity to ultraviolet laser radiation, as needed to irradiate the fusion fuel capsule. Even with smoothing by spectral dispersion, as has been implemented at the NIF, existing Nd–glass lasers have nonuniformity of beam intensity at the target that is substantially greater than that of KrF lasers. These nonuniformities lead to Rayleigh–Taylor instabilities in direct drive systems. Finally, repetition rates for large Nd–glass lasers are four to five orders of magnitude too low to produce useable energy.

KrF lasers have overcome the drawbacks of Nd–glass lasers for inertial confinement fusion. These lasers show efficiencies of an order of magnitude or more than Nd–glass lasers, without the need for frequency doubling or tripling. As well, KrF lasers using induced spatial incoherence show significantly reduced beam nonuniformity, compared to Nd–glass lasers. Finally, repetition rates of 10 Hz have been demonstrated experimentally for more than 30 000 shots. The basic drawback of KrF lasers is the insufficient power of present devices.

New technologies

Two approaches to achieve progress in inertial confinement laser fusion would be to overcome the shortcomings of Nd–glass lasers or to increase the available power from KrF lasers and there are active research programs worldwide in both of these areas.

A major factor in the low efficiency of Nd–glass lasers, as well as the low repetition rate, comes from the use of xenon flash tubes to create the population inversion. In recent years, the use of laser diodes to pump Nd–lasers has provided a potentially viable option that would deal with both of these issues. Low power diode pumped solid state lasers, or DPSSLs, are in common use for a variety of applications. The most common use of DPSSLs is as green laser pointers. In these devices an infrared GaAlAs laser diode operating at 808 nm is used to pump a Nd-doped material, which then lases at 1064 nm. The Nd-doped material is most commonly Nd-doped yttrium aluminum garnet (Nd:YAG) or a Nd-doped yttrium orthovanadate ($Nd:YVO_4$) crystal. The infrared Nd-laser output is then frequency doubled, typically using a nonlinear optical crystal, typically potassium titanyl phosphate (KTP), thereby providing 532 nm green laser light.

The application of high power pulsed DPSSLs for laser fusion is an attractive possibility. DPSSLs have much higher efficiency than typical flash tube pumped Nd-lasers, in the range of 10%–20% or higher. This higher efficiency reduces heating and provides the opportunity for higher repetition rates. The Mercury laser system at Lawrence Livermore National Laboratory is a prototype DPSSL laser using Nd-doped ytterbium-strontium fluorapatite crystals and has been used to investigate this possibility. Mercury has run continuously at 10 Hz for more than 8 h (3×10^5 shots) providing 50 J per 15 ns pulse. Initial experiments have been conducted using infrared Nd-laser output but experiments using frequency doubling and tripling are in progress. Other investigations of DPSSLs as drivers for inertial confinement fusion have included the Polaris laser at the Helmholtz Institute in Jena, Germany, the HALNA laser at the Institute for Laser Engineering, Osaka University, Japan and the Lucia laser at Laboratoire pour l'Utilisation des Lasers Intenses, École Polytechnique, Palaiseau, France.

Research on the development of high energy KrF lasers is underway at a number of locations. At the Lawrence Livermore National Laboratory, the Rapier laser system has been developed and produces 800 J per pulse. Other notable KrF laser developments have been the 3 kJ Super Ashuna laser at the Electrotechnical Laboratories of the Ministry of International Trade and Industry, Japan and the 10 kJ Aurora laser at Los Alamos National Laboratory in New Mexico.

Despite the progress that has been made in inertial confinement fusion using a laser driver over more than 50 years of research, it is clear that there is still a long way to reach the goal of a viable commercial reactor. Lasers with sufficient energy, efficiency and repetition rate, as well as economical and convenient fuel target design, are prerequisites for successful implementation of this technology. It is also likely that the implementation of suitable advanced ignition techniques, such as fast ignition or shock ignition, will be necessary to achieve sufficient gain to exceed commercial breakeven conditions.

References and suggestions for further reading

Betti R and Hurricane O A 2016 Inertial confinement fusion with lasers *Nat. Phys.* **12** 435–48

Boehly T R *et al* 1997 Initial performance results of the OMEGA laser system *Opt. Commun.* **133** 495–506

Campbell J H and Suratwala T I 2000 Nd-doped phosphate glasses for high-energy/high-peak-power lasers *J. Non-Cryst. Solids* **263 and 264** 318–41

Carlson R L, Carpenter J P, Casperson D E, Gibson R B, Godwin R P, Haglund R F Jr, Hanlon J A, Jolly E L and Stratton T F 1981 Helios: a 15 TW carbon dioxide laser-fusion facility *IEEE J. Quantum Electron.* **17** 1662–78

Guo L *et al* 2015 Uranium hohlraum with an ultrathin uranium–nitride coating layer for low hard X-ray emission and high radiation temperature *New J. Phys.* **17** 113004

He X T *et al* 2016 The updated advancements of inertial confinement fusion program in China *J. Phys.: Conf. Ser.* **688** 012029

Hurricane O A *et al* 2014 Fuel gain exceeding unity in an inertially confined fusion implosion *Nature* **506** 343–48

Jing L F *et al* 2018 Comparison of three hohlraum configurations with six laser entrance holes for indirect-drive inertial confinement fusion *Nucl. Fusion* **58** 096017

Kirkwood R K *et al* 2013 A review of laser–plasma interaction physics of indirect-drive fusion *Plasma Phys. Control. Fusion* **55** 103001

Landen O L *et al* 2012 Progress in the indirect-drive National Ignition Campaign *Plasma Phys. Control. Fusion* **54** 124026

Los Alamos National Laboratory 1979 *Laser Fusion Program at Los Alamos* Report LASL-79-29

Meier W R, Dunne A M, Kramer K J, Reyes S, Anklam T M and the LIFE Team 2014 Fusion technology aspects of laser inertial fusion energy (LIFE) *Fusion Eng. Des.* **89** 2489–92

Miquel J-L, Lion C and Vivini P 2016 The laser mega-joule: LMJ & PETAL status and program *J. Phys.: Conf. Ser.* **688** 012067

Moses E I 2010 Advances in inertial confinement fusion at the National Ignition Facility (NIF) *Fusion Eng. Des.* **85** 983–86

Obenschain S, Lehmberg R, Kehne D, Hegeler F, Wolford M, Sethian J, Weaver J and Karasik M 2015 High-energy krypton fluoride lasers for inertial fusion *Appl. Opt.* **54** F103–22

Sangster T *et al* 2007 Overview of inertial fusion research in the United States *Nucl. Fusion* **47** S686–95

Seka W, Jacobs S, Rizzo J, Boni R and Craxton R 1980 Demonstration of high efficiency third harmonic conversion of high power Nd-glass laser radiation *Opt. Commun.* **34** 469–73

Sethian J D *et al* 2010 The science and technologies for fusion energy with lasers and direct-drive targets *IEEE Trans. Plasma Sci.* **38** 690–703

Sethian J D *et al* 1994 Overview of the Nike KrF Laser Program *Fusion Technol.* **26** 717–21

Sethian J D *et al* 2002 The Electra KrF Laser program *Proc. 13th IEEE Pulsed Power Conf.* p 232–36

Wieneke S and Gerhard C *Lasers in Medical Diagnosis and Therapy Basics, Applications and Future Prospects* (Bristol: IOP Publishing)

Wolford M, Sethian J, Myers M, Hegeler F, Giuliani J and Obenschain S 2013 KrF laser development for fusion energy *Plasma Fusion Res.* **8** 3404044

Wolford M F, Sethian J D, Myers M C, Hegeler F, Giuliani J L and Obenschain S P 2013 Krypton fluoride (KrF) laser driver for inertial fusion energy *Fusion Sci. Technol.* **64** 179–86

Zohuri B 2017 *Inertial Confinement Fusion Driven Thermonuclear Energy* (Cham: Springer)

IOP Publishing

Energy from Nuclear Fusion

Richard A Dunlap

Chapter 6

Other approaches to fusion

6.1 Introduction

While magnetic confinement and inertial confinement fusion have received the vast majority of funding and scientific attention, there are a number of other approaches to achieving laboratory nuclear fusion that have occurred over the past half century or more. Some of these approaches are scientifically sound and have clearly demonstrated the existence of fusion reactions. While many of these have provided very useful fundamental scientific information, they have, unfortunately, failed to produce results that are consistent with viable sources of future energy. Other alternative approaches to fusion have been shown to be based on questionable scientific principles or have, in the worst cases, been shown to be hoaxes. Probably the greatest fusion hoax was perpetrated by Austrian-born German physicist Ronald Richter (1909–1991). In 1951, Richter convinced Argentine president Juan Perón to fund a project to develop a nuclear fusion power reactor. While Richter's 'research' lacked any scientific credibility, it brought nuclear fusion energy to the public's attention and also, in a way, motivated Layman Spitzer's work on fusion. When Spitzer heard of Richter's proposed reactor, he was convinced that it would not work and consequently invented the stellarator, as described in chapters 3 and 4.

A few of the experiments that are discussed in this chapter may be scientifically valid and may overcome the fundamental technical difficulties that prevent them from providing useful energy. The present chapter is, by no means, complete in its coverage of alternative approaches to nuclear fusion. However, it tries to cover those experiments that are scientifically interesting or are historically significant, as well are those that show some possibility of producing useful energy and those that have been most effective in attracting the interest of the general public.

doi:10.1088/978-0-7503-3307-8ch6

6.2 Electrostatic confinement

The fusor

In 1924 French physicist Camille Gutton (1872–1963) first observed the multipactor effect. This effect can occur in radio frequency vacuum tubes when secondary electron emission is in resonance with an alternating electric field. This effect can give rise to exponential electron multiplication. About a decade after Gutton's discovery, Philo T Farnsworth (1906–1971) made an extensive study of the multipactor effect. Farnsworth, who is best known as the inventor of the television and video camera, observed that the multipactor effect gave rise to the accumulation of electric charge in certain regions of the vacuum tube.

In the 1960s Farnsworth was inspired by the concept of the multipactor effect and how it might be of use in the construction of a controlled fusion device. At that time, virtually all fusion research aimed at utilizing nuclear fusion as a viable source of energy, considered the use of magnetic fields for the purpose of plasma confinement. Farnsworth considered the option of using electric fields as a means of confinement. This approach to fusion is referred to as inertial electrostatic confinement (or IEC). Similar to a particle accelerator, IEC utilizes an electric field to accelerate ions. It is the kinetic energy of the ions that allows them to penetrate the Coulomb barrier and achieve fusion. Even fairly modest accelerating voltages can achieve fusion. The kinetic energy of a singly charged ion accelerated across a potential of V volts is

$$E = eV \tag{6.1}$$

Relating this to the kinetic energy of a particle at a temperature T;

$$E = k_B T \tag{6.2}$$

means that the effective temperature of an accelerated particle is

$$T = \frac{qV}{k_B} \tag{6.3}$$

where k_B is the Boltzmann constant. Even a modest accelerating potential of (say) 10 kV, corresponds to a temperature of 1.2×10^8 K, typical of plasma temperatures in magnetic confinement devices. There is also the advantage that the ions are both monoenergetic and their velocity is directional, unlike the Maxwellian energy distribution and random directions of ions in a plasma.

The device that Farnsworth developed in 1964 is known as a fusor. Robert L Hirsch (b. 1935) refined Farnsworth's design and the resulting device is generally referred to as a Hirsch–Farnsworth fusor or (sometimes) a Farnsworth–Hirsch fusor. A diagram of the Hirsch–Farnsworth fusor as proposed by Hirsch is shown in figure 6.1. The device consists of two spherical cages, as shown in the figure, which are enclosed in a vacuum chamber. The outer cage is the anode and is held at a positive potential relative to the inner cage, which is the cathode. The sequence of events leading to fusion in the fusor is shown in figure 6.2.

Unfortunately, the Hirsch–Farnsworth fusor is not very efficient and is not capable of producing a net energy output. The main reason for this is that the

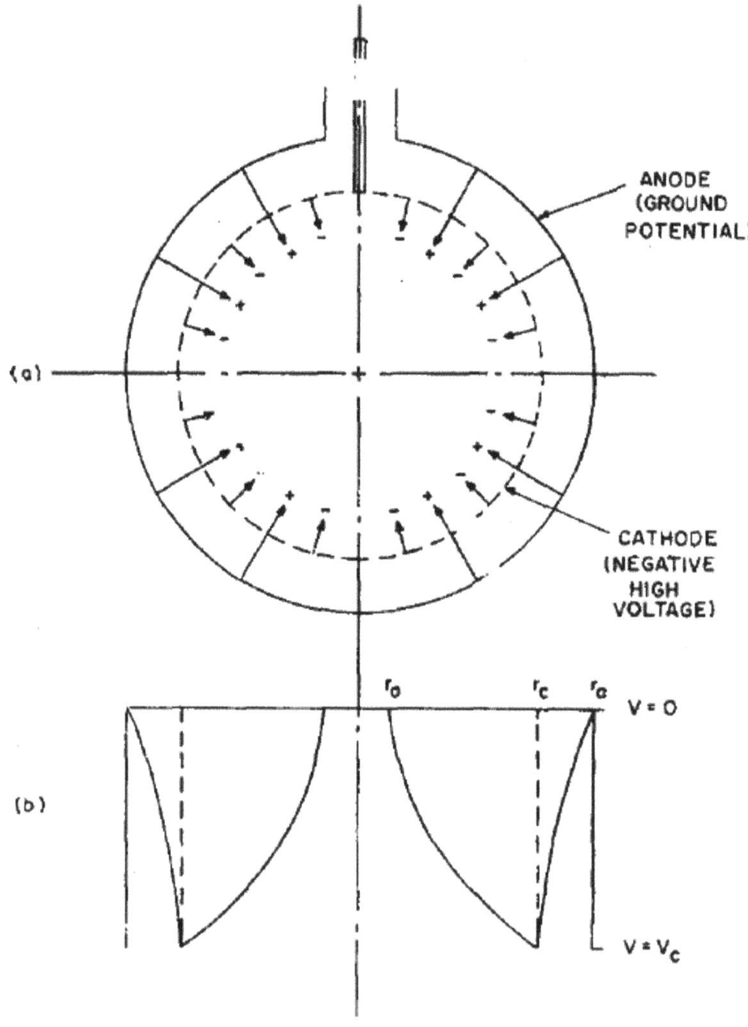

Figure 6.1. Diagram of a Hirsch–Farnsworth fusor from Hirsch's original paper; (a) device geometry and (b) electric potential. Reproduced from Hirsch (1967) with the permission of AIP Publishing.

positive ions must pass through the grid of the inner cathode cage. The cage can be made as open as possible, but ultimately some of the ions will impact on the wires of the cage and be lost.

An interesting variation of the Hirsch–Farnsworth fusor was actually proposed about five years before Farnsworth's work on the fusor by William Elmore (1909–2003), James Tuck (1910–1980) and Kenneth Watson (b. 1921), although it was not constructed at the time. In this device, known as the Elmore–Tuck–Watson fusor, the polarities of the cages are reversed, with the outer cage acting as the cathode and the inner cage acting as the anode. In this device electrons are injected through the outer cage and are accelerated toward the positive potential of the inner cage. Many of the electrons will pass through the inner cage creating a negatively charged region

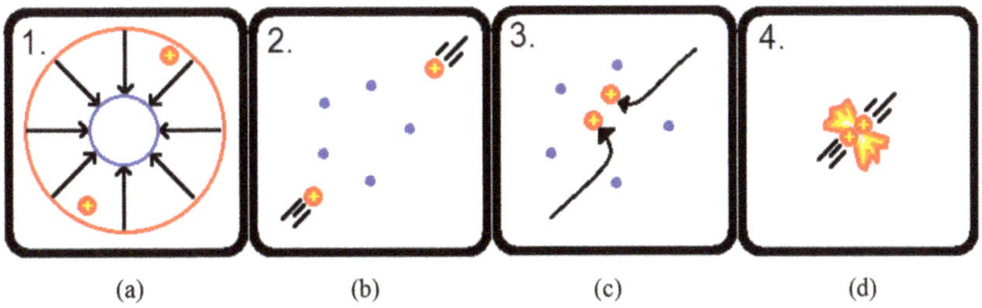

Figure 6.2. Sequence of events leading to fusion in a Hirsch–Farnsworth fusor; (a) Positive ions (i.e. the fusion fuel) are injected into the device from the outside, (b) the positive ions are accelerated toward the inner cage by the electrostatic potential, (c) ions miss the grid of the inner cage and enter into the central portion of the fusor, and (d) some ions will collide and fuse. This Hirsch–Farnsworth Fusor Events image has been obtained by the author from the Wikimedia website https://commons.wikimedia.org/wiki/File:Fusor_Mechanism.png where it was made available by WikiHelper2134 under a CC BY-SA 3.0 license. It is included within this book on that basis. It is attributed to WikiHelper2134.

near the center of the device. Since the electrons achieve high velocities, they may make multiple passes through the anode cage. Positive ions can now be injected inside the inner anode cage and will be repealed by the positive anode and attracted to the negatively charged region of electrons, thereby being confined within the central region of the device. These ions, at high velocities (i.e. kinetic energies) may collide and fuse. An interesting aspect of fusors that distinguishes them from devices such as the tokamak or inertial confinement fusion devices, as presented in chapters 4 and 5, respectively, is the fact that electrons and ions are injected into the device independently, leading to the feature that the net plasma is not necessarily neutral. In fact, in order to maintain the negative potential at the center of the Elmore–Tuck–Watson fusor, it is essential that the electron density is greater than the positive ion density. Unfortunately, as in the Hirsch–Farnsworth fusor, conductive losses due, in this case, to electrons impacting on the inner wire cage make the output insufficient to produce net useable energy.

Although it seems unlikely that fusors, as described above, will ever contribute to commercial energy production, they have been used successfully as neutron sources. Neutrons are used for the production of radioisotopes, for example in medical diagnostics and treatment. While commercial fusors produce a smaller neutron flux than large-scale neutron generators, such as particle accelerators, they are much more economical and typically produce larger neutron fluxes than radioisotope sources. They also have the advantage, compared to radioisotope sources and reactors, that they can readily be turned on and off.

The polywell

As noted above, the main difficulty in producing net useable energy from a fusor is the conductive loses due to charges impacting on the wires of the inner cage. The polywell avoids this problem by utilizing magnetic fields to trap electrons which, in turn, create a negative potential well to attract positively charged ions. This concept was first suggested by the Russian physicist Oleg Lavrentiev (1926–2011) in 1975. Around

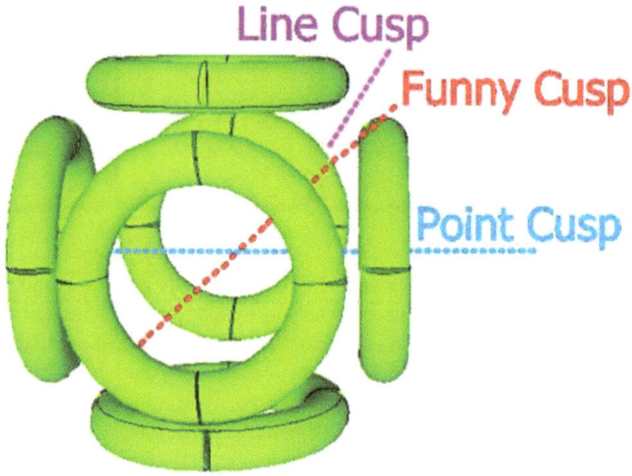

Figure 6.3. Basic arrangement of magnetic coils in a polywell. This Magnetic Coils in a Polywell image has been obtained by the author from the Wikimedia website https://commons.wikimedia.org/wiki/File:Cusps_inside_the_Polywell.png where it was made available by WikiHelper2134 under a CC BY-SA 3.0 license. It is included within this book on that basis. It is attributed to WikiHelper2134.

1983 the concept was refined by the American physicist Robert W Bussard (1928–2007). Bussard proposed a device based on a hexahedral arrangement of magnetic coils as illustrated in figure 6.3. Much of the early research of the polywell concept was done by the company Energy/Matter Conversion Corporation, Inc. (aka EMC2) that was founded by Bussard in 1985. The figure shows opposing magnetic fields positioned on opposite sides of a six-sided cube. The fields cancel at the center of the device leading to a three-dimensional region of zero magnetic field which is surrounded by an increasing magnetic field in all directions. A two-dimensional representation of the magnet field geometry through the center of the polywell is illustrated in figure 6.4. The shape of the magnetic field lines inside the polywell is, of course, dependent on the geometry of the magnetic coils. Point cusps, as shown in figure 6.4, are lines along the directions through the center of opposite magnets (what a solid state physicist might call the (100) direction). The line cusps occur along the lines defined by two adjacent magnets (i.e. the (110) direction) and the funny cusps occur along lines defined by three adjacent magnets (i.e. the (111) direction). The resulting field geometry is roughly spherical with holes along the cusps and has been referred to as 'Wiffleball' confinement by Bussard.

In a functioning polywell the magnetic coils as shown in figure 6.4 are enclosed in a vacuum chamber as illustrated in figure 6.5. As well, the magnetic coils are covered with metallic shells, referred to as 'magrids', and are held at a high positive potential. Electrons are injected into the polywell as indicated in figure 6.5 along the point cusps and are concentrated at the center of the device by the magnetic fields thereby forming a 'virtual cathode'. Electrons may escape through the cusps but are attracted back into the device by the positive potential on the magrids. This approach reduces the conductive losses that are present in fusors. The number of

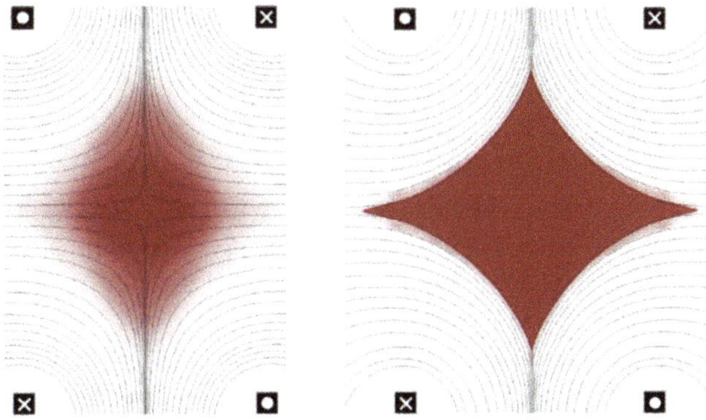

Figure 6.4. Magnetic field geometry at the center of a polywell for (a) low β and (b) for high β, as discussed in the text. Reprinted from Park *et al* (2015).

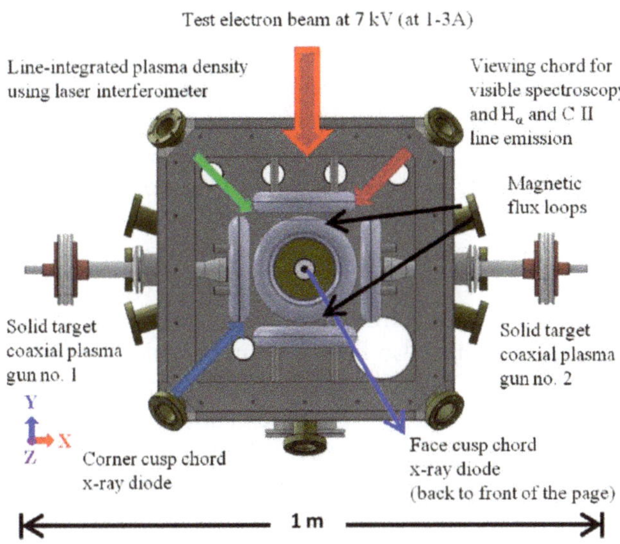

Figure 6.5. Experimental design of a polywell fusion device showing the vacuum enclosure, the locations of the six magnetic coils inside the enclosure and ports for ion injection and experimental observations. Reprinted from Park *et al* (2015).

passes electrons make through the polywell before being lost is an important consideration in assessing the viability of the device for net energy production. Neutral gas containing the fusion nuclei is injected at the corners of the polywell and is ionized by interactions with the electrons. The positive ions thus formed are attracted to the virtual cathode and gain sufficient energy that fusion through ion collisions may occur.

The question of plasma neutrality is an important aspect of the polywell that warrants consideration. As noted above, it is essential that the total negative electron charge in the region of the center of the polywell is greater than the total positive ion charge in the same region. This is to maintain the net negative potential well that is needed to confine the ions. It is important to realize that the net overall charge of the central region of the polywell depends on the difference in electron and ion densities, not on the ratio of the two densities. This means that the greater the particle densities, the closer the plasma will be to electrically neutral. It is the electrons that are confined by the magnetic fields which produce the potential well that accelerates the ions. This is in contrast to the much larger confinement energy required in devices such as the tokamak where the fields are used directly to confine the ions as well as the electrons.

The behavior of the polywell depends on the operating conditions. One of the important parameters in determining the performance of the device is the parameter β, which is defined as the ratio of the plasma pressure to the pressure exerted on the plasma by the applied magnetic field. This is expressed as (see chapter 4)

$$\beta = \frac{nk_{\mathrm{B}}T}{\left(B^2/2\mu_0\right)} \tag{6.4}$$

where n is the number density of particles, k_{B} is the Boltzmann constant, T is the temperature in Kelvin, B is the magnetic flux density and μ_0 is the vacuum permeability. This expression may be defined independently for electrons and for ions, as the number density and temperature of these two particles in the polywell can be substantially different. In the assessment of the operation of the polywell it is the electron β that defines the confinement properties. Bussard estimated that the number of electron passes, as discussed above, is between a few tens for the low β regime of the polywell to a few thousand for the high β regime. Thus, high β devices have attracted the most interest as they are less susceptible to electron losses.

The choice of fusion fuel has also been a point of concern for polywell research. The simplest reaction (i.e. d–d fusion) has been the subject of virtually all experimental work. The d–d reaction is convenient because the materials are easy to work with and are relatively inexpensive. While d–t fusion has a greater cross section for typical operating conditions, tritium is more difficult to work with as it is radioactive and is also substantially more expensive than deuterium (having to be artificially prepared). Proton–boron (i.e. p–^{11}B) fusion,

$$\mathrm{p} + {}^{11}\mathrm{B} \rightarrow 3\,{}^{4}\mathrm{He} \tag{6.5}$$

has been an interesting possibility for future polywell machines. As noted in chapter 2, p–^{11}B fusion is aneutronic. This means that virtually no neutrons are emitted during the fusion process. This has the advantage that energy loss to neutrons is not a concern. As well, neutron activation of structural materials leading to radioactive contamination is not a problem and the shielding of the reactor that is necessary for operator safety is minimal. Finally, the by-product of the p–^{11}B reaction is three alpha particles. Extracting energy from the alpha particles is straightforward as they are charged and will readily interact with matter to transfer their energy. They are also amenable to direct energy conversion as discussed in more detail in the next chapter, and this approach can have higher efficiencies than thermal conversion.

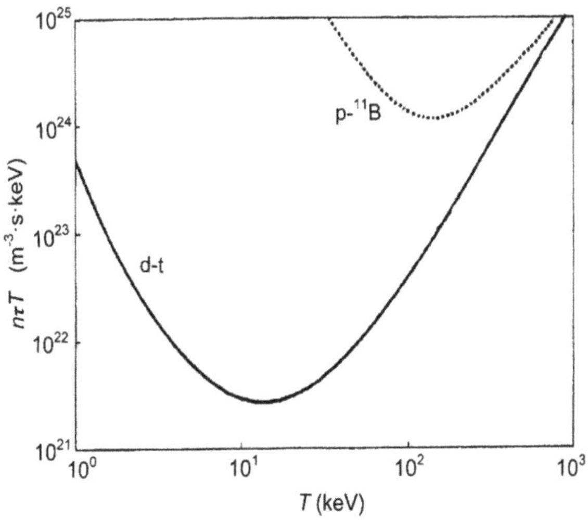

Figure 6.6. Triple product as a function of plasma temperature need to achieve ignition for d–t and p–^{11}B fusion.

Proton–boron fusion, however, is not without its own potential drawbacks. Because of the higher Coulomb barrier associated with the charged boron nucleus (compared to the hydrogen nucleus), conditions to achieve a viable fusion reaction are more demanding for p–^{11}B fusion than for d–t fusion. Figure 6.6 shows the triple product necessary to achieve ignition for d–t and p–^{11}B fusion. Clearly, the conditions for p–^{11}B fusion require higher temperatures, densities and/or confinement times. As well, the minimum in the triple product occurs at a substantially higher temperature. This higher temperature gives rise to greater energy loss through bremsstrahlung. Bremsstrahlung occurs predominantly as a result of the behavior of the free electrons in the plasma. This situation is compounded for p–^{11}B fusion because each fusion pair creates six free electrons in the plasma, compared to two for each fusion pair in d–t fusion. While bremsstrahlung losses are generally not a concern for d–t fusion, they are typically two orders of magnitude greater for p–^{11}B fusion and must be dealt with accordingly.

A substantial fraction of polywell research has been conducted by EMC2, although in recent years there has also been growing activity in university and industrial laboratories. Although much progress has been made in polywell development, it is still very far from the production of a viable power reactor. Experimental studies of confinement have shown encouraging results, as have extensive theoretical and modeling studies. Proponents of polywell technology have suggested much faster development times compared to large scale magnetic fusion and lower costs due to the much more modest infrastructure requirements.

6.3 Accelerator based fusion

Fusion is very easy to achieve in particle accelerators, as the energies required to overcome the Coulomb barrier between positively charged nuclei is very modest compared to typical accelerator energies. The study of fusion reactions using particle accelerators, beginning with the first observation of laboratory fusion by Mark Oliphant in 1934, have provided considerable information concerning the basic physics of fusion reactions. Generally speaking, accelerator based fusion experiments fall into four basic categories beam–target fusion, beam–beam fusion, muon catalysis and heavy ion drivers for inertial confinement fusion, as described in the present section.

Beam–target fusion

Beam–target fusion refers to a technique where one of the fusing nuclei (in the form of an ion) is accelerated in a particle accelerator while the other nucleus is part of a stationary target (typically a solid containing the relevant nuclei). The main difficulty with this approach is that, although it is straightforward to accelerate one of the nuclei to the necessary energy to achieve fusion, the Coulomb scattering cross section is typically several orders of magnitude larger than the fusion cross section. This means that most of the ion's energy is lost to the emission of bremsstrahlung or to the ionization of atoms in the target. While this approach is not suitable as a potential candidate as a viable energy source because of the low efficiencies for fusion energy production, it has found applications as a neutron source, much as fusors have. Neutron generators containing deuterium and tritium gas may be constructed so that the deuterons and tritons may be accelerated and collide against solid hydride targets containing deuterium and tritium. The resulting fusion reactions release neutrons, as described in chapter 2, which can be used for various measurement technologies. These devices are in common use by the petroleum industry for mapping oil deposits.

Beam–beam fusion

Beam–beam fusion refers to a technique where two or more beams of accelerated ions collide with the purpose of undergoing fusion reactions. The fusor and the polywell, as described in the previous section, are, technically, beam–beam devices, although the terminology is more commonly used to describe devices that are based on more conventional particle accelerator technology. We can make an estimate of the amount of energy that can be generated in such a situation by looking in detail at a specific example of beam–beam fusion. We consider the possibility of p–^{11}B fusion for two colliding beams. The number of fusions per unit time, $\mathrm{d}N/\mathrm{d}t$, is given by

$$\frac{\mathrm{d}N}{\mathrm{d}t} = \sigma L \tag{6.6}$$

where σ is the fusion cross section and L is the beam luminosity measured in particles per unit area per unit time. The power generated is merely the energy per fusion reaction, E, times the fusion rate,

$$P = E\frac{\mathrm{d}N}{\mathrm{d}t} = E\sigma L \tag{6.7}$$

For the present example, the energy per p–^{11}B fusion is 8.7 MeV (1.4×10^{-12} J) and the fusion cross section is 9×10^{-29} m^2. As an extreme example we consider the maximum luminosity of the large hadron collider near Geneva, 2×10^{38} m^{-2} s^{-1}, equation (6.7), therefore, gives a power output of

$$P = (1.4 \times 10^{-12}\ \mathrm{J}) \times (9 \times 10^{-29}\ \mathrm{m}^2) \times (2 \times 10^{38}\ \mathrm{m}^{-2}\ \mathrm{s}^{-1}) = 25\ \mathrm{mW} \qquad (6.8)$$

This simple example clearly indicates the unsuitability of beam–beam fusion as a practical source of energy.

Muon-catalyzed fusion

In 1935 the Japanese physicist Hideki Yukawa (1907–1981) predicted (Yukawa 1935) the existence of a subatomic particle, which he referred to as a meson, with a mass intermediate between the mass of the electron and the mass of the neutron and proton, as a mechanism for mediating the interactions in the nucleus (i.e. the strong interaction, see chapter 2). A year later Carl David Anderson (1905–1991) and Seth Henry Neddermeyer (1907–1988), while studying cosmic radiation, observed a particle with a mass of 105.7 MeV/c^2, intermediate between the masses of the electron and the proton. This particle was, at the time, generally believed to be the meson predicted by Yukawa. In reality, the particle discovered by Anderson and Neddermeyer was the muon, which was a lepton and, as such, was related to the electron. The meson, as predicted by Yukawa, i.e. the pi meson (now often called the pion) with mass of 140 MeV/c^2, was discovered in 1947 by Cecil Powell (1903–1969) and co-workers. It is, however, the muon, and its relationship to the electron that is of relevance here.

In 1913 Niels Bohr (1985–1962) proposed his semi-classical model of the atom. This model gives the radius of the ground state electron orbit in a hydrogen atom as

$$r = \frac{\hbar^2}{ke^2 m_e} = 5.29 \times 10^{-11}\ \mathrm{m} \qquad (6.9)$$

where $\hbar$ is the reduced Planck's constant, k is the Coulomb constant, e is the electronic charge and m_e is the electron mass. The distance between the nuclei of the two hydrogen atoms in a covalently bonded hydrogen molecule is slightly greater than the Bohr radius or about 7.5×10^{-11} m. For a singularly ionized hydrogen molecule, H_2^+, the separation is somewhat larger, about 1.1×10^{-10} m.

By the late 1940s it was known that the muon observed by Anderson and Neddermeyer was a lepton with properties related to those of the electron. The concept of muon-catalyzed nuclear fusion was first suggested in the late 1940s by the Russian physicist Andrei Sakharov (1921–1989) and the British physicist Frederick Charles Frank (1911–1998) (see Frank 1947). They suggested that a negatively charged muon could replace the electron in a molecular hydrogen ion. Since the mass of the muon, 105.7 MeV/c^2, is 207 times the electron mass, 0.511 MeV/c^2, Bohr's relation in equation (6.9) shows that the separation of the hydrogen nuclei in a muonic molecular hydrogen ion would be a factor of 207 smaller than in a normal molecular hydrogen ion, or about 5×10^{-13} m. If one, or both, of the hydrogen nuclei in the molecular hydrogen ion are deuterons or tritons, then there is a high probability of the nuclei fusing because of their proximity. This process is referred to as muon-catalyzed fusion.

The first experimental evidence of muon-catalyzed fusion came in 1954 when Luis Walter Alvarez (1911–1988) and co-workers were analyzing bubble chamber results of experiments involving muons. They found evidence for the p–d fusion reaction

$$p + d \rightarrow {}^{3}He + \gamma \tag{6.10}$$

accompanied by the release of about 5.4 MeV of energy. In 1957, Canadian-American physicist John David Jackson (1925–2016) published a detailed theoretical description of muon-catalyzed fusion and concluded that this technique would not provide a viable method of energy production (Jackson 1957). In order to understand the reasons for Jackson's conclusion and the knowledge that has been gained since that time, we need to look at the technique in more detail.

The muon exists in both charge configurations, negative (μ^-) and positive (μ^+), analogous to the electron and positron. Both μ^- and μ^+ have identical masses and are unstable with a lifetime of 3.2×10^{-6} s. It is, obviously, the negative muon that participates in muon-catalyzed fusion. The μ^- decay process is

$$\mu^- \rightarrow e^- + \bar{\nu}_e + \nu_\mu \tag{6.11}$$

where ν_μ is the muon neutrino. An analogous decay occurs for the positive muon. Muons can be created in the laboratory as by-products of the decay of negative pions by the process

$$\pi^- \rightarrow \mu^- + \bar{\nu}_\mu \tag{6.12}$$

with a lifetime of 2.6×10^{-8} s. The pions are typically created in collisions involving high energy hadrons and both negative and positive pions are created leading to both negative and positive muons. The ratio of negative pions/muons to positive pions/muons should be maximized in order to optimize muon-catalyzed fusion efficiency. This can be achieved by proper choice of the particles involved in the collisions to produce pions and is discussed further below.

Because of the very low cross sections involved, muon-catalyzed p–d fusion as observed by Alvarez is certainly not an option for practical energy production. d–d fusion, as has been promoted for IEC devices, is an attractive option for muon-catalyzed fusion for many of the same reasons as it is attractive for electrostatic confinement fusion approaches. However, most research has suggested that d–t fusion is a highly preferable alternative to d–d fusion for several reasons:

- higher fusion rate (by about two orders of magnitude);
- greater energy per fusion (17.6 MeV fusion^{-1} compared to ~3.5 MeV fusion^{-1}, see chapter 2);
- lower sticking fraction (see below).

Therefore, the present discussion concentrates primarily on muon-catalyzed d–t fusion experiments.

A negative muon which is incident on a mixture of deuterium and tritium (of approximately equal proportions) will, most likely, undergo the series of processes

$$\mu + t + d \rightarrow t\mu + d \rightarrow dt\mu \tag{6.13}$$

That is, the muon will replace the electron in a tritium atom thereby forming a muonic tritium atom which is a factor of about 200 smaller than the original tritium atom. This muonic tritium atom will bond with a deuteron and form a bound dtμ system where the triton and deuteron are very close together. An alternative, although less likely process to dtμ formation is through the formation of a muonic deuterium atom followed by the transfer of the muon to a tritium atom and then the binding of the muonic tritium atom with a deuteron. This process is

$$\mu + t + d \rightarrow t + d\mu \rightarrow t\mu + d \rightarrow dt\mu \tag{6.14}$$

The process as described above is illustrated in figure 6.7. The figure shows that the binding of the muonic tritium atom with a deuteron generally involves a deuteron that is part of a deuterium molecule (D_2). That is

$$t\mu + D_2 \rightarrow [(dt\mu)d2e] \tag{6.15}$$

where the muon is a component of a muonic dt molecular ion as shown in figure 6.7.

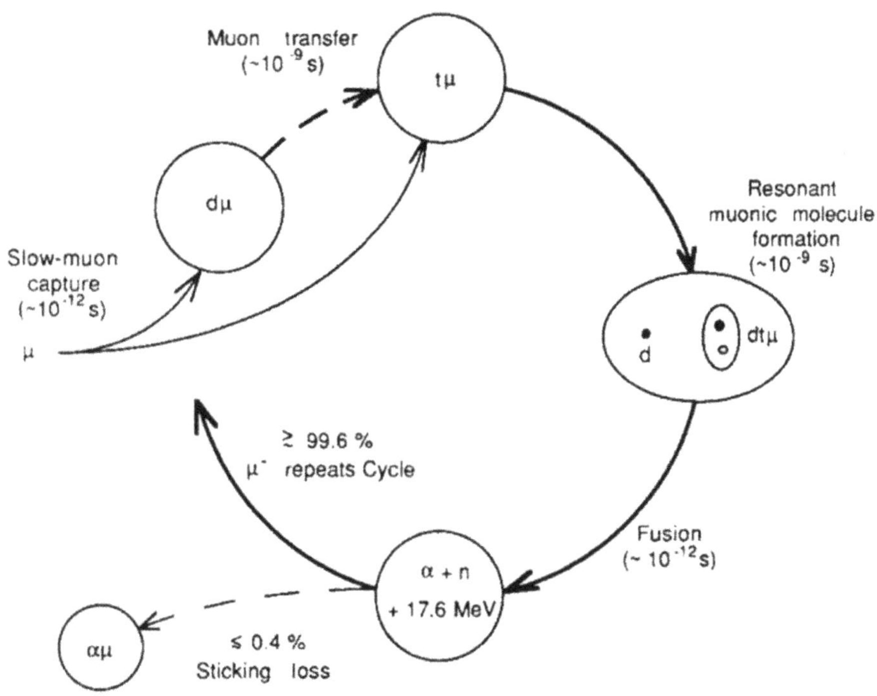

Figure 6.7. Processes occurring when a muon is incident on a mixture of deuterium–tritium (approximately 50%/50%). Reprinted with permission from Macmillan Publishers Ltd, Jones (1986).

The proximity of the deuteron and triton in the muonic dt molecular ion results d–t fusion according to the reaction

$$d + t \rightarrow \alpha + n \tag{6.16}$$

which releases 17.6 MeV of energy as shown in figure 6.7. The fusion process for the muonic dt molecular ion is, therefore,

$$dt\mu \rightarrow \alpha + n + \mu \tag{6.17}$$

where the muon is released and can go on to catalyze additional d–t fusions. The ability of the muon to catalyze more than one fusion reaction is an essential component of the potential viability of this process for energy production. The creation of a muon in an accelerator is an energy intensive process, typically requiring on the order of 5 GeV to produce one negative muon. There is, however, a limit to the number of fusion reactions a single muon can catalyze. One obvious limit results from the lifetime of the muon (2.2×10^{-6} s) after which the muon decays to an electron. As seen in figure 6.7, the time scale for the muon-catalyzed process is very short compared to the muon lifetime.

A more serious problem that limits the number of fusions that a muon can catalyze is the process of sticking. After the fusion event in equation (6.17) occurs, the muon may not be released as a free particle. Rather it may stick to the alpha particle that is produced in the fusion process,

$$dt\mu \rightarrow \alpha\mu + n \tag{6.18}$$

In general, this means that the muon is no longer available to catalyze fusion reactions. In a few cases, the muon may be stripped from the alpha particle as a result of Coulomb interactions with other particles in the d–t mixture. The process involving a particle, x, is of the form

$$\alpha\mu + x \rightarrow \mu + \alpha + x \tag{6.19}$$

By the 1960s research on muon-catalyzed fusion generally concluded that the sticking fraction was too high to make this approach attractive as a possible method for energy production. However, in the 1980s there was a resurgence of interest in muon-catalyzed fusion. Much of this resulted from the work of the American physicist Steven Earl Jones (b. 1949). At that time there was considerable interest in determining how fuel parameters such as temperature, pressure, density and tritium fraction affected the sticking fraction, or more relevantly, the average number of fusions per muon.

Figure 6.8 shows some experimental results for the number of fusions per muon compared to a theoretical prediction. It is clear that control of d–t density can result in substantially larger numbers of fusions per muon that had been suspected on the basis of previous theoretical studies. The number of fusions per muon required to achieve breakeven, that is, the point at which net energy is produced, is uncertain. The requirements for breakeven depend on the number of fusions per muon and also on the energy necessary for the production of negative muons. Typical estimates of

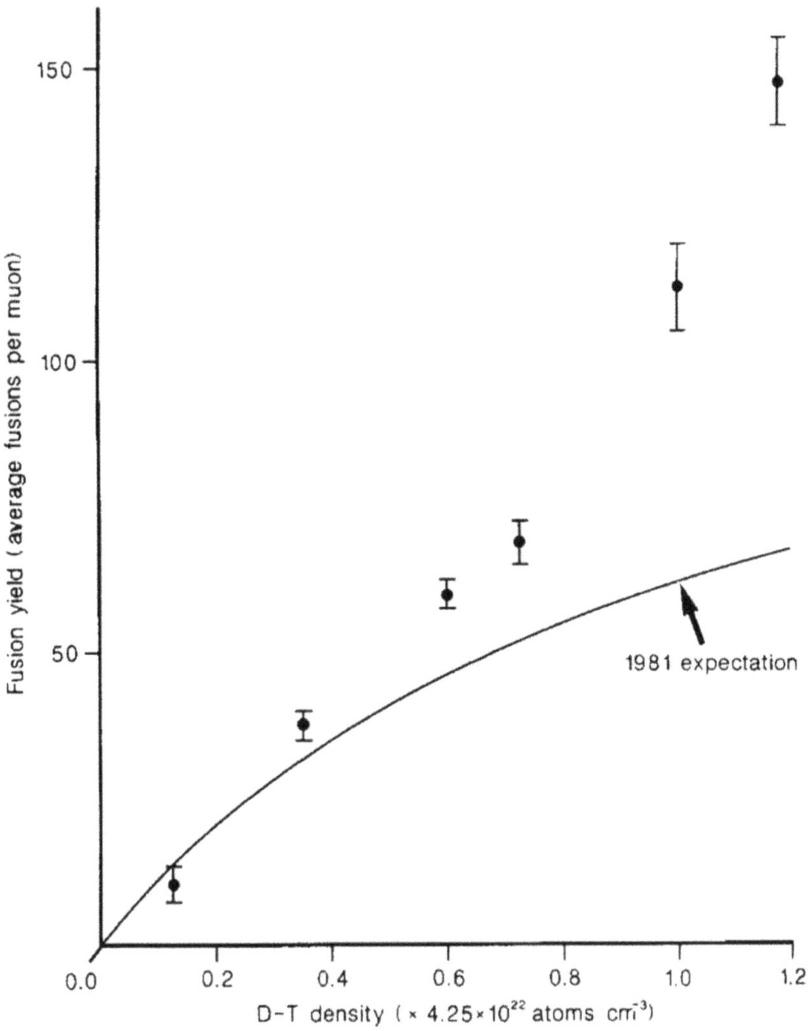

Figure 6.8. Experimentally observed number of muon-catalyzed d–t fusions per muon as a function of d–t density and a comparison with a 1981 theoretical prediction. Reprinted with permission from Macmillan Publishers Ltd, Jones (1986).

the number of fusions per muon that are needed using current accelerator technologies are in the range of 1000. So, although the results in figure 6.8 are encouraging, there is still a substantial way to go before muon-catalyzed fusion is commercially viable. Section 6.4 deals with hybrid fusion–fission reactors and muon-catalyzed fusion may be an appropriate fusion technology to integrate into this approach. Models have suggested that only 100–300 fusions per muon might be needed in this case to achieve breakeven.

A final point of interest concerning muon-catalyzed fusion is the design of a possible d–t fusion power reactor. A possible design is illustrated in figure 6.9. Here

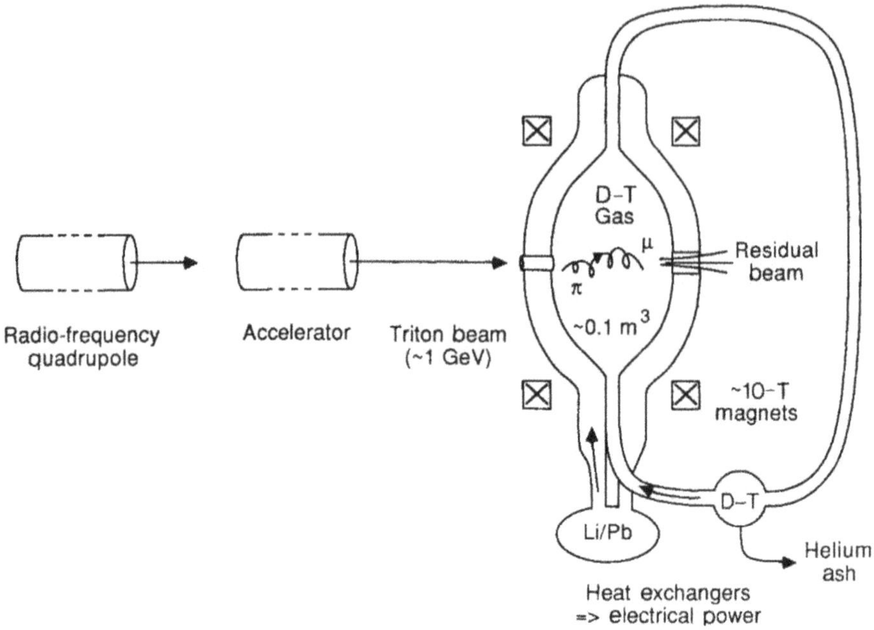

Figure 6.9. Potential design of a muon-catalyzed d–t fusion power reactor as described in the text. Reprinted with permission from Macmillan Publishers Ltd, Jones (1986).

the fuel in the core of the reactor consists of a mixture of deuterium and tritium. Muons are generated *in situ* in the core by bombarding the fuel with a high energy beam of tritons from an accelerator. Pions (both negative and positive) are produced in the core by a variety of hadronic interactions between the triton beam and the deuterons and tritons in the fuel. These pions in turn decay to muons. The ratio of negative to positive pions, and subsequently muons, is improved by using an incident beam of tritons, rather than deuterons or protons, as the ratio of neutral neutrons to positive protons is greater.

Alpha particles, known as helium ash, are separated from the d–t fuel, which is then reinjected into the core. A liquid eutectic Li–Pb mixture is used to carry heat away from the reactor core and transfer this heat through heat exchangers to steam turbine generators. The presence of Li also has the advantage that emitted neutrons, as in equation (6.17), can be used to breed additional tritium fuel by the reactions

$$n + {}^{6}\mathrm{Li} \rightarrow \alpha + t \tag{6.20}$$

and

$$n + {}^{7}\mathrm{Li} \rightarrow n + \alpha + t \tag{6.21}$$

where natural lithium is 7.5% ^{6}Li and 92.5% ^{7}Li. The reaction in equation (6.20) is exothermic with an energy release of 4.78 MeV, while the reaction in equation (6.21) is endothermic and requires an additional 2.47 MeV to proceed. Neutrons emitted by fusion reactions are typically of sufficient energy to permit the reaction in equation (6.21) to proceed. The utilization of lithium to breed tritium for d–t fusion reactors is considered in more detail in the next chapter.

Heavy ion drivers

Chapter 5 has overviewed some of the research activity in the area of inertial confinement fusion using laser drivers. It is clear that there are a number of concerns that need to be dealt with in order to make this approach to commercial energy production feasible. Specifically, the low efficiency for returning energy to the fusion fuel for subsequent energy production and the low repetition rate for lasers with sufficient power to approach breakeven conditions. While future technical developments may be capable of resolving these issues, the current state of technology is far from achieving these goals. Heavy ions may provide a driver that mitigates problems associated with low efficiency and low repetition rate. As well, heavy ion accelerators may exhibit improved long-term durability compared with laser drivers. Since the mid-20th century, there has been extensive research on particle accelerators, largely for the purpose of understanding the properties of subatomic particles and their interactions. Much of this work has dealt with accelerating small numbers of electrons, protons or light nuclei, often to as high an energy as possible. Although the most suitable potential ion drivers for inertial confinement fusion are fairly large numbers of heavy ions accelerated to rather moderate energies, the knowledge gained from traditional particle experiments forms a good basis for investigating heavy ion inertial confinement fusion.

The earliest work using particle beams as drivers for inertial confinement fusion involved the use of electrons. Research along these lines was conducted at Sandia National Laboratories in Albuquerque, New Mexico and the Kurchatov Institute Moscow in the 1970s. Because of the electron's small mass, the energy per unit space charge is limited. The large mass associated with heavy ions provides a greatly increased energy per unit space charge compared to electrons, and more recent work along these lines has concentrated on these heavy particles as drivers for inertial confinement fusion.

Typical ions that may be suitable for heavy ion fusion include Xe, Cs, Bi and Pb. Pulses consisting of large bunches of heavy ions are used to compress and ignite the target. Typical pulse characteristics might be 10^{15} Bi ions (corresponding to a charge of about 10^{-3} C for singly charged ions) accelerated to 10 GeV. This pulse corresponds to a total energy of about 3 MJ, compatible with pulse energy requirements for inertial confinement fusion based on the results of laser fusion experiments. Based on the analysis for repetition rates for laser fusion experiments as discussed in the previous chapter, ion beam pulses at about 10 Hz would suffice for the design of a commercial facility. This pulse rate is well within the capabilities of existing particle accelerators.

Two fundamentally different linear accelerator (linac) designs for heavy ions may be considered for inertial fusion facilities—radio frequency (rf) accelerators and induction accelerators. Radio frequency accelerators have typical efficiencies of a few percent, while induction accelerators can have efficiencies of 30%–50%. Thus, in terms of returning fusion energy output to future fusion events and achieving breakeven, induction accelerators are the preferable choice. This high efficiency exceeds that which can currently be achieved by sufficiently powerful laser systems and may open up the possibility (from an energy efficiency standpoint) to indirect drive as well as direct drive approaches.

The general design of a potentially viable heavy ion inertial confinement fusion device consists of a several components, as follows:

- source of multiple singly charged heavy ion beams;
- injector to introduce multiple beams into accelerator;
- linear accelerator to accelerate ions to GeV energies;
- buncher to separate ion beam into packets;
- storage ring to store packets of ions;
- pulse compressors to make ion pulses of appropriate duration;
- focusing;
- fuel target.

The accelerator is typically a linear accelerator consisting of an initial stage with electrostatic quadrupole focusing and a final stage with magnetic quadrupole focusing. To achieve appropriate GeV energies, a typical accelerator would be about 3 km in total length. The fuel target may be a simple d–t fuel capsule with an ablation layer, as illustrated in figure 5.13, for a direct drive system or a fuel capsule enclosed in a hohlraum, as illustrated in figure 5.14, for an indirect drive system.

A common approach has been to use a technique analogous to fast ignition for laser fusion, where one ion beam pulse is used for compression and a second pulse is used for ignition. In principle, there is also the possibility of combining a heavy ion driver with a laser driver, where an initial ion pulse is used to compress the fuel and a subsequent laser pulse is used to produce ignition.

From an experimental standpoint, heavy ion inertial confinement fusion is in its very early stages of development. To a large extent this situation is the result of much less funding being directed toward this fusion approach compared to laser fusion or, particularly, magnetic confinement fusion. However, at present there are research activities in heavy ion fusion at the University of California at Berkeley, the Princeton Plasma Physics Laboratory, GSI Helmholtzzentrum für Schwerionenforschung in Darmstadt, Germany and the Institute for Theoretical and Experimental Physics in Moscow. Although induction accelerators are preferable from an efficiency standpoint, existing accelerator facilities, which include radio frequency accelerators, are used for experimental studies. It is also to be seen if financial considerations, particularly infrastructure costs, make this fusion approach a viable option.

6.4 Hybrid fusion–fission

Hybrid fusion–fission reactors were first considered in the 1950s. In the 1970s they were advocated by the German-American physicist Hans Bethe (1906–2005). This technology was promoted as a means of utilizing fusion energy in a beneficial way even when breakeven was not achieved. The general concept has some similarities with fast breeder reactors (see chapter 1). In the hybrid fusion–fission reactor a fusion reactor, e.g. d–t reactor, is surrounded by a blanket of fission fuel. The requirements for the quality of the fission fuel are much less demanding than those for a conventional thermal neutron fission reactor or a traditional fast breeder reactor. Fission fuel may be natural uranium ore or waste fuel from a conventional thermal neutron reactor. It is important to note that the neutrons emitted from d–t fusion have an energy of about 14 MeV, compared to neutrons from a typical fission reaction with an energy of about 2 MeV. Thus, these very high energy fusion neutrons can cause reactions that the much lower energy neutrons from fission cannot. For example, the fusion neutrons can induce fission in non-fissile ^{238}U. Specifically, three important reactions that the fusion neutrons can induce are:

(1) The neutron can be captured by a fissile or non-fissile nucleus and the available energy will be above the fission threshold and fission will be induced. This fission process, as discussed in chapter 2, will liberate two or three additional neutrons with about 2 MeV of energy.
(2) The neutron may interact with nuclei and rather than inducing fission it may release one or two neutrons from the nucleus with sufficient energy to induce fissions in other nuclei.
(3) The neutron or one of the lower energy neutrons that are released during induced fission events can be captured by a non-fissile nucleus and breed fissile material, as described in chapter 2. As naturally occurring ore or waste reactor fuel may contain non-fissile ^{238}U or ^{232}Th, the breeding reactions are

$$n + {}^{238}U \rightarrow {}^{239}U + \gamma \rightarrow {}^{239}Np + e^- + \bar{\nu}_e \rightarrow {}^{239}Pu + e^- + \bar{\nu}_e \tag{6.22}$$

and

$$n + {}^{232}Th \rightarrow {}^{233}Th + \gamma \rightarrow {}^{233}U + e^- + \bar{\nu}_e \tag{6.23}$$

The net result of the above processes is that the neutron emitted during one d–t fusion event which produces 17.6 MeV can go on to induce two or even more fissions, each of which liberates about 200 MeV of energy. This process, known as fission multiplication, produces much more energy than the fusion process itself and is capable of using low grade fission fuel that would not be acceptable for conventional thermal or fast breeder reactors without expensive reprocessing. As well, fissile material that is produced in the fissile blanket by breeding reactions can be separated from the remaining fissile material and used in conventional thermal neutron reactors. However, the approach as described above is somewhat overly optimistic. As noted above for the muon-catalyzed fusion approach, it is necessary to utilize some of the neutrons that are produced to breed additional tritium fuel for the fusion reaction. Overall, calculations have suggested that, on the average, about one fission per fusion is typical.

Although the one fission per fusion that can be expected from a fusion–fission hybrid will produce a substantial amount of energy compared to the fusion energy alone, the most significant contribution of hybrid reactors is the production of fissile ^{233}U or ^{239}Pu which can be used in so-called satellite thermal neutron fission reactors. The number of satellite fission reactors that can be operated from the fissile fuel generated in a hybrid reactor depends on a number of criteria. These criteria include the specific design of the fusion and fission components of the reactor and whether there is ^{238}U or ^{232}Th, or both, in the breeding blanket. Research conducted at Argonne National Laboratory (Chang *et al* 1977) suggested that a single hybrid fusion–fission reactor could fuel between 4 and 16 satellite thermal fission reactors. Thus, in a sense, whether the fusion reactor achieves breakeven, or even if the hybrid reactor generates net power itself, is not as relevant as the energy that can be produced in its satellite reactors.

The situation for breeding fissile fuel in a hybrid reactor may be compared to the case for a conventional fast neutron breeder reactor. Studies of breeder reactors have shown that one breeder reactor can supply 0.7–2.7 satellite thermal neutron reactors. Thus, the actual production of net power from the breeder reactor is an important or even essential component of the overall energy efficiency. This difference in requirements for fissile fuel production from a hybrid fusion–fission reactor compared to a fast breeder reactor emphasizes an important advantage of the hybrid approach. The reactor designed to breed fissile fuel can be separate from reactors that are intended to generate electricity. This approach facilitates the implementation of security measures for reactors containing large quantities of fissile materials (i.e. weapons grade material). It also means that new reactors could serve merely to provide fuel for an existing fission reactor infrastructure.

Somewhat contrary to the results from Argonne discussed above, a study from Los Alamos National Laboratory (Barrett and Hardie 1980) did not find such a substantial difference between hybrid fusion–fission reactors and fast breeder reactors for fissile fuel production and suggested that the fast breeder route would be preferable, at least for the present, because of their greater degree of current technical development.

Over the years since around 1980, there has been occasional renewal of interest in fusion–fission hybrid reactors, partly because of the desire to utilize progress made on pure fusion reactors without the need to actually achieve breakeven. At present, however, interest in pursuing this approach further appears to be minimal.

6.5 Some new approaches to magnetic confinement

Magneto-inertial fusion

Magneto-inertial fusion is a fairly generic term that applies to a variety of somewhat different fusion approaches that, as the name implies, combine some aspects of magnetic confinement and some aspects of inertial confinement. The categories of magneto-inertial fusion can be quite broad and are often not very well defined, although, in general, fusion is pulsed, and the event is initiated by suddenly squeezing the plasma to achieve appropriate condition. In most cases, fusion conditions fall between those of magnetic confinement fusion and inertial confinement fusion, in terms of plasma density and confinement time, while still (hopefully) achieving conditions that satisfy a relevant criterion such as the Lawson criterion or the triple product.

A common approach to magneto-inertial fusion is magnetized target fusion. The concept of magnetized target fusion is to utilize a magnetic field to contain and heat a plasma at a density that is intermediate between the density achieved in magnetic confinement fusion devices (e.g. tokamaks) and inertial confinement fusion devices. Unfortunately, as in the case of magnetic confinement fusion, substantial external magnets are needed to confine the plasma and as in the case of inertial confinement fusion, the output is pulsed. However, by using a suitable compromise between plasma density and confinement time, the adverse effects of plasma instabilities may be minimized. Various techniques have been utilized to compress the plasma to conditions that are appropriate for fusion. As infrastructure is somewhat more modest than that utilized in major tokamak, stellarator or laser fusion experiments, magnetized target fusion has been a popular approach at industrial fusion laboratories.

General Fusion
General Fusion is a Canadian based company in Burnaby, British Columbia that is attempting to initiate deuterium–tritium fusion in a magnetized target device. A general schematic of their experimental set-up is shown in figure 6.10. The basic device consists of a small spherical tokamak confinement vessel, about 3 m in diameter, which is surrounded by a number of hydraulic rams, as shown in the figure. A liquid lithium–lead mixture is injected into the core tangentially at the

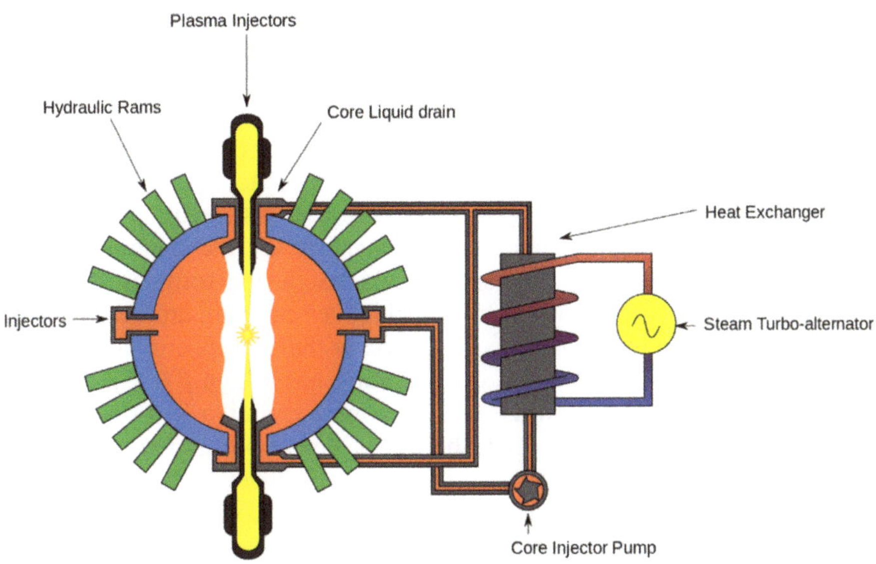

Figure 6.10. Magnetized target fusion experiment conducted by General Fusion. The magnetic field coils are shown in blue and the lithium–lead blanket is shown in orange. This Magnetized Target Fusion image has been obtained by the author from the Wikimedia website https://commons.wikimedia.org/wiki/File:General_Fusion_Reactor.svg where it was made available by Evan Mason under a CC BY-SA 3.0 license. It is included within this book on that basis. It is attributed to Evan Mason.

equator so that it rotates inside the chamber. The centrifugal force experienced by the rotating liquid pushes it outward along the walls of the vessel. A deuterium-plasma is then injected into the core of the reactor from the top and bottom. Once the plasma is inserted inside the rotating liquid lithium–lead layer, hydraulic rams, which are distributed over the surface of the spherical vessel, are used to uniformly compress the liquid, thereby reducing the volume of the d–t plasma in the center. The large number of hydraulic rams distributed over the surface of the spherical vessel act, more-or-less, like the large number of incident laser beams typically used in inertial confinement fusion experiments and help to minimize the difficulties associated with Rayleigh–Taylor instabilities. General Fusion has aimed at a compression ratio of about 1000:1 in order to achieve fusion conditions with a triple product of the order of

$$n\tau T \sim 5 \times 10^{21}\ \mathrm{keV\ s\ m^{-3}} \tag{6.24}$$

After being heated by energy from fusion reactions in the core of the reactor, the rotating lithium–lead mixture exits from the vessel at the poles. The lithium–lead blanket serves, as discussed for previous reactor designs, to transfer heat to a heat exchanger for electricity production and to breed new tritium from lithium for reactor fuel.

Recent results reported by General Fusion have indicated a final plasma density of 1.16×10^{26} ions $\mathrm{m^{-3}}$, a temperature of 24.6 keV and a confinement time of 6.93×10^{-6} s. These conditions give a triple product of $\sim 2 \times 10^{22}$ keV s $\mathrm{m^{-3}}$, consistent with requirements for fusion. Tritium breeding ratios of ~ 1.5 have been achieved.

Field reversed configuration

Compact toroids are devices that produce a self-stable toroidal plasma that do not require magnet coils which run through the center of the toroid. This configuration simplifies construction and may significantly reduce infrastructure costs. The most common types of devices that fall into this category are the spheromak and the reversed field configuration. In the spheromak design, the toroidal field and the poloidal field have similar strengths. In the field reversed configuration, the toroidal is small or absent. The result of these differences is that the spheromak has a low β (i.e. $\beta \sim 0.1$) while the field reversed configuration has a high β (i.e. $\beta \sim 1$). These devices also differ in geometry, where the spheromak is similar to a smoke ring (i.e. the axial dimension is much less than the radial dimension) and the field reversed configuration is more like a hollow sausage (i.e. axial dimension much greater than the radial dimension). A diagram of the field reversed configuration is shown in figure 6.11. A toroidal electric current is induced in a hollow cylindrical plasma as illustrated. This toroidal current creates a poloidal field that is reversed in direction to the externally applied magnetic field, hence the terminology used for this device. At a particular radius inside the device, the external poloidal field and the induced field cancel each other to create a toroidal region of zero net field which confines the plasma between increasing magnetic fields for larger and smaller radii.

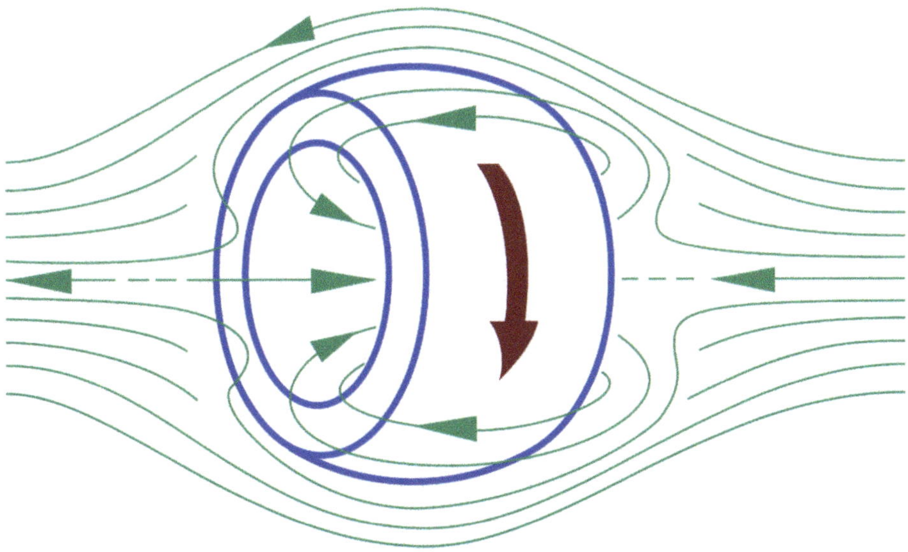

Figure 6.11. A field reversed configuration showing the induced toroidal current (brown arrow) and the poloidal field lines (green arrows). This Reversed Field Configuration image has been obtained by the author from the Wikimedia website https://commons.wikimedia.org/wiki/File:Field-Reversed_Configuration.svg where it was made available by Tokamac under a CC BY-SA 3.0 license. It is included within this book on that basis. It is attributed to Tokamac.

Helion Energy

Helion Energy in Redmond, WA takes its name from the general term for the ^{3}He nucleus. The company is developing a field reversed configuration fusion device, which they refer to as 'The Fusion Engine', to fuse deuterium and ^{3}He according to the reaction

$$d + {}^3\text{He} \rightarrow {}^4\text{He} + p \tag{6.25}$$

This reaction has an energy release of 18.3 MeV per fusion. The obvious advantage of this reaction is that it is aneutronic and eliminates the concern of neutron damage to reactor components and the loss of useable energy in the form of neutron kinetic energy.

Helion Energy's approach to achieving fusion conditions is to create two field reversed configuration plasmas and to accelerate these toward each other where they merge and create conditions amenable to fusion. Each plasma (or plasmoid) has a mass of approximately 0.2 mg and a diameter of less than 0.1 m and is accelerated using an inductive plasma accelerator. The inductive plasma accelerator (or inductive plasmoid accelerator) uses a high-current pulse, created by the discharge of a capacitor, to energize an electromagnetic coil in the vicinity of a plasmoid. The magnetic field pulse that is produced induces a current in the plasma which opposes the field from the coil and the resulting interaction accelerates the plasmoid away from the coil. It has been suggested that this approach could be adapted to provide a means of spacecraft propulsion. Helion Energy uses this method to create a region where the two plasmoids merge to create conditions of increased density and ion energy.

Helion Energy anticipates that this approach will be pulsed with plasmoids being fired at a rate of about 1 Hz, followed by extraction of fusion energy. As all of the fusion by-products of the d–^{3}He reaction are charged particles, Helion Energy plans to utilize direct energy conversion (see section 7.2) to convert the fusion energy into electricity. The availability of ^{3}He, as discussed in section 7.4, is a major concern for this approach.

Experimental results along with computer simulations have indicated that Helion Energy's device is capable of accelerating plasmoids to a velocity of 300 m s^{-1} and creating merged plasmas with ion densities greater than 10^{22} m^{-3} and temperatures greater than 800 eV.

TAE Technologies

TAE Technologies was founded as Tri-Alpha Energy in 1998 in Foothill Ranch, California. They are undertaking field reversed configuration experiments similar to those of Helion Energy, with a couple of important differences. Firstly, they are using fusion based on the p–^{11}B reaction, as described in some detail above for the polywell. Secondly, the stability of the plasma that is formed by the merger of two plasmoids is stabilized by neutral beam injection. The potential disadvantages of the more demanding fusion conditions of the p–^{11}B reaction compared to the d–t reaction may be offset (at least partially) by longer confinement times and densities that result from the neutral beam injection.

An illustration of TEA Technologies' field reversed configuration device design is shown in figure 6.12. Their most recent device is nick-named 'Norman' after one of the company's co-founders, Canadian physicist Norman Rostoker (1925–2014).

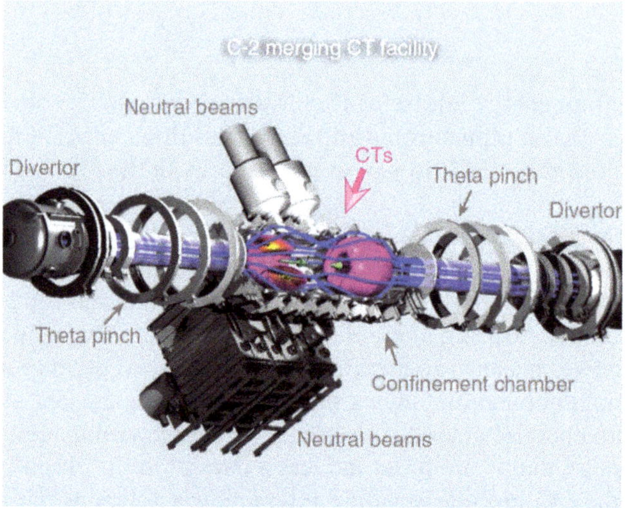

Figure 6.12. C-2 CT-merging FRC experiment at TAE Technologies. Reprinted with permission from Macmillan Publishers Ltd, Guo *et al* (2015). Copyright 2015.

6.6 *z*-Pinch machines

The pinch machine, as described in section 3.6, was one of the very earliest devices to attempt to produce controlled nuclear fusion. The *z*-pinch machine is an inertial confinement fusion device in which a pulse of high current is passed through a wire immersed in a region of fusion fuel. The current pulse causes ohmic heating of the wire which ionizes the fuel atoms to create a plasma. The plasma is confined and compressed by the magnetic field that is created by the current flowing through the wire and the plasma itself. If the density and temperature of the plasma reaches the necessary conditions, fusion will occur.

A major difficulty with the *z*-pinch machine is the formation of kink instabilities (see section 4.4). Ideally the current flowing through the plasma, and hence the magnetic field will be axially symmetric. If any transverse perturbations occur, a 'kink' or bend will be formed in the plasma. On the outside curvature of the bend the magnetic field lines will be further apart and on the inside curvature of the bend the field lines will be closer together. These differences in magnetic field will cause the kink to grow until the plasma is forced against the walls of the chamber. The magnitude of this instability is proportional to the amount of current flowing through the plasma. While the move in research direction to devices such as the stellarator and the tokamak has been motivated by the need to eliminate kink instabilities, work has also continued in several research facilities on other approaches to minimize kink instabilities while maintaining a linear geometry.

The most extensive work on the *z*-pinch approach has been conducted at Sandia National Laboratories in Albuquerque, New Mexico. An approach that Sandia has used to minimize instabilities is the use of a circular array of parallel wires rather than a single conductor. This has the benefit that each wire carries a smaller current and the magnetic field is the combined field from a large number of currents. An arrangement utilizing two coaxial arrays is shown in figure 6.13. When a current

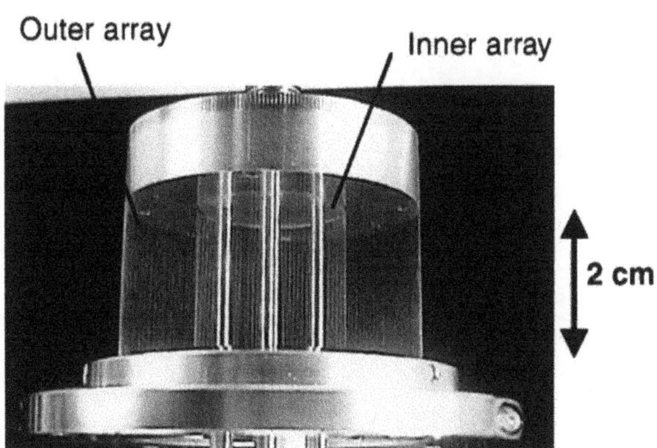

Figure 6.13. Photograph of a nested wire array used at the *z*-pinch machine at Sandia National Laboratories. The outer array consists of 240 wires and is 40 mm in diameter, while the inner array consists of 120 wires and is 20 mm in diameter. Reprinted figure 1 with permission from Deeney *et al* (1998). Copyright (1998) by the American Physical Society.

pulse is directed through the array, an imploding plasma is created. This plasma generates an intense pulse of x-rays, as described in the previous chapter for indirect drive laser fusion experiments.

Early experiments at the Sandia *z*-pinch facility dealt largely with the study of materials under extreme condition. More recently, experiments related to nuclear weapons design, as well as tests of power production by inertial confinement fusion have been undertaken. The x-rays produced by the imploding array of current carrying wires and the plasma that they create can be used to irradiate a d–t fusion fuel capsule in the same manner that x-rays from a hohlraum are utilized in laser fusion experiments. These studies are in their fairly early stages (compared to laser fusion investigations) but Sandia has estimated that a viable 300 MW commercial reactor facility could be constructed using this technique with a repetition rate of 0.1 Hz.

6.7 Magnetized Liner Inertial Fusion (MagLIF)

Magnetized Liner Inertial Fusion (MagLIF) is a technique which combines some aspects of inertial confinement laser fusion and the *z*-pinch approach. Figure 6.14 shows a basic diagram of a MagLIF device. The fusion fuel, either a d–d or d–t mixture in experimental investigations thus far, is enclosed in a centimeter sized metal cylinder. This cylinder is referred to as a hohlraum, by analogy with the similar component in inertial confinement fusion experiments (see chapter 5). A magnetic field applied along the axial direction of the hohlraum (i.e. the *z*-direction) is used to pre-magnetize the fuel plasma. Fusion is initiated by two actions—irradiating the fuel with a laser and running a large current through the hohlraum.

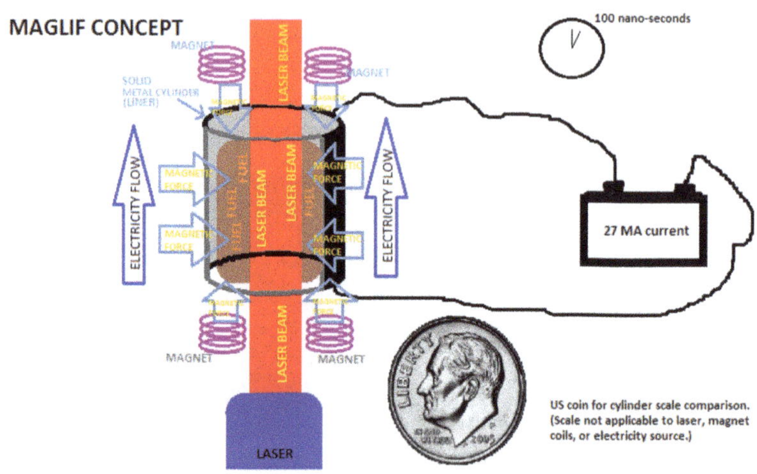

Figure 6.14. Diagram of a MagLIF fusion device. The United States 10 cent coin (17.9 mm diameter) is shown to indicate the scale of the hohlraum but not the magnets, laser or electronics. This MagLIF image has been obtained by the author from the Wikimedia website https://en.wikipedia.org/wiki/File:MAGLIF_crude_conceptual_drawing.png where it was made available by Johan Sosa under a CC BY-SA 3.0 license. It is included within this book on that basis. It is attributed to Johan Sosa.

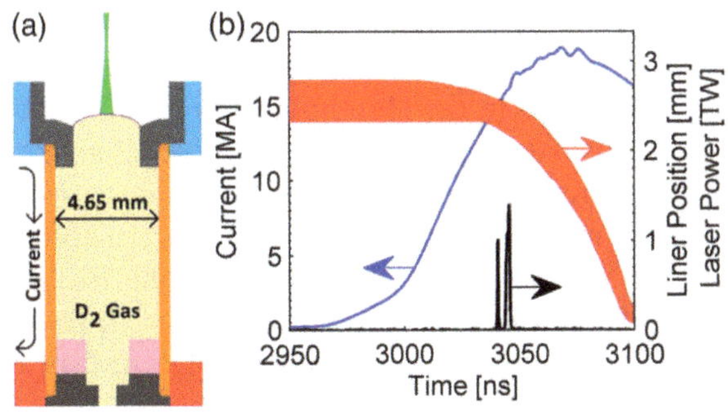

Figure 6.15. (a) Diagram of hohlraum used at Sandia National Laboratories for MagLIF experiment; beryllium hohlraum (orange), aluminum cushions (gray), nylon cushion (pink), current electrodes (red/blue) and laser beam (green). (b) Results of experiments on d–d fusion showing current pulse (blue), simulated implosion trajectory of hohlraum and laser power (black). Reprinted figure 2 with permission Gomez *et al* (2014). Copyright (2014) by the American Physical Society.

The laser beam will preheat the fuel to a high temperature and the current through the hohlraum will act like the current in a z-pinch machine. This current, in the z-direction, will interact with the induced magnetic field, in the θ-direction, to produce a force in the radial direction, as shown in figure 6.14,

$$f_r = \boldsymbol{J}_z \times \boldsymbol{B}_\theta \tag{6.26}$$

where f_r is the force per unit volume, $\boldsymbol{J}_z$ is the current density and $\boldsymbol{B}_\theta$ is the magnetic field. This force will cause the hohlraum to implode and will compress the fusion fuel.

The most significant MagLIF experiments and simulations have been conducted at the z-pinch facility at Sandia National Laboratories beginning in the early 2010s. Typical results from the work at Sandia are shown in figure 6.15. Here the timing of the laser pulse and the current pulse and the resulting implosion of the hohlraum are illustrated. Experiments have concentrated on the effects of containment field, laser power and current. Studies utilizing both d–d and d–t fuels have shown fusion neutrons and simulations along with experimental data have suggested that facilities at Sandia should be capable of achieving scientific breakeven.

6.8 Cold fusion

The term 'cold fusion' could, in principle, apply to a variety of techniques where nuclei are fused in an environment where the fuel is not raised to a high temperature. In fact, it is terminology that was often used to refer to muon catalyzed fusion prior to the late 1980s. However, since 1989 the term cold fusion has been used almost exclusively to refer to nuclear fusion in electrochemical cells or related techniques. In more recent years, researchers who have continued to consider the possibility of cold fusion have often used the term 'low-energy nuclear reactions' or LENR to describe such phenomena to distance this work from the reputation that it acquired 30 years ago.

Although the field of cold fusion is generally associated with the work that began in 1989, as described below, the basic ideas that were the genesis of this research date from more than 60 years earlier. Around 1927 two research groups reported results that suggested the existence of cold fusion. Chemists Friedrich Paneth (1887–1958) and Kurt Peters (1897–1978) from the Humboldt-Universität zu Berlin reported the transformation of hydrogen that was absorbed in palladium powder into helium suggesting solid state catalysis of nuclear fusion at room temperature. The use of palladium powder was of interest in these experiments as the absorption of hydrogen gas by palladium metal was first reported in 1866 by the British chemist Thomas Graham (1805–1869). It is known that at atmospheric pressure the maximum composition of hydrogen in palladium is about $PdH_{0.7}$ and provided a means of storing elemental hydrogen for commercial applications. Paneth and Peters later retracted their claims indicating that the helium that they had detected was from the background concentration in the atmosphere. At about the same time, Swedish scientist John Tandberg reported the fusion of hydrogen into helium in the palladium electrode of an electrochemical cell. Because Paneth and Peters had retracted their claim of nuclear fusion in palladium power, Tandberg's results were not considered as seriously as they might have been otherwise. In 1931 Harold Urey (1893–1981) discovered deuterium and with Chadwick's discovery of the neutron in 1932, the nuclear structure of this heavy isotope of hydrogen was established. For a number of years following the discovery of deuterium, Tandberg continued studies of nuclear fusion in electrochemical cells containing heavy water (D_2O).

The Fleischmann–Pons experiment

The most notable observation of cold fusion came in April 1989 when Martin Fleischmann (1927–2012) from the University of Southampton and Stanley Pons (b. 1943) from the University of Utah reported fusion of deuterium nuclei in an electrochemical cell (see Fleischmann, Pons and Hawkins 1989). This experiment was basically the same as had been conducted earlier by Tandberg, although Fleischman and Pons were not aware of this earlier work. The basic design of the electrochemical cell used by Fleischman and Pons is illustrated in figure 6.16. The cell consists of a container filled with deuterated water (D_2O) with an electrolyte (in this case 0.1 M LiOD) added in order to increase electrical conductivity.

Much has been written about the results of the Fleischmann and Pons cold fusion experiment. Most accounts of this work concentrate on the history, publishing chronology, political implications and the reaction of the scientific community. Here we concentrate on an assessment of the validity of the physics presented in this work and some subsequent related investigations. First, however, we briefly review the nuclear fusion reactions that are expected between deuterons.

When two deuterons are fused, an alpha particle is formed according to the reaction

$$\mathrm{d} + \mathrm{d} \rightarrow {}^{4}\mathrm{He} + \gamma \ (23.8\ \mathrm{MeV}) \qquad (6.27)$$

This process is exothermic and releases a substantial amount of energy (23.8 MeV). This excess energy leaves the ^{4}He nucleus (alpha particle) that is formed in a highly excited state. Perhaps, the most obvious way in which this energy is released is

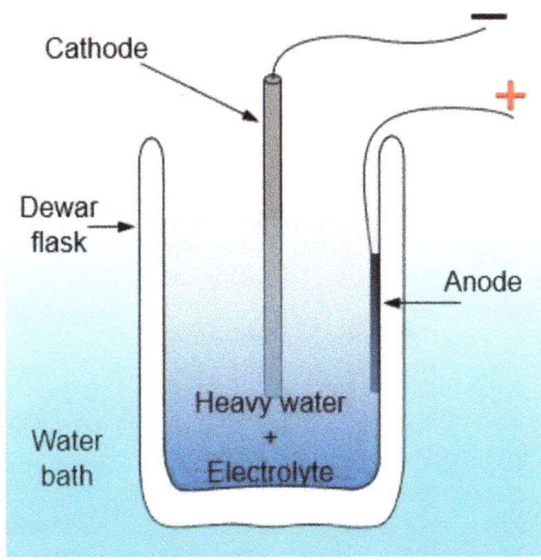

Figure 6.16. Electrochemical cell for diffusing deuterium into palladium metal. This Electrochemical Cell image has been obtained by the author from the Wikimedia website https://en.wikipedia.org/wiki/File:Cold_fusion_electrolysis.PNG where it was made available by Theresa Knott under a CC BY-SA 3.0 license. It is included within this book on that basis. It is attributed to Theresa Knott.

through the emission of a 23.8 MeV gamma-ray to the ground state of the ^{4}He nucleus, as indicated in equation (6.27). However, one should note that the energy associated with this excited state of the ^{4}He nucleus is well above the neutron and proton separation energies. Therefore, rather than gamma decay to the ground state of ^{4}He, the excited ^{4}He nucleus is more likely to undergo neutron or proton emission as given in the following equations:

$$\mathrm{d} + \mathrm{d} \rightarrow {}^4\mathrm{He} \rightarrow {}^3\mathrm{He} + \mathrm{n}\ (3.27\ \mathrm{MeV}) \tag{6.28}$$

$$\mathrm{d} + \mathrm{d} \rightarrow {}^4\mathrm{He} \rightarrow \mathrm{t} + \mathrm{p}\ (4.03\ \mathrm{MeV}) \tag{6.29}$$

Equations (6.27) through (6.29) are illustrated graphically in figure 6.18. The figure also shows the measured branching ratios for the different reactions. It is seen from the figure that approximately half of the d–d fusions ultimately produce ^{3}He and a neutron while the other half produce ^{3}H (triton) and a proton, only about 1 in 10^9 fusions produces a ^{4}He nucleus that decays by gamma decay to the ^{4}He ground state. The information in equations (6.27) through (6.29) and in figure 6.17, help us to understand the results of the Fleischmann–Pons experiment.

In Fleischmann and Pons' initial report of cold fusion, they cited the following evidence to support their claim:

- The generation of excess heat as evidenced by calorimetric measurements.
- Gamma rays at an energy of 2.2 MeV from the formation of a deuteron with neutrons generated according to equation (6.28) and protons generated according to equation (6.29),

$$\mathrm{n} + \mathrm{p} \rightarrow \mathrm{d} + \gamma \tag{6.30}$$

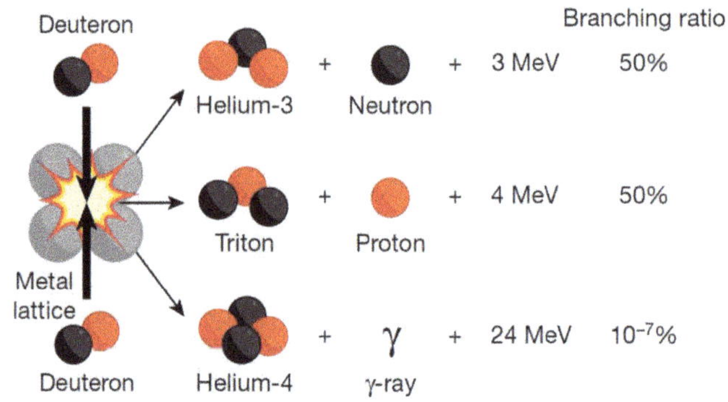

Figure 6.17. Possible outcomes of d–d fusion with branching ratios. Reprinted with permission from Macmillan Publishers Ltd, Berlinguette *et al* (2019), copyright (2019).

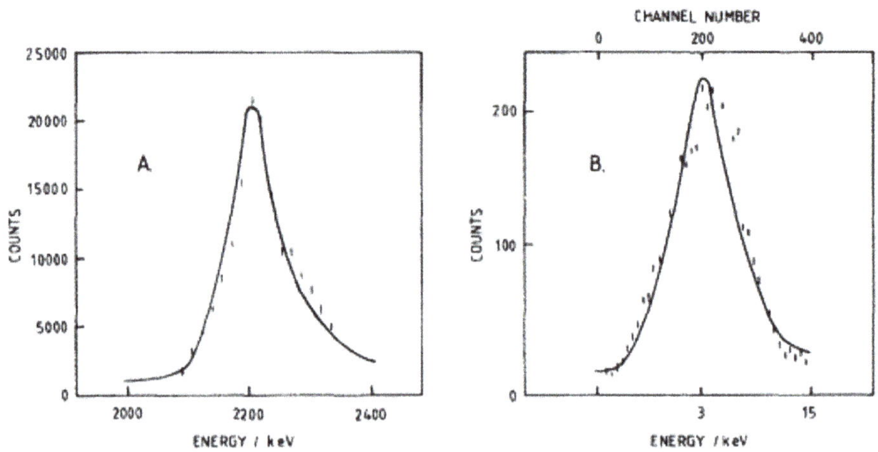

Figure 6.18. (a) Gamma-ray spectrum and (b) electron spectrum from d–d cold fusion reactions as reported by Fleischmann, Pons and Hawkins in 1989. Reprinted from Fleischmann *et al* (1989). Copyright (1989) with permission from Elsevier.

The gamma ray spectrum from Fleischmann and Pons' original experiment is illustrated in figure 6.18(a).

- Neutrons from equation (6.28). Neutron fluxes of approximately three times background were observed.
- Tritium from equation (6.29). The tritium was detected by the observation of electrons resulting from its β^- decay;

$$t \rightarrow {}^{3}He + e^{-} + \bar{\nu}_e \tag{6.31}$$

These results are illustrated in figure 6.18(b).

After the initial report of Fleischmann and Pons, there was considerable activity in the field of cold fusion. Some research was aimed at providing a viable physical explanation of cold fusion. Other work provided an analysis of the Fleischmann–

Pons results. The vast majority of the effort on cold fusion that followed the report by Fleischmann and Pons was aimed at repeating and validating their results.

The general hypothesis that describes cold fusion is that deuterium nuclei that are absorbed into a metal host are forced closer together than they normally are in deuterium compounds. It is speculated that these deuterons are in sufficiently close proximity for a sufficient period of time to satisfy the necessary fusion conditions. However, theoretical calculations have not been able to demonstrate the existence of conditions that are compatible with fusion.

There has been considerable critique of the results reported by Fleischmann and Pons that has clearly indicated that their results are not consistent with nuclear fusion and are not self-consistent themselves. Some of the points that refute the validity of Fleischmann and Pons' clams are discussed below.

In a press conference prior to the publication of their original report, Fleischmann and Pons showed a gamma-ray spectrum with a peak at 2.5 MeV (the incorrect energy for d–d fusion as given by equation (6.30)). In the publication that followed, the peak appeared at the correct energy of 2.2 MeV, as in figure 6.18(a). It is interesting to note, however, that the incorrect value of 2.5 MeV was given in the text of the publication. It is also noteworthy that the gamma-ray energy spectrum presented by Fleischmann and Pons lacked the Compton edge that is characteristic of all such gamma spectra.

The neutron flux as reported by Fleischmann and Pons is not consistent with the thermal energy production. If the d–d fusion processes followed branching ratios as given in figure 6.17, then the heat generated would correspond to a significantly greater neutron flux. It has been estimated that 1 W of thermal output, as would be roughly consistent with the author's claims, would correspond to about 10^{12} neutrons per second and this would constitute a lethal dose of neutrons for the researchers in the laboratory. While Fleischmann and Pons suggested that other fusion processes were occurring that had branching ratios that were substantially different than those in figure 6.17, there is no theoretical explanation for why this would be the case.

While the data in figure 6.18(b) suggest the production of tritium in accordance with equations (6.27) and (6.29), the quantity of tritium as evidenced by the gamma-ray flux, is not consistent with the reported thermal energy and the neutron flux in the context of the branching ratios.

Finally, reproducibility is an essential component for validating any experimental results. In view of the potential significance of the results reported by Fleischmann and Pons, there was considerable activity worldwide to try to reproduce these results. While some experiments tended to support Fleischman and Pons, at least in the observation of some aspects of their experimental results, i.e. either heat production, gamma-ray production, neutron production or tritium production, as well as the observation of ^{4}He, other experiments were unable to support any of the original claims. Even in those cases where experimental results seemed to confirm the existence of cold fusion, results tended to be inconsistent and unreliable. By the end of 1989, the general consensus of the scientific community was that there was no valid evidence for d–d cold fusion. Initial positive evidence was generally attributed to poor experimental techniques, inappropriate data analysis, overly optimistic conclusions or, unfortunately, in some cases data fabrication.

Ongoing cold fusion experiments

Despite the rejection of early studies of cold fusion by the scientific community, some research activity in this area has continued around the world. While some studies have reported some evidence of cold fusion, e.g. excess heat or tritium production, the most careful and detailed experiments have yielded negative results. Although recent studies of electrochemical cells similar to those of the original Fleischmann–Pons experiment have not shown evidence of nuclear fusion, such studies have sometimes provided important information about the formation and properties of metal hydrides.

Some recent studies have taken a somewhat different approach to inserting deuterons into a metallic host. Rather than using an electrolytic cell, as in the case of the Fleischmann–Pons experiments, deuterons are implanted in a metal target, such as palladium, by ionizing deuterium gas and accelerating the resulting ions using an electrostatic potential. The potential is established between an anode consisting of a stainless steel cage and a palladium cathode wire. The device is enclosed in an evacuated chamber into which low pressure deuterium gas can be introduced. A similar approach has been used in the past to study low energy nuclear reactions that are relevant to astrophysical phenomena. It is more likely that cold fusion experiments along these lines will provide information on the fundamental physics of such low energy nuclear reactions than to lead to a viable source of energy. The current status of such experiments has recently been reviewed by Berlinguette *et al* (2019).

6.9 Sonofusion

Sonofusion, which is also referred to as bubble fusion, first became widely known in 2002 when it was reported by Rusi Taleyarkhan (b. 1953) from Oak Ridge National Laboratories (Taleyarkhan *et al* 2002). An early report of this phenomenon by Hugh G Flynn (1982) went unnoticed until later. An extensive and informative overview of the history of sonofusion has been presented by Seife (2009).

Sonofusion is an extension of the well-known phenomenon sonoluminescence. Sonoluminescence was first observed by H Frenzel and H Schultes at the Universität zu Köln in 1934. In this early experiment, ultrasound was utilized in a tank of photographic developer with the hope of improving the development process. The ultrasound introduced bubbles in the developer which then emitted light which was observed as small dots on the photographic film.

A detailed description of the process by which sonoluminescence occurs did not begin to be formulated until the 1960s. While the general features of the bubbles formed in liquids by their interaction with ultrasound are known, the details of the process by which light is emitted is still not fully understood. However, the general steps by which light is produced may be described as follows: a cavity may exist with the liquid as a naturally occurring bubble or it may be introduced by the interaction of the ultrasound with the liquid by a process known as cavitation. Cavitation commonly occurs in liquids when rapid changes of pressure introduce vapor filled cavities. This may happen, for example, around ship propellers when the liquid pressure is not sufficient to maintain the liquid in its liquid state. In many laboratory experiments, the interaction between a single bubble and a standing acoustic wave has been investigated. The changes in the liquid pressure resulting from the sound wave cause the bubble to alternately expand and contract. During the contraction phase the vapor in the bubble heats and the high temperature results in the emission of a burst of light. The sequence of events resulting in light emission are summarized in figure 6.19. While several theoretical models of light emission during sonoluminescence have been proposed, there is no general consensus concerning the detail of this process.

It is difficult to experimentally measure the temperature inside the bubble at the time of maximum compression. However, a number of investigations seem to be consistent with a central temperature in the range of 10 000–20 000 K. Speculation during the 1990s that the internal temperature of the bubble could reach values in the range of 10^6 K, prompted renewed interest in sonoluminescence from the standpoint of a possible approach to nuclear fusion. The 2002 report by Taleyarkhan and co-workers claimed the first experimental evidence of d–d fusion initiated by ultrasound in a liquid cell.

A diagram of the basic sonofusion experiment conducted by Taleyarkhan and co-workers at Oak Ridge National Laboratory is shown in figure 6.20. The experiment consists of a cell containing deuterated acetone (C_3D_6O) and an acoustic transducer that emits ultrasound. As noted in the figure, an external pulsed neutron source has been utilized as a means of bubble nucleation. Also shown in the figure are a scintillation detector for gamma ray and neutron measurements and a photomultiplier tube for the observation of sonoluminescence events. As the external neutron source was pulsed, the nucleation of bubbles followed by their implosion through the interaction with the ultrasound, was also expected to be pulsed and correlated in time with the neutron pulses.

Several observations were reported by Taleyarkhan and co-workers that they interpreted as evidence for d–d sonofusion as follows:

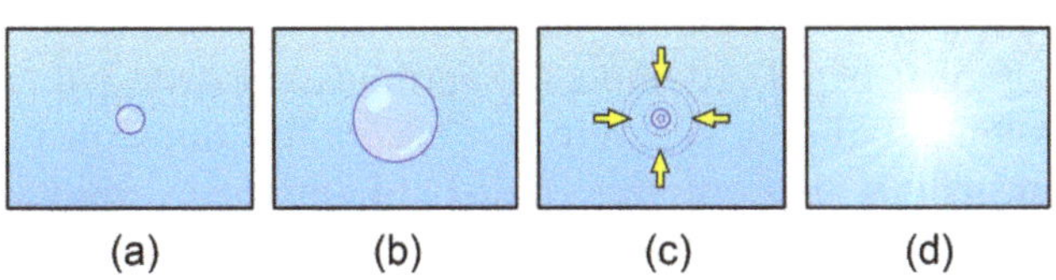

Figure 6.19. Sequence of processes leading to sonoluminescence; (a) formation of bubble, (b) quick expansion of bubble, (c) sudden collapse of bubble (short compared to expansion phase) and (d) production of light as a result of high temperature and pressure in the bubble. This Sonoluminescence image has been obtained by the author from the Wikimedia website https://commons.wikimedia.org/wiki/File:Sonoluminescence.png where it was made available by Dake under a CC BY-SA 3.0 license. It is included within this book on that basis. It is attributed to Dake.

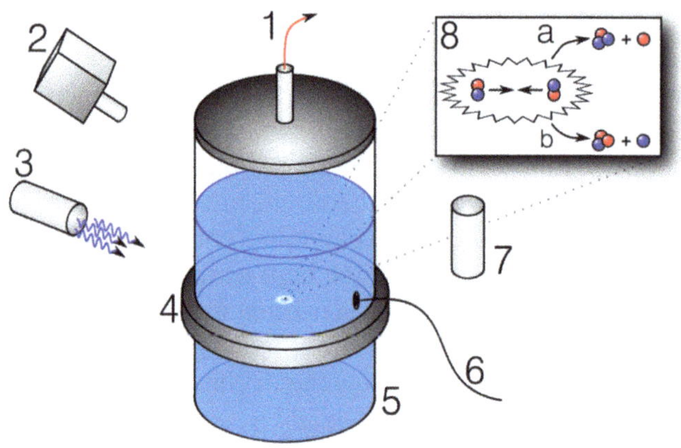

Figure 6.20. Basic diagram of apparatus for observing sonofusion; (1) connection to vacuum pump, (2) scintillation detector, (3) source of neutrons, (4) acoustic transducer, (5) tank of fluid, (6) microphone, (7) photomultiplier tube, (8a) d + d → ^{3}He + n fusion event and (8b) d + d → t + p fusion event. This Sonofusion Apparatus image has been obtained by the author from the Wikimedia website https://commons.wikimedia.org/wiki/File:Sonofusion.svg where it was made available by Silver Spoon under a CC BY 3.0 license. It is included within this book on that basis. It is attributed to Silver Spoon.

- Observation of gamma-rays as expected by the decay of radioactive tritium given in equation (6.29).
- Neutrons with energies in the neighborhood of 2.5 MeV, as expected for d–d fusion, with fluxes of 10^{3}–10^{6} neutrons per second.
- Temporal correlations between the observed d–d fusion neutrons and the neutron pulses provided by the external neutron generator.

Additional evidence provided by the authors included control experiments using non-deuterated acetone (C_3H_6O) which showed no evidence of tritium production. As well, computer simulations predicted temperatures in the range of 10^{6}–10^{7} K inside the imploding bubbles, consistent with the requirements for d–d fusion.

While Taleyarkhan's experimental techniques seem to deal with a significant number of objections that had been voiced more than a decade earlier concerning Fleischmann and Pons' cold fusion experiments, these new table-top fusion results were still viewed with considerable skepticism. Even before the publication of Taleyarkhan's original paper, attempts to reproduce positive d–d fusion results at Oak Ridge National Laboratory, were not successful. Subsequent research at a number of prestigious laboratories failed to produce results that verified Taleyarkhan's claims. These research laboratories included the University of California at Los Angeles, the University of Göttingen (Georg August Universität Göttingen) and the University of Illinois. In addition to the lack of confirming results, a major criticism of Taleyarkhan's work was the use of the external neutron source for bubble nucleation and its possible contribution to the observed neutrons that were attributed to d–d fusion reactions.

In 2006 Taleyarkhan, who had subsequently moved to Purdue University, published additional results indicating d–d fusion from an experiment which did not use an external neutron source but in which bubbles were formed directly as a result of the ultrasound (Taleyarkhan *et al* 2006). To a large extent, this work, and that reported previously in 2002, was discredited in the eyes of the scientific community by published experimental work that purported to be independent confirmation for the existence of sonofusion but turned out to be conducted in Taleyarkhan's laboratory and under Taleyarkhan's supervision. Overall, sonofusion appears to be a phenomenon that, much like cold fusion, was the result of a combination of poor experimental techniques, overly optimistic data interpretation and unfortunately, even some unethical scientific methods.

6.10 The energy catalyzer

The energy catalyzer (often referred to as the E-Cat) is a low temperature fusion device that was developed and promoted by Italian entrepreneur Andrea Rossi (b. 1950) and Italian physicist Sergio Focardi (1932–2013). Rossi and Focardi first publicized their device in 2011, however, the history of the field began about two decades earlier with the research of the Italian biophysicist Francesco Piantelli at the University of Bologna. Piantelli was involved in research on the absorption of hydrogen in nickel with relationship to techniques used for processing fats. One aspect of this work was the observation of anomalous heat generated by the nickel hydride system. Sergio Focardi became interested in this phenomenon and began collaborating with Piantelli. The result of this collaboration was a publication in 1994 (Focardi *et al* 1994) which reported excess heat when hydrogen was diffused into nickel. They suggested that this heat could be produced by p–d nuclear fusion where the deuterium was the naturally occurring component of elemental hydrogen, although their experiments did not observe any neutron or gamma-ray emission. Subsequent research by this same group (Battaglia *et al* 1999) claimed to have observed neutron emission from a similar experiment where hydrogen was diffused into nickel. However, measured neutron flux appeared to be inconsistent with heat production in the context of known nuclear fusion processes. In 1995 Piantelli obtained a European patent for a device based on this work that purported to generate excess thermal energy.

In the late 2000s, Focardi joined with Andrea Rossi to commercialize a fusion device based on the work he had conducted with Piantelli. Figure 6.21 shows a basic diagram of the Energy Catalyzer or E-Cat that they developed. Early versions of the device contained a nickel foil or solid cylinder, as shown here, while later versions contained nickel powder. The nickel sample also includes an added catalyst (of unknown composition, as Rossi has never revealed the details of this material) and is heated using a resistive heater. Hydrogen gas is introduced into the system and, according to the inventors, the nickel (in the presence of the catalyst) causes the hydrogen to ionize and the proton is absorbed into the nickel. The proton

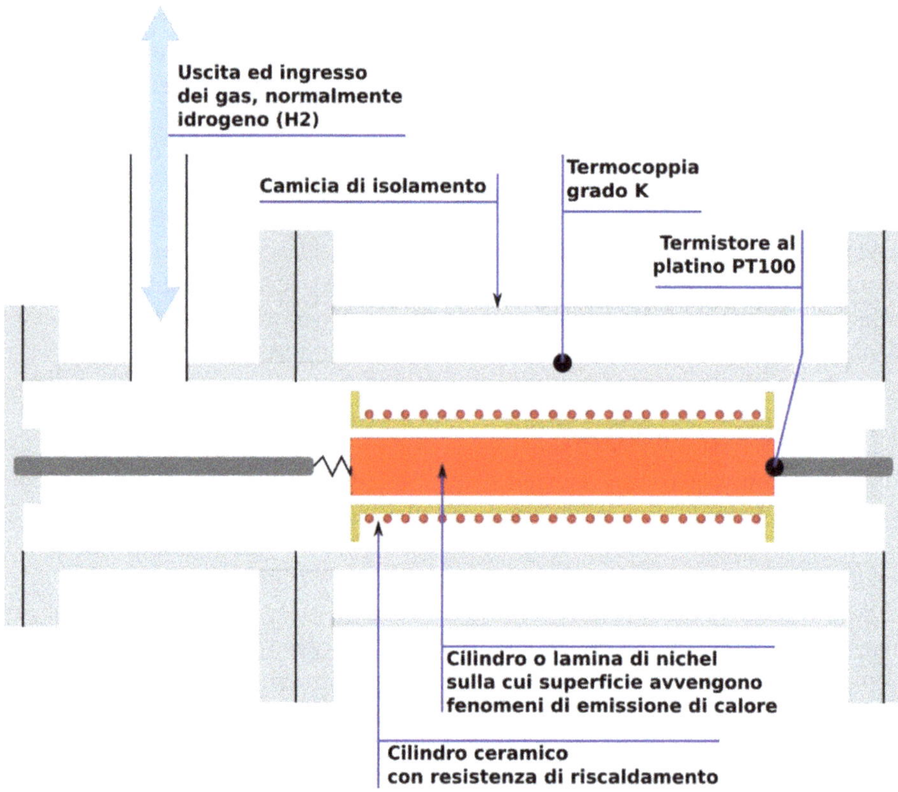

Figure 6.21. Diagram of the Energy Catalyzer based on the work of Piantelli and Focardi. (uscita ed ingresso dei gas normalmente idrogeno = normal hydrogen gas inlet and outlet; camicia di isolamento = sheilding; termacoppia grado = thermocouple; termistore al platino = platinum thermistor; cilindro o lamina di nichel sulla cui superficie avvengono fenomeni di emissione di calore = nickel cylinder or nickel foil on whose surface heat emission phenomena occur; cilindro ceramico con resistenza di riscaldamento = ceramic cylinder with resistance heater). This Energy Catalyzer image has been obtained by the author from the Wikimedia website https://commons.wikimedia.org/wiki/File:Piantelli_Focardi_schema_reattore_01_it.jpg where it was made available by Abassign under a CC BY-SA 3.0 license. It is included within this book on that basis. It is attributed to Abassign.

purportedly fuses with a nickel nucleus, thereby creating a copper nucleus and energy is released during this exothermic process.

Beginning in 2011 Rossi and Focardi provided demonstrations of their energy catalyzer, showing that it generates more energy than it consumes. However, some suspicion has existed that quantities that have been measured in these demonstrations might not be a true representation of the actual operating conditions of the device but that external power could be diverted to appear as generated power. It may also be noted that attempts by other research groups to reproduce Rossi and Focardi's results have not been successful. Part of the difficulty in reproducing these results stems from the lack of transparency in how the device actually operates.

The inventors have failed to provide a description of the device in any peer-reviewed journal. Rather, they have published incomplete descriptions in the *Journal of Nuclear Physics*. Despite its rather impressive title, this journal is merely a non-refereed internet blog created by Rossi for dissemination of selected information.

In order to better analyze the operation of the energy catalyzer, it is relevant to consider the physics of its operation, to the extent that it can be inferred from the information available. We begin with an analysis of the fuel used in the energy catalyzer. Natural occurring nickel consists of five isotopes with abundances as shown in table 6.1. The fusion reaction between a proton and a nickel nucleus may be written as

$$^{A}\mathrm{Ni} + \mathrm{p} \rightarrow {}^{A+1}\mathrm{Cu} + Q \tag{6.32}$$

where Q is the energy associated with the reaction. This energy may be expressed in terms of the masses of the reactants as

$$Q = [m(^{A}\mathrm{Ni}) + m(^{1}\mathrm{H}) - m(^{A+1}\mathrm{Cu})]c^2 \tag{6.33}$$

where the atomic masses of the various isotopes are used in order to properly account for electron masses. The energy associated with proton fusion reactions for the naturally occurring isotopes of nickel are given in table 6.1. In all cases, fusion is an exothermic process, indicating that the reaction will occur if the nuclei tunnel through the Coulomb barrier and fuse together. The table also shows the stability of the resulting copper isotopes. In all cases the unstable copper isotopes decay by β^+ decay,

$$^{A+1}\mathrm{Cu} \rightarrow {}^{A}\mathrm{Ni} + e^+ + \nu_e \tag{6.34}$$

with a small branching ratio to electron capture;

$$e^- + {}^{A+1}\mathrm{Cu} \rightarrow {}^{A}\mathrm{Ni} + \nu_e \tag{6.35}$$

A clear indication that excess power is being produced by the energy catalyzer as a result of nuclear fusion reactions would include the following:

1. Annihilation radiation (gamma rays at 511 keV) from annihilation of electrons with positrons produced by the reaction in equation (6.34) with for fusion involving ^{58}Ni, ^{60}Ni and ^{61}Ni.

Table 6.1. Naturally occurring isotopes of nickel, their natural abundances and fusion products as described in the text.

^{A}Ni	Natural abundance	$^{A+1}$Cu	Q (MeV)	$\tau_{1/2}(^{A+1}\mathrm{Cu})$ or natural abundance
^{58}Ni	68.077%	^{59}Cu	3.417	81.5 s
^{60}Ni	26.223%	^{61}Cu	4.800	3.33 h
^{61}Ni	1.14%	^{62}Cu	5.867	9.74 m
^{62}Ni	3.634%	^{63}Cu	6.122	69.17%
^{64}Ni	0.926%	^{65}Cu	7.453	30.83%

2. Characteristic gamma radiation from the de-excitation of excited Cu nuclear states for ^{63}Cu and ^{65}Cu for fusion involving ^{62}Ni and ^{64}Ni.
3. A quantity of elemental copper in the spent fuel that is consistent with the amount of nickel in the original fuel and the quantity of energy that was produced by nuclear fusion as measured by the thermal energy release.
4. Copper content in the spent fuel with abundances of the stable isotopes that are consistent with the abundances of nickel isotopes in the original fuel.
5. Trace radioactivity in the spent fuel from unstable copper isotopes in proportion to their formation and lifetime.

Experimental evidence for the existence of nuclear fusion in the energy catalyzer, or lack thereof, comes from measurements made during the supposed energy production by the catalyzer and independent analysis of spent fuel samples. With regard to the points itemized above, this evidence can be summarized as follows:

1, 2. During demonstrations of the energy catalyzer provided by Rossi and Focardi during which excess energy was supposedly measured, nuclear counting measurements failed to show the presence of any ionizing radiation.

3. An analysis of a sample of spent fuel provided by the inventors to an independent laboratory, showed the presence of 10% copper in the sample. This quantity was much larger than the amount expected on the basis of the nickel content of the original fuel and the quantity of energy supposedly produced by fusion reactions.
4. An analysis of a sample of spent fuel containing copper has shown abundances of copper isotopes that were consistent with the natural abundances of copper isotopes and not the abundances that would be expected on the basis of by-products of proton–nickel fusion reactions. Rossi, at one point, countered this inconsistency on his internet blog by indicating that the nickel fuel used in the energy catalyzer was artificially enriched in heavier isotopes using an undisclosed process. In order for this explanation to be valid, the relative abundances of ^{62}Ni and ^{64}Ni would also have to be changed to yield a copper by-product with the proper isotopic abundances.
5. No trace radioactivity has been observed in spent reactor fuel. Although Rossi's explanation for point (4) above might reduce the occurrence of radioactivity in spent fuel. Changing the isotopic abundances of nickel in order to satisfy points (3), (4) and (5), if that is even possible, would be expensive and would not have any obvious advantage in terms of energy production.

In view of the above points, some recent analyses of the operation of the e-cat (Bryce 2019, Hassani 2019) have cast doubt on the viability of this device as a potential power source. The scientific validity of the claims by Rossi and Focardi have also been disputed by theoretical astrophysicist Ethan R Siegel and nuclear physicist Peter Thieberger (Seigel and Thieberger 2019) in a recent review of the operation of the e-cat.

6.11 Summary

We have seen a few, quite diverse, alternative approaches to the construction of a viable nuclear fusion reactor for power production. In some cases, the methodology may be scientifically valid and has been shown to produce nuclear fusion, although it may be uncertain if the approach can be made energetically favorable enough to achieve breakeven. At present none of these approaches is at that point. Their shortcomings may indicate that additional technical development is needed, or may be the result of limitations placed on the technique by fundamental physics. Some approaches have been pursued primarily by industrial research organizations because they involve infrastructure that is substantially less complex and expensive than the large-scale approaches such as tokamaks and laser fusion and which, if successful could lead to viable power reactors on a shorter time scale than these larger efforts. Approaches that fall into this category and which have shown some promising results include the polywell and magnetized target fusion. Despite the fact that such approaches are not currently, and may never be, at a point where they are will yield viable commercial power reactors, many have been beneficial in other ways. For example, fusors have been used as an economical neutron source for medical purposes. Other alternative fusion approaches, for example muon catalysis, have been important for the advancement of our fundamental understanding of nuclear and particle properties and interactions.

References and suggestions for further reading

Alvarez L W *et al* 1957 Catalysis of nuclear reactions by μ mesons *Phys. Rev.* **105** 1127–28

Barrett R J and Hardie R W 1980 The Fusion–Fission Hybrid as an Alternative to the Fast Breeder Reactor Los Alamos Scientific Laboratory (Technical Report LA-8503-MS)

Battaglia A, Daddi L, Focardi S, Gabbani V, Montalbano V, Piantelli F, Sona P G and Veronesi S 1999 Neutron emission in Ni–H systems *Nuovo Cimento* **112A** 921–31

Berlinguette C P, Chiang Y M, Munday J N, Schenkel T, Fork D K, Koningstein R and Trevithick M D 2019 Revisiting the cold case of cold fusion *Nature* **570** 45–51

Bethe H A 1979 The fusion hybrid *Phys. Today* **32** 44–51

Bryce I 2019 Rossi's E-Cat: Exposé of a claimed cold fusion device *Skeptical Inquirer* **43** 40–48 https://web.archive.org/web/20190808190536/ https://skepticalinquirer.org/2019/05/rossis-e-cat-expose-of-a-claimed-cold-fusion-device/

Bussard R W 1991 Some physics considerations of magnetic inertial-electrostatic confinement: a new concept for spherical converging-flow fusion *Fusion Sci. Technol.* **19** 273–93

Chang Y I, Till C E, Rudolph R R, Deen J R and King M J 1977 Alternative Fuel Cycle Options Performance Characteristics and Impact on Nuclear Power Growth Potential Argonne National Laboratory (Report RSS-TM-4)

Deeney C, Douglas M R, Spielman R B, Nash T J, Peterson D L, L'Eplattenier P, Chandler G A, Seamen J F and Struve K W 1998 Enhancement of x-ray power from a *Z* pinch using nested-wire arrays *Phys. Rev. Lett.* **81** 4883–86

Elmore W C, Tuck J L and Watson K M 1959 On the inertial electrostatic confinement of a plasma *Phys. Fluids* **2** 239–46

Fleischmann M, Pons S and Hawkins M 1989 Electrochemically induced nuclear fusion of deuterium *J. Electroanal. Chem. Interfacial Electrochem.* **261** 301–8 (note: this paper was originally published with the authorship Fleischmann M and Pons S. In an *erratum* published in *J. Electroanal. Chem. Interfacial Electrochem.* **263** 187–88, Fleischmann and Pons added the name of University of Utah Graduate Student, Marvin Hawkins, to the list of authors of this paper)

Flynn H G 1982 Method of generating energy by acoustically induced cavitation fusion and reactor therefor *U.S. Patent* 4333796 filed 1978, issued 08 June 1982

Focardi S, Habel R and Piantelli F 1994 Anomalous heat production in Ni–H systems *Nuovo Cimento* **107A** 163–67

Frank F C 1947 Hypothetical alternative energy sources for the 'second meson' events *Nature* **160** 525–27

Gomez M R *et al* 2014 Experimental demonstration of fusion-relevant conditions in Magnetized Liner Inertial Fusion *Phys. Rev. Lett.* **113** 155003

Guo H Y *et al* 2015 Achieving a long-lived high-beta plasma state by energetic beam injection *Nat. Commun.* **6** 6897

Hamilton T 2011 *Mad Like Tesla—Underdog Inventors and Their Relentless Pursuit of Clean Energy* (Toronto: ECW Press)

Hassani S 2019 *Why E-Cat is a hoax Skeptical Inquirer* **43** 36–41 https://skepticalinquirer.org/2019/01/why-e-cat-is-a-hoax/

Hirsch R L 1967 Inertial-electrostatic confinement of ionized fusion gases *J. Appl. Phys.* **38** 4522–34

Jackson J D 1957 Catalysis of nuclear reactions between hydrogen isotopes by μ^--mesons *Phys. Rev.* **106** 330–39

Jones S E 1986 Muon-catalysed fusion revisited *Nature* **321** 127–33

Krivit S and Winocur N 2004 *The Rebirth of Cold Fusion: Real Science, Real Hope, Real Energy* (Los Angeles, CA: Pacific Oaks Press)

Mahaffey J 2017 *Atomic Adventures: Secret Islands, Forgotten N-Rays, and Isotopic Murder—A Journey Through the Wild World of Nuclear Science* (Cambridge: Pegasus Books)

Miley G H and Murali S K 2014 *Inertial Electrostatic Confinement (IEC) Fusion—Fundamentals and Applications* (New York: Springer)

National Research Council of the National Academies 2013 *An Assessment of the Prospects for Inertial Fusion Energy* (Washington, DC: The National Academies Press) available at http://www.nap.edu/catalog.php?record_id=18289

Parisi J and Ball J 2019 *The Future of Fusion Energy* (London: World Scientific)

Park J Y, Krall N A, Sieck P E, Offermann D T, Skillicorn M, Sanchez A, Davis K, Alderson E and Lapenta G 2015 High-energy electron confinement in a magnetic cusp configuration *Phys. Rev.* X **5** 021024

Seife C 2009 *Sun in a Bottle* (New York: Penguin Books)

Siegel E and Thieberger P 2019 The physics of why the e-Cat's cold fusion claims collapse Science Blogs (05 December 2011) available at https://scienceblogs.com/startswithabang/2011/12/05/the-nuclear-physics-of-why-we

Slough J, Votroubek G and Pihl C 2011 Creation of a high-temperature plasma through merging and compression of supersonic field reversed configuration plasmoids *Nucl. Fusion* **51** 053008

Taleyarkhan R P, West C D, Cho J S, Lahey R T, Nigmatulin R I and Block R C 2002 Evidence for nuclear emissions during acoustic cavitation *Science* **295** 1868–73

Taleyarkhan R P, West C D, Lahey R T, Nigmatulin R I, Cho J S, Block R C and Xu Y 2006 Nuclear emissions during self-nucleated acoustic cavitation *Phys. Rev. Lett.* **96** 03430

Yukawa H 1935 On the interaction of elementary particles *Proc. Phys. Math. Soc. Japan* **17** 139–48

Chapter 7

Is fusion useful?

7.1 Introduction

The previous chapters have provided an overview of the technologies that are under development to make use of energy from nuclear fusion reactions. The progress toward useful energy production has been considered. The final chapter of the present book considers the utilization of energy produced by nuclear fusion and the ways in which it can help to fulfill our future energy needs. The analysis of the usefulness of fusion energy must include a consideration of a number of factors including:

- the conversion of fusion energy into electricity;
- the environmental impact of fusion energy;
- the availability of resources for fusion energy production;
- the cost of fusion energy production;
- the scientific and technical challenges of developing fusion energy.

The present chapter provides an overview of these factors and how they may influence the future development and use of nuclear fusion energy.

7.2 Electricity generation

The viable application of fusion energy requires that the energy can be converted into a form that suits our requirements. In general, electricity that can be connected to the grid is a goal of most large-scale energy production. The present chapter begins with a consideration of the ways in which energy from nuclear fusion can be converted into useable electricity.

The energy that is liberated during a fusion reaction appears, firstly, as the kinetic energy of the by-products of the reaction. The energy that is associated with charged particles that are given off during the reaction can readily be used to generate electricity. This can be done in one of two ways. These charged particles transfer energy via Coulomb interactions with atoms resulting in the production of thermal energy which can be converted into electricity through the use of a heat engine. Alternately, the kinetic

doi:10.1088/978-0-7503-3307-8ch7

energy associated with charged particles can be converted into electricity directly through charge collection. These two different approaches are discussed below.

Thermal generation

Thermal generation of electricity by means of a heat engine is the most common method employed in coal-fired and nuclear fission generating stations. Heat produced by combustion of a chemical fuel or nuclear reactions is used to boil water to produce steam. The steam is used to operate a turbine which drives a generator to produce electricity. A diagram of a nuclear fusion reactor which uses this approach to generating electricity was shown in figure 4.27. In this case energy is provided by a magnetic confinement fusion reactor. A similar design could incorporate an inertial confinement laser fusion reactor.

In an evaluation of the net energy (i.e. electricity) production of a fusion reactor, the energy conversion efficiency, from fusion energy to electrical energy, is relevant. For thermal generation, it is the thermodynamic efficiency of the heat engine that is most relevant. This is readily approximated by the Carnot efficiency (in percent) as

$$\eta = 100 \times \left[1 - \frac{T_{\text{cold}}}{T_{\text{hot}}}\right] \tag{7.1}$$

where T_{hot} is the temperature of the hot reservoir and T_{cold} is the temperature of the cold reservoir (both in Kelvin). It is important to make T_{hot} as high as practical and to make T_{cold} as low as practical in order to optimize the thermodynamic efficiency. Minimizing T_{cold} requires the use of an appropriate cold reservoir. Further details of power plant cooling, as well as its potential environmental effects, is discussed below in the section on cooling requirement. Modern coal-fired and nuclear generating stations typically have efficiencies around 40%. This means that slightly less than half of the energy produced by a nuclear fusion reactor would be converted into useable electrical energy.

Direct energy conversion

An alternative to the thermal generating approach outlined above, another possible way of converting energy from a nuclear fusion reactor into electricity is through direct conversion. The potential designs as described below utilize the charged nature of fusion reaction by-products and are, therefore, most suitable for reactions where the majority of the fusion energy is carried by particles with charge. Although work toward the development of a direct energy converter is in its fairly early stages, several different approaches for suitable designs have been proposed, and in some cases, initial experimental devices have been tested. Several basic categories of direct energy converters are discussed below.

Electrostatic converters

The concept of electrostatic energy converters was first promoted at Lawrence Livermore National Laboratory in the 1960s by American physicist Richard F Post

(1918–2015). The conversion of the charged particle kinetic energy to electricity involves the following steps:

- collimating charged particles into a beam;
- separating positively and negatively charged particles;
- separating particles according to their energy;
- collecting the charged particles using collectors that are part of an electric circuit.

A generalized schematic of the design of an electrostatic direct energy converter is illustrated in figure 7.1. The stages of energy conversion as described above can be seen in the diagram. The beam of particles from the plasma is first expanded and then collimated by passing through an opening in the shield. The positive charges (ions) and negative charges (electrons) are separated by repelling the electrons from the suppressor. Finally, the positive ions deposit charge on a series of collector plates as they decelerate. The so-called venetian blind direct energy converter has collector plates arranged in a geometry that is similar to the window blinds bearing the same name. An example of this type of design is shown in figure 7.2. Various electrostatic direct energy converter designs are suitable for fusion by-products in the range of energies from about 100 keV to about 800 keV. Typical electrostatic direct energy converters have efficiencies around 60%.

Induction converters

In 1963 Soviet physicist Lev Andreevich Artsimovich (1909–73) proposed a method of direct energy conversion for pulsed fusion cycles where the magnetic field used for plasma confinement is inductively coupled to a coil. This approach is suitable, for example, for a pinch machine as described in chapter 3. The basic sequence of events corresponding to energy conversion for a fusion cycle in a pinch machine is as follows:

- fusion fuel is introduced into the device and is ionized to create a plasma;
- the plasma is compressed by a magnetic field;
- this compression heats the plasma and a fusion reaction is initiated;

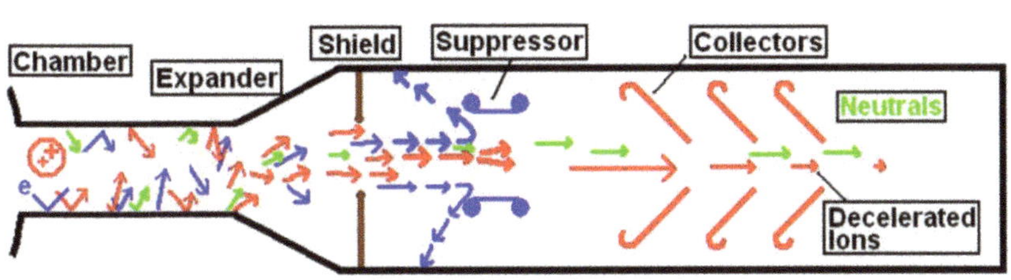

Figure 7.1. Generalized diagram of an electrostatic direct energy converter. Negatively charged electrons are shown in blue, positively charged ions are shown in red and neutral particles are shown in green. This Electrostatic Direct Energy Converter image has been obtained by the author from the Wikimedia website https://commons.wikimedia.org/wiki/File:Direct_Converter.jpeg where it was made available by WikiHelper2134 under a CC BY-SA 3.0 license. It is included within this book on that basis. It is attributed to WikiHelper2134.

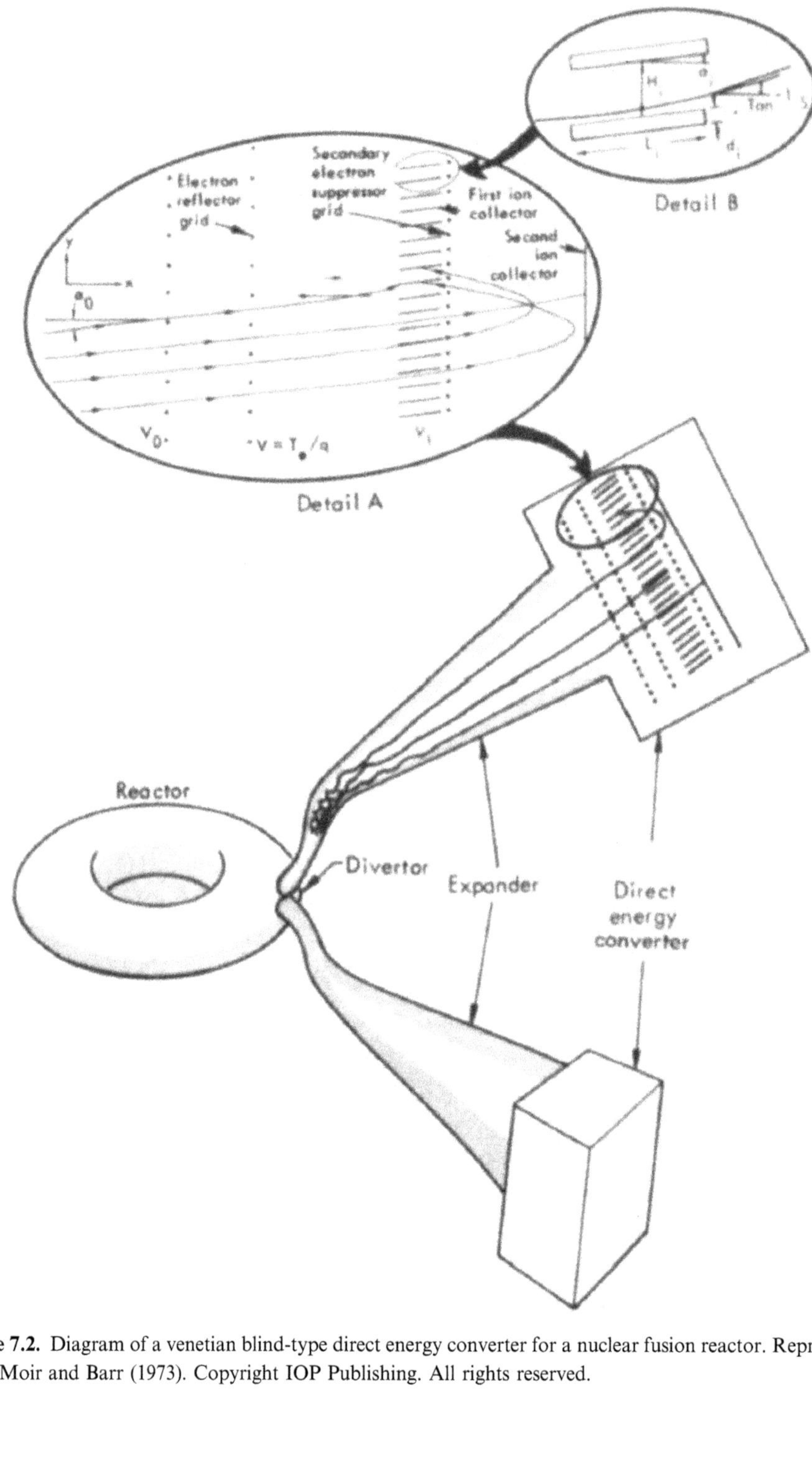

Figure 7.2. Diagram of a venetian blind-type direct energy converter for a nuclear fusion reactor. Reproduced from Moir and Barr (1973). Copyright IOP Publishing. All rights reserved.

- the charged particles resulting from the fusion process are driven outward expanding the magnetic field lines;
- the changing magnetic field induces a current in a coil which provides electrical energy to an external circuit.

This approach has been studied by Oliphant *et al* (1973) who estimated the overall efficiency of conversion from fusion energy to electricity at 62%.

Inverse cyclotron converters

The linear motion of charge particles produced in fusion reactions may be converted into circular motion by the application of a magnetic field. Electrons can be removed from the particle beam along the lines described above for electrostatic energy converters. The energy of the remaining ion beam, E, can be controlled, i.e. decreased, by the application of an electric field parallel to the beam path. The application of an orthogonal magnetic field with flux density, B, will cause singularly charged ions of mass m_{ion} to undergo cyclotron orbits with an angular frequency, ω;

$$\omega = \frac{eB}{m_{\text{ion}}} \tag{7.2}$$

If the ion energy and magnetic flux density are chosen such that the radius of the cyclotron orbit, given by

$$r = \sqrt{\frac{2E}{m_{\text{ion}}\omega^2}} \tag{7.3}$$

is compatible with the spacing of charge collector plates, then, as for the venetian blind converter, the kinetic energy of the ions will be converted to an electrical potential on the plates. While traditional electrostatic energy converters are useful for charged particles with energies in the 100s of keV range, the inverse cyclotron converter is useful for much higher energy charged particles. The inverse cyclotron converter was originally proposed as a means to extract energy from 14.7 MeV protons from d–^{3}He fusion (see section 2.5 and section 7.4, below) and has a theoretical maximum efficiency of about 90%.

As noted above, direct conversion of nuclear fusion energy to electricity is a field that is in the very early stages of development. Some of the approaches are specifically aimed at aneutronic fusion technologies, as described in some detail below, as the fusion energy appears as primarily kinetic energy of charged particles. Direct energy conversion is less attractive for d–t fusion, for example, where a significant fraction of fusion energy is given off as kinetic energy of neutrons. It is likely that initial plans to commercialize fusion energy will rely upon traditional, and well tested, thermal approaches to energy conversion, particularly since early commercialization is likely to utilize the d–t fusion reaction. These are the proposed plans for the DEMO reactor described in chapter 4. Direct energy conversion methods may play a role in fusion energy technology at some point in the future, possibly in conjunction with thermal methods.

7.3 Environmental and safety considerations

Reactor safety considerations

All methods of producing useable energy have some drawbacks. While energy production by nuclear fusion is relatively benign, compared to some other methods such as nuclear fission, it is important to carefully consider the potential adverse effects that can result from the possible widespread use of fusion reactors.

While fusion power has been promoted as a safe alternative to fission power, a careful comparison of the two approaches is informative. A comparison of safety concerns for fission and fusion reactors can be summarized as follows:

- fission reactions are easy to start and hard to stop;
- fusion reactions are hard to start and easy to stop.

Typically, in a magnetic confinement fusion reactor, fuel appropriate for a few minutes of operation may be injected into the reactor at a time. In a hypothetical 1000 MW d–t fusion reactor this would correspond to a few grams of d–t mixture, containing a gram of so of radioactive tritium. This amount of radioactive material can be contrasted with about 75 tons of uranium contained in the core of an operational 1000 MW thermal neutron fission reactor. Lack of reactor control or appropriate cooling can result in an uncontrolled fission reaction in a typical fission reaction, leading to accidents such as those that have occurred in the past at Three Mile Island, Chernobyl and Fukushima. In a fusion reaction, typically, any increase in fusion energy production will lead to excess loss of heat through the containment walls and cessation of fusion reactions. Thus, it is straightforward to see that fusion reactors are not susceptible to the kind of runaway accidents that can occur in fission reactors that endanger lives and the environment.

In the event of the release of any reactor fuel to the environment, fusion fuels are, for the most part, relatively non-toxic and do not pose any significant health risk. Tritium, as used for d–t reactors is radioactive and decays by β^- decay as illustrated in figure 7.3. This decay has a relatively short half life (12.3 years) compared to fuels used in fission reactors and emits only low energy electrons.

Radioactive waste

The problem of radioactive waste disposal is a significant deterrent to long term, large scale fission reactor use. Radioactive waste from fission reactors can be

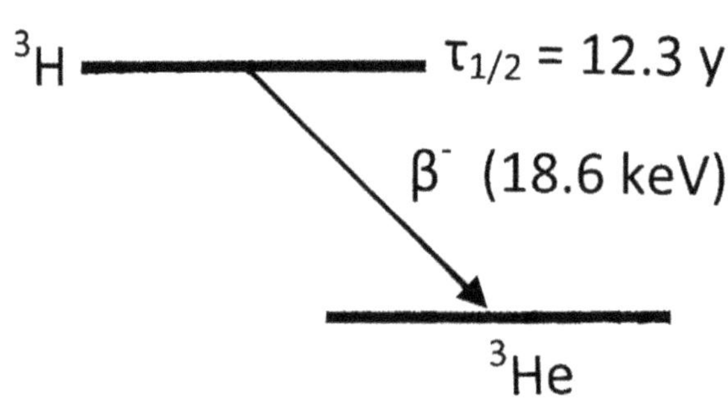

Figure 7.3. Decay scheme of ^{3}H.

classified according to the level of activity, the nature of the radioactive nuclides and their lifetime. Intermediate level radioactive waste comes largely from material that has been activated by neutron irradiation, as described in detail below. High level radioactive waste consists of the spent fuel from the reactor core and this represents the most severe challenge for waste disposal as it is highly radioactive and many of the nuclides are long-lived. Table 7.2 shows the principal longer lived fission products present in spent fuel from a reactor utilizing ^{235}U as a fuel. The fission yield is the per cent (by moles) of the particular nuclide in the spent fuel. In addition, radioactive actinides are produced by neutron capture (see below) by uranium fuel (or other actinide fuels). Figure 7.4 shows the total calculated activity for spent fuel from different types of fission reactors. A wide variety of short-lived nuclides contribute to the high initial activity shown in the figure. For intermediate times, i.e. a few years to a few hundred years, ^{90}Sr and ^{137}Cs are the major contributors to the activity. This is seen by an inspection of the half lives and fission yields in table 7.1. For longer times (i.e. $>10^3$ years) the major contributors to the radioactivity are ^{99}Tc and actinides produced from neutron capture by the reactor fuel.

By comparison, the direct by-products of fusion reactions that are likely candidates for future fusion power reactors, are shown in table 7.2. All nuclear

Table 7.1. Principal fission products from a thermal neutron fission reactor using ^{235}U fuel with half lives of more than 10 years.

Nuclide	Fission yield	Half life
^{129}I	0.9%	15.7 My
^{107}Pd	0.16%	6.5 My
^{93}Zr	6.30%	1.53 My
^{79}Se	0.05%	327 ky
^{126}Sn	0.02%	230 ky
^{99}Tc	6.05%	211 ky
^{151}Sm	0.42%	90 y
^{137}Cs	6.09%	30.17 y
^{90}Sr	5.75%	28.9 y
^{85}Kr	0.27%	10.78 y

Table 7.2. Potential fusion reactions and their by-products. Note additional comments in the section on aneutronic fusion below.

Reaction	By-products
d + t	α + n
d + d	^{3}He + n
d + d	t + p
d + ^{3}He	α + p
p + ^{11}B	3α

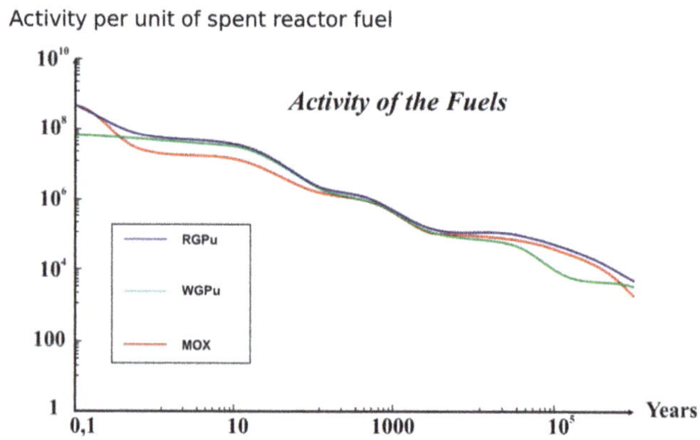

Figure 7.4. Total activity of spent fuel from a fission reactor. Curves are for the following spent fuels: RGPu = thorium with reactor-grade plutonium, WGPu = thorium with weapons-grade plutonium and MOX = mixed oxide fuel with no thorium. This Spent Fuel Activity image has been obtained by the author from the Wikimedia website https://commons.wikimedia.org/wiki/File:Activitytotal1.svg where it was made available by Teijo1 under a CC BY-SA 3.0 license. It is included within this book on that basis. It is attributed to Teijo1.

by-products, except tritium from d–d fusion, are stable nuclides and are non-toxic. Tritium is useful as a further fusion fuel, so waste disposal is not a concern.

Neutron activation

Radioactive materials that result from neutron irradiation of reactor materials is an important concern for fission reactors. These materials represent intermediate level radioactive materials, as they are typically lower activity and shorter lived than the most troublesome high-level nuclides from spent fuel. While neutron activated material is a major concern at the time of decommissioning of aging reactors, it is also a factor for worker's safety as well as maintenance and repair protocol.

Neutron activation of construction materials is also a concern for fusion reactors that produce neutrons as a reaction by-product. These reactions include both d–t fusion and d–d fusion. As with neutron activated materials in fission reactors, radioactive materials that are produced by neutron flux in fusion reactions are a factor for worker safety during reactor operation. In some experimental fusion reactor facilities, maintenance is performed by robots to minimize human radiation exposure. For commercial fusion reactors decommissioning must include provisions for neutron activated material disposal. The present section overviews some of the materials that are components of a fusion reactor infrastructure that may become radioactive during normal reactor operation and the methods by which this potential problem can be mitigated.

Neutrons produced by fusion reactions such as

$$d + d \rightarrow {}^{3}He + n \tag{7.4}$$

and

$$d + t \rightarrow {}^{4}He + n \tag{7.5}$$

can interact with reactor materials by a variety of different nuclear reactions. Most commonly it is the (n,γ) reaction, where n is the incident particle and γ is the emitted particle, involving thermal neutrons that occurs:

$${}^{A}_{Z}E^{N} + n \rightarrow {}^{A+1}_{Z}E^{N+1} + \gamma \tag{7.6}$$

where, in order to properly account for all particles, the proton number, Z, neutron number, N, and the total number of nucleons, $A = N + Z$, are indicated by the subscripts and superscripts for each nuclide. The (n,γ) reaction can also occur for high energy neutrons. As well, other neutron induced reactions can produce radioactive by-products. These include (but are not limited to) the following:

$$(n,2n) \qquad {}^{A}_{Z}E^{N} + n \rightarrow {}^{A-1}_{Z}E^{N-1} + 2n \tag{7.7}$$

$$(n,p) \qquad {}^{A}_{Z}E^{N} + n \rightarrow {}^{A}_{Z-1}E^{N+1} + p \tag{7.8}$$

$$(n,d) \qquad {}^{A}_{Z}E^{N} + n \rightarrow {}^{A-1}_{Z-1}E^{N} + d \tag{7.9}$$

and

$$(n,\alpha) \qquad {}^{A}_{Z}E^{N} + n \rightarrow {}^{A-3}_{Z-2}E^{N-1} + \alpha \tag{7.10}$$

The element, E, for each nuclide is determined by the number of protons, Z. The relationship of the number of neutrons, N, and the number of protons, Z, for the by-products of these various neutron reactions is shown graphically in figure 7.5. The by-products of these neutron reactions can also undergo further interactions with reactor neutrons.

Nuclei produced by neutron reactions are often in excited states and decay by emitting gamma rays. Nuclei may also be unstable against β decay, or in some cases α decay, and may further decay, emitting ionizing radiation. The severity of the risk associated with neutron activated materials depends on a number of factors including:

- the magnitude of the neutron flux;
- the amount of the nuclide ${}^{A}E$ present;
- the neutron absorption cross section of the ${}^{A}E$ nuclei;
- the decay scheme of the by-product of the neutron activation (if it is unstable);
- the lifetime of nuclides in the by-product decay scheme.

Nuclides that are of importance in a consideration of neutron activation may occur as natural components of structural materials, such as Fe in steel, or they may occur as impurities, such as many 3d transition metals in iron. The most common structural materials that can become radioactive as a result of neutron irradiation are steel (structural steel or stainless steel) or concrete and the details of the

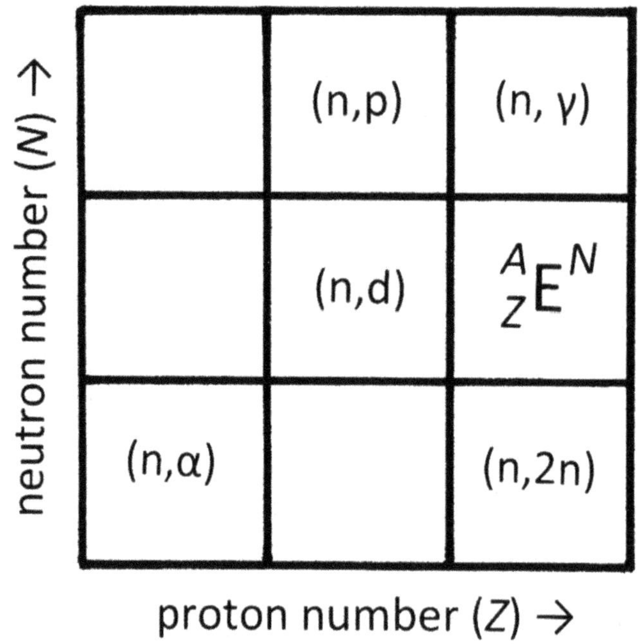

Figure 7.5. Graphical representation of neutron and proton numbers for different neutron activation processes.

radioactive materials that are produced depends on the composition of the material. While the composition of structural steels and concrete may be designed to minimize the adverse effects of neutron activation, the major consideration in the design of materials is the suitability of their physical and chemical properties for the purpose for which they are intended. Table 7.3 summarizes some radioactive nuclides that can be produced by neutron irradiation of steel (or stainless steel) and concrete. While a significant number of other radioactive nuclides are also produced, those given in the table are generally the most important as they occur in significant amounts for typical reactor materials and they have sufficiently long half-lives that disposal is a concern. The role of the various elements given in the table below in the composition of reactor structural materials, specifically stainless steel and concrete, are discussed below.

Stainless steel

Nitrogen, since it comprises about 80% of the Earth's atmosphere, commonly appears as an impurity in manufactured materials, including both steels and concrete. However, in the case of stainless steel, nitrogen is often intentionally incorporated into the alloy as it improves corrosion resistance and increases mechanical strength.

Iron, of course, is the major component of stainless steel, comprising typically 65%–80% of the alloy. Usually, the remainder of stainless steel is chromium, minimum ~11%, which is the principal compositional element that provides rust

Table 7.3. Common radioactive nuclides and their properties that are formed by neutron activation of elements frequently found in steel/stainless steel and concrete. Data adapted from Evans *et al* (1984). For principal radiation, β^- refers to electrons emitted as part of the β^- decay process, x-rays refer to x-rays emitted from atomic processes in the progeny nuclide and electrons refer to Auger electrons emitted from atomic process in the progeny nuclide. ec = electron capture.

Nuclide	Half-life (years)	Decay mode	Principal radiation	Precursor	Natural abundance (%)	Principal production reaction	Occurrence
^{14}C	5730	β^-	β^-	^{14}N	99.63	(n,p)	Steel/concrete
^{41}Ca	1.03×10^5	ec	X-rays	^{40}Ca	96.94	(n,γ)	Concrete
^{55}Fe	2.7	ec	X-rays, electrons	^{54}Fe	5.9	(n,γ)	Steel
^{59}Ni	8.0×10^4	ec	X-rays, electrons	^{58}Ni	68.1	(n,γ)	Steel
^{60}Co	5.3	β^-	β^-, γ	^{59}Co	100	(n,γ)	Steel
^{63}Ni	100	β^-	β^-	^{62}Ni	3.6	(n,γ)	Steel
^{94}Nb	2.0×10^4	β^-	β^-, γ	^{93}Nb	100	(n,γ)	Steel
^{133}Ba	10.5	ec	γ	^{132}Ba	0.1	(n,γ)	Concrete
^{152}Eu	13.5	β^-, ec	β^-, γ	^{151}Eu	47.8	(n,γ)	Concrete
^{154}Eu	8.6	β^-	β^-, γ	^{153}Eu	52.2	(n,γ)	Concrete

inhibition to the alloy. Nickel is commonly a component of stainless steel as it aids with reducing corrosion.

Cobalt is of particular concern in terms of the production of radioactive material by neutron activation. ^{59}Co is 100% naturally abundant and has a large thermal neutron cross section (19 b) for the (n,γ) reaction. The resulting ^{60}Co decays by β^- decay to stable ^{60}Ni thereby emitting gamma radiation (primarily) at 1.17 MeV and 1.33 MeV. ^{60}Co has a half-life of 5.3 years, so it is both highly radioactive and sufficiently long-lived to be of concern for disposal. Cobalt is a common impurity in other 3d transition metals, but it is also intentionally added to stainless steel to increase strength and creep resistance at high temperatures.

Finally, the addition of niobium to stainless steel improves weldability and increases strength and creep resistance.

Thus, the elemental components of stainless steel that are the most problematic as far as radioactivity induced by neutron activation is concerned, are all frequently added to such alloys to improve their properties. As the environment surrounding structural stainless steel in both fission and fusion reactors can be very demanding, the design of alloys with suitably robust properties can be challenging. Parallel to the actual development of viable fusion reactors, work will be undertaken at the International Fusion Materials Irradiation Facility (IFMIF) to develop suitable structural materials that minimize the production of problematic radioactive materials by neutron activation. Three possible approaches to this problem are under active consideration. Firstly, the development of steels which contain a minimum of the elements that are most susceptible to neutron activation, but which retain the necessary physical and chemical properties. Secondly, the development of vanadium alloys which are low activation materials. These alloys typically contain (in addition to vanadium) chromium, titanium and silicon, all of which minimize the potential of undesirable radioactive materials produced by neutron irradiation. Finally, the development of silicon carbide composites which may be suitable low activation materials for fusion reactor construction. Development of appropriate materials along these lines is a very significant materials science task and one which can proceed alongside fusion reactor research itself.

Concrete

Concrete consists of an aggregate material (usually sand with particles of a range of different sizes) bonded together with cement. Cement, in this sense, begins as calcium oxide, CaO, which is mixed with water, and undergoes the following reaction,

$$CaO + H_2O \rightarrow Ca(OH)_2 \tag{7.11}$$

This cement, mixed with the aggregate material, is hardened by first allowing excess water to evaporate (a process referred to as *setting*). Setting is followed by the carbonation reaction,

$$Ca(OH)_2 + CO_2 \rightarrow CaCO_3 + H_2O \tag{7.12}$$

which hardens the concrete. Thus, calcium is an integral component of concrete and, as such, will contribute to radioactivity produced by neutron activation in a reactor setting. Radioactive ^{41}Ca is of particular importance from a medical perspective. If ^{41}Ca is ingested, it can displace calcium in bones and will remain in the body for several decades, thereby introducing adverse biological effects.

Barium, in the form of barite ($BaSO_4$), can be used as a high-density aggregate in concrete and, when exposed to reactor neutrons, will produce radioactive ^{133}Ba from naturally occurring ^{132}Ba. Although the natural abundance of ^{132}Ba is small, the large amount of barium that is sometimes used as a component of concrete, can lead to significant radioactivity.

Although the by-products of the reactions in a fusion reactor are, in themselves, not radioactive, neutron activated material must be considered as factor in fusion reactor safety. Overall studies have suggested that total hazardous radioactive by-products from fusion reactors will be about 10^{-4} of that produced per unit energy by current fission reactors (Zarnstorff and Goldston 2017). The aneutronic fusion reactions d–^{3}He and p–^{11}B, which were introduced in chapter 2, and which were discussed further with respect to alternative fusion technologies, are appealing options. These reactions are considered further in section 7.4 in the context of the availability of fusion reactor fuels.

Radiation damage

Materials which are subject to irradiation by high energy subatomic particles can suffer from radiation damage. In the case of nuclear reactors (both fission and fusion) neutrons emitted during energy production can cause radiation damage in reactor components. In addition to producing radioactive nuclides by (e.g.) the (n,γ) reaction, neutrons can interact with nuclei of atoms in the crystalline structure of reactor materials and create defects by (e.g.) knocking atoms out of their lattice positions and thereby creating a lattice vacancy. Over time, continued creation of lattice defects can lead to a degradation of the structural properties of the reactor materials.

Radiation damage caused by neutrons is a well-known problem in fission reactors where reactions liberate neutrons with an energy around 2 MeV. This problem is exacerbated in fusion reactors (i.e. those utilizing d–t fusion) as the fusion neutrons have an energy of around 14 MeV and can introduce substantially more damage to reactor structural materials. As well, reactions such as the (n,p) reaction given in equation (7.8) give rise to the formation of protons (or ^1H) within the crystalline lattice. Hydrogen accumulation in metals (for example steels) leads to embrittlement and potential mechanical failure.

The degradation of reactor structural materials, particularly steels, is a factor that must be incorporated into an economic analysis of the facility including scheduled maintenance and facility life expectancy.

Cooling requirements

The energy associated with the nuclear reactions in a fusion reactor is manifested as kinetic energy of the reaction by-products. In a commercial power reactor this kinetic energy is ultimately converted into electricity. In current approaches to producing electricity from nuclear reactions, the particle kinetic energy is converted into heat, which is then converted into mechanical energy by a heat engine and used to drive a generator. The thermodynamic efficiency of this process can be as high as about 40%. Alternatively, as discussed above, some future fusion reactors could make use of direct conversion of the kinetic energy of charged particles into electricity, with potentially higher efficiencies. In either case, fusion energy that is not converted into electricity must be expelled as heat to the environment.

Traditionally, for coal-fired generating stations, as well as many nuclear fission stations, natural bodies of water, i.e. the ocean, a large lake or a river, are used for cooling. This approach is referred to as 'once through' cooling as water taken from the source is used for cooling and then returned to the source. The quantity of water required for cooling is easily estimated as follows. The heat, Q, that must be removed from the reactor is given as

$$Q = \left[\frac{1}{\eta} - 1\right]E \tag{7.13}$$

where E is the electrical energy generated and η is the efficiency expressed as a fraction. The heat removed by the cooling water is expressed as

$$Q = Cm\Delta T \tag{7.14}$$

where C is the specific heat of water ($C = 4180\ \mathrm{J\ kg^{-1}\ °C^{-1}}$), m is the water mass and ΔT is the change in water temperature. Combining equations (7.13) and (7.14), we can write the flow rate of cooling water, ϕ (volume per unit time), in terms of the electrical power generated, P (energy per unit time), as

$$\phi = \frac{P}{C\rho\Delta T}\left[\frac{1}{\eta} - 1\right] \tag{7.15}$$

where ρ is the density of water (10^3 kg m^{-3}). The change in water temperature through the heat exchanger is, of course, determined by the intake temperature and the exhaust temperature. The intake temperature depends largely on the location of the plant and the season. As an example, we consider a facility where the intake temperature averages around 15 °C. The exhaust temperature is largely regulated by government policies, which aim to minimize, as much as practical, environmental effects of power production. This temperature might typically be around 30 °C, thereby giving $\Delta T = 15$ °C. For a 1000 MW (electric) generating station operating at 40% efficiency, equation (7.15) will give a cooling water flow rate of about 24 m^3 s^{-1}. This amounts to close to 8×10^8 m^3 per year.

This use of substantial cooling water from natural sources has the undesirable effect of introducing thermal pollution into the environment, and this can have adverse effects on wildlife, specifically fish populations. Sufficient quantities of water may not be available in all locations, and, in locations where it is available, its use for power plant cooling may compete with requirements for (e.g.) agriculture. Droughts which have been exacerbated in many regions by global climate change, make water availability uncertain.

To minimize the environmental consequences of excess heat production, as well as excessive water consumption, by thermal generating stations, many contemporary facilities utilize cooling towers, as illustrated in figure 7.6, in order to dissipate heat to the environment. In fact, some locations mandate the use of cooling towers rather than natural bodies of water. As well, this approach is more universally applicable, and allows for the construction of facilities where natural water cooling is not available. Two designs for cooling towers are in current use, wet cooling towers and dry cooling towers. Wet cooling towers use excess heat to evaporate water while dry cooling towers transfer heat directly to the atmosphere using a design analogous to an automobile radiator. Although cooling towers alleviate much of the adverse effects associated with transferring heat to natural water bodies, some localized atmospheric effects, such as increased temperature and humidity can occur, particularly with wet cooling towers.

It is seen, therefore, the concerns about thermal pollution and water use that are relevant for thermal generating stations that burn coal or other fossil fuels and nuclear fission generating stations are also relevant for future nuclear fusion generating stations. These concerns must be factored into any assessment concerning the long-term viability of fusion power.

Security considerations

Fission reactors pose a security risk because they use and contains materials from which nuclear weapons (i.e. fission bombs) can be constructed. Principal among

Figure 7.6. Cooling towers at the Dukovany Nuclear Power Station in the Czech Republic. This Dukovany Nuclear Power Station image has been obtained by the author from the Wikimedia website https://commons.wikimedia.org/wiki/File:Nuclear.power.plant.Dukovany.jpg where it is stated to have been released into the public domain. It is included within this book on that basis.

these materials are fissile nuclides, such as ^{235}U and ^{239}Pu, which can be separated from the remaining fuel and from which fission weapons can be constructed directly. As well, fission reactors contain fertile materials, such as ^{238}U, the largest component of the fuel in most thermal reactors. Such nuclides can be used to breed fissile material using neutron irradiation by such processes as

$$n + {}^{238}U \rightarrow {}^{239}U + \gamma \rightarrow {}^{239}Np + e^{-} + \bar{\nu}_e \rightarrow {}^{239}Pu + e^{-} + \bar{\nu}_e \tag{7.16}$$

Fusion reactors do not utilize either fertile or fissile materials that could be used in the construction of a fission weapon. However, there are two possible security concerns associated with fusion reactors. Firstly, they are (assuming they are based on d–t or d–d fusion reactions) a source of significant neutron flux. Thus, if significant quantities of fertile material (e.g. ^{238}U) is available, then fissile material can be bred according to the reaction in equation (7.16) using fusion neutrons. This is the same approach as discussed in chapter 6 for a hybrid fusion–fission reactor where fissile material bred using fusion neutrons may become available for possible weapons applications.

A second security concern related to fusion reactors stems from the utilization of tritium (for d–t reactors). Tritium could be used to boost the yield of a fission weapon by using the fission reaction to induce fusion in the tritium, as described in chapter 3. This risk is minimal, as the amount of tritium fuel that would need to be physically present in a facility at any given time is small.

Both of these possibilities are much less of a security concern than that associated with the large amount of fertile and fissile material available at any time in the core of a fission reactor. In addition, it is much easier to monitor the use of a fusion reactor compared to a fission reactor to ensure that fissile material is not being bred from fertile material, as no fertile material should be present in the facility. Finally, in the event of possible security risks it may be useful to be able to safely shut down the facility immediately and even remotely. This would be possible for a fusion reactor but is not possible for existing fission reactors.

7.4 Fusion resources

The long-term sustainability of fusion energy requires the availability of the necessary resources. While there seems to be a general belief that nuclear fusion would provide a virtually unlimited source of energy for humanity, it is important to look at this assumption analytically. In order to consider fusion reactor resources in detail we need to look at the specific reactions that have been considered in this book and which could, in principle, be implemented in a commercial fusion reactor.

d–t Fusion resources

Firstly, we consider the d–t reaction, as this will most likely be the reaction suitable for the first commercialization of nuclear fusion energy. While deuterium is a natural component of hydrogen, tritium is, as discussed in previous chapters, produced artificially through neutron reactions with lithium.

The natural abundance of deuterium varies between about 0.0026% and 0.0184% by atomic percent of hydrogen on earth. The lower abundance of deuterium occurs in trace amounts of hydrogen gas, while hydrogen in sea water contains, on the average, 0.0156% deuterium. The use of sea water as a source of deuterium is, therefore, advantageous both for the relatively high concentration, as well as the great amount of resource available. Although we would presumably never be in a position of requiring all of the deuterium atoms that are present in the oceans, for reasons as described below, a calculation of the total deuterium availability, emphasizes the vastness of this resource. The oceans on the Earth contain about 1.35×10^{18} m^3 of water, corresponding to a mass of about 1.35×10^{21} kg or 7.5×10^{22} moles of water molecules. This quantity of water contains about 4.67×10^{16} kg or 2.34×10^{19} moles of deuterium atoms. At present there is considerable commercial production of deuterium in the form of heavy water (D_2O). This is used primarily for heavy water fission reactors such as the CANDU (Canadian Deuterium Uranium) reactor, but also has numerous uses in medical and scientific research. Current production levels are around 1000 tons of D_2O per year, corresponding to about 2×10^5 kg y^{-1} of deuterium.

For d–t fusion, however, it is the availability of lithium, that is used to produce tritium, that is the limiting factor. Tritium is produced from the naturally occurring isotopes of lithium, ^{6}Li and ^{7}Li. Neutrons produced by d–t fusion,

$$d + t \rightarrow \alpha + n \quad (17.6\ \text{MeV}) \tag{7.17}$$

breed tritium from lithium contained in a breeder blanket, as discussed in section 4.4, by the reactions,

$$n + {}^{6}\text{Li} \rightarrow \alpha + t \quad (4.78\ \text{MeV}) \tag{7.18}$$

and

$$n + {}^{7}\text{Li} \rightarrow n + \alpha + t \quad (-2.47\ \text{MeV}) \tag{7.19}$$

where the energy associated with each reaction is given. Natural lithium consists of 7.5% ^{6}Li and 92.5% ^{7}Li. In evaluating the usefulness of these breeding reactions, it is also important to consider the cross sections for the reactions. The neutron cross section is a function of energy. Measurements show that the reaction in equation (7.18) has a resonant peak at about 250 keV with a maximum cross section of 3.29 b (Komoda and Igarasi 1978). On the other hand, Smith *et al* (1981) have shown that the cross section for equation (7.19) is relatively constant at 0.372 b for neutrons between 7 MeV and 9 MeV. So, while the reaction in equation (7.19) makes use of the largest fraction of natural lithium, as well as generating an additional neutron, which can be used for further breading reactions, it is also endothermic, requiring at least 2.47 MeV of kinetic energy from the incident neutron, and, in general, has a relatively small cross section. The breeding reaction in equation (7.18), although only making use of a fraction of available lithium, can utilize neutrons of any energy, has a larger cross section and produces excess energy that can contribute to the total reactor heat output though the kinetic energy of the charged particle reaction by-products. In this way fusion energy that might otherwise be lost as kinetic energy of uncharged neutrons can be, at least in part, recovered. The fraction of energy carried by the neutron for d–t fusion is very substantial. For the reaction in equation (7.17) conservation of energy and momentum require that the ratio of energy carried by the alpha particle, E_α, and the neutron, E_n, is

$$\frac{E_\alpha}{E_n} = \frac{m_n}{m_\alpha} \tag{7.20}$$

where m_α and m_n are their respective masses. For d–t fusion equation (7.20) gives $E_\alpha \sim 3.7$ MeV and $E_n \sim 14.6$ MeV. So, the ability to extract some of the neutron energy is highly advantageous.

It is also relevant to consider other potential uses of elemental lithium. Table 7.4 summaries the use of world lithium production in 2019. It is clear that batteries (specifically rechargeable batteries) utilize the largest fraction of lithium and this fraction has been increasing consistently over recent years. While it is possible that future rechargeable batteries may make use of other chemistries, it is clear that there is a substantial need for lithium for non-fusion reactor applications. As non-fusion lithium uses are generally not isotope specific, a reasonable approach to lithium utilization would be to separate the ^{6}Li and ^{7}Li components of natural lithium. The ^{6}Li would be suitable for tritium breeding for d–t fusion reactor fuel, while the ^{7}Li

Table 7.4. Distribution of world lithium use in 2019. Total 2019 world lithium use is estimated at 5.23×10^7 kg. Data are from https://pubs.usgs.gov/periodicals/mcs2020/mcs2020-lithium.pdf.

Use	Percent world lithium use
Batteries	65
Ceramics/glass production	18
Grease	5
Polymer production	3
Casting mold flux powders	3
Air treatment	1
Other	5

could fulfill all other lithium needs. The economic impact of this approach is discussed below in section 7.5.

The worldwide resources of lithium are somewhat difficult to estimate. Lithium is a fairly common element in the Earth's crust but in most cases occurs in very low concentrations in rocks and brine deposits and is not economically viable. At present estimated lithium resources associated with operational commercial mines is 1.35×10^{10} kg. Estimated total commercially viable terrestrial lithium is 3.95×10^{10} kg or about 5.64×10^{12} moles of lithium atoms. If we utilize only the ^{6}Li component of lithium for tritium breeding, the total commercially viable terrestrial lithium corresponds to 2.96×10^9 kg or 4.94×10^{11} moles of ^{6}Li. This may be compared to 2.34×10^{19} moles of deuterium atoms available from sea water and it is clear that the availability of lithium is, by far, the limiting factor for d–t fusion fuel.

For the long-term viability of d–t fusion energy we can look to the oceans as a source of lithium. Sea water contains between 0.14 ppm and 0.25 ppm of lithium, for total resource of 2.3×10^{14} kg. Following along the lines of the above discussion, this corresponds to 1.73×10^{13} kg or 2.88×10^{15} moles of ^{6}Li atoms, four orders of magnitude less than the available deuterium.

If resources for d–t fusion reactors are fully exploited, then the available fusion energy at (say) 40% conversion to electricity is,

$$\begin{aligned}(0.4) \times (6.02 \times 10^{23}\,\mathrm{mol^{-1}}) \times (2.88 \times 10^{15}\,\mathrm{mol}) \times (17.6\,\mathrm{MeV})\\ \times (1.6 \times 10^{-13}\,\mathrm{J\,MeV^{-1}}) = 1.95 \times 10^{27}\,\mathrm{J}\end{aligned} \tag{7.21}$$

Returning to our estimate in equation (1.1) from chapter 1 of projected future energy consumption of 1.7×10^{21} J y^{-1} giving a longevity for energy from d–t fusion as

$$(1.95 \times 10^{27}\,\mathrm{J})/(1.7 \times 10^{21}\,\mathrm{J\,y^{-1}}) = 1.1 \times 10^6\,\mathrm{y} \tag{7.22}$$

A similar calculation using only terrestrial lithium resources gives a longevity of about 240 years. The value given in equation (7.22) is, no doubt, an overestimate in some ways, as it is unlikely that all of the lithium in all of the oceans could be extracted. Although predictions concerning our energy needs a million years from now cannot be made with any degree of confidence, the conclusion that we can draw from the above discussion is that d–t fusion reactors could provide energy for humanity on a timescale that is long compared to nuclear fission or any non-renewable chemical source.

d–d Fusion resources

The information presented in the above section allows us to readily calculate the longevity of d–d fusion energy. The two deuterium fusion reactions and the energy release in each are,

$$d + d \rightarrow {}^{4}He \rightarrow {}^{3}He + n \quad (3.27\ MeV) \tag{7.23}$$

and

$$d + d \rightarrow {}^{4}He \rightarrow t + p \quad (4.03\ MeV) \tag{7.24}$$

The branching ratios for these reactions are approximately equal, as shown in figure 6.17, meaning that the average energy per d–d fusion will be 3.65 MeV.

The world's supply of deuterium (as available from sea water) is 4.67×10^{16} kg or 2.34×10^{19} moles of deuterium atoms. Since two deuterium nuclei are required for each d–d fusion the total energy available from d–d fusion will be (using 40% conversion as above)

$$\begin{aligned}(0.4) \times (6.02 \times 10^{23}\ mol^{-1}) \times (1.17 \times 10^{19}\ mol) \times (3.65\ MeV) \\ \times (1.6 \times 10^{-13}\ J\ MeV^{-1}) = 1.65 \times 10^{30}\ J\end{aligned} \tag{7.25}$$

The longevity of this energy source will be

$$(1.65 \times 10^{30}\ J)/(1.7 \times 10^{21}\ J\ y^{-1}) = 9.7 \times 10^{8}\ y \tag{7.26}$$

That is, infinite, for all practical purposes.

Resources for other fusion reactions

Other fusion reactions, specifically those which are aneutronic, have been discussed in chapter 6 with regard to their potential suitability in some alternative fusion technologies. Two fusion reactions are, in particular, of interest in this respect. These are d–^{3}He fusion,

$$d + {}^{3}He \rightarrow {}^{4}He + p \quad (18.3\ MeV) \tag{7.27}$$

and p–^{11}B fusion,

$$p + {}^{11}B \rightarrow 3\alpha \quad (8.7\ MeV) \tag{7.28}$$

The present section considers available resources for fusion reactors utilizing these two reactions.

d–^{3}He fusion resources

The availability of deuterium, as discussed above, is very substantial. It is the availability of ^{3}He that is the limiting factor for the reaction in equation (7.27).

^3He is a stable isotope of helium with a natural abundance of 0.000 14%, the remainder of natural helium being ^4He. Most ^3He on earth occurs as a dilute gas in the atmosphere, and is, therefore difficult and expensive to extract. The total amount of ^3He is estimated (Takada *et al* 1988) to be about 4×10^6 kg, or 1.3×10^9 moles. Following the above discussion, the longevity of energy from the fusion reaction in equation (7.27) is

$$(0.4) \times (6.02 \times 10^{23}\,\mathrm{mol^{-1}}) \times (1.3 \times 10^{9}\,\mathrm{mol}) \times (18.3\,\mathrm{MeV}) \\ \times (1.6 \times 10^{-13}\,\mathrm{J\,MeV^{-1}})/(1.7 \times 10^{21}\,\mathrm{J\,y^{-1}}) = 0.54\,\mathrm{y} \tag{7.29}$$

This is clearly not a useful contribution to our future energy needs.

As an aside, it is interesting to note that it has been suggested (Takeda *et al* 1988) that ^3He could be used as a means of breeding tritium for d–t fusion reactors according to the reaction,

$$\mathrm{n} + {}^3\mathrm{He} \rightarrow {}^3\mathrm{H} + \mathrm{p} \quad (0.76\,\mathrm{MeV}) \tag{7.30}$$

This is an interesting possibility as the reaction is exothermic and can occur for low energy neutrons. As well, the thermal neutron cross section for the reaction in equation (7.30) is about 60 times that for the reaction in equation (7.18). Unfortunately, the lack of significant ^3He resources makes this approach unviable. ^3He is presently used in very small quantities for several scientific applications. Most commonly ^3He is used as a cryogen in helium dilution refrigerators. Neutron detectors containing ^3He utilize the reaction in equation (7.30) where the resulting charged proton can be readily detected.

p–^{11}B fusion resources

^{11}B is a naturally occurring isotope of boron with a natural abundance of 80.1%, the remainder of natural boron is ^{10}B. Boron is a common element in the Earth's crust and is a component of numerous widely distributed minerals. Commercially produced boron comes largely from the minerals colemanite, kernite and ulexite. Viable boron resources on earth are estimated to be in excess of 10^{12} kg (https://en.wikipedia.org/wiki/Boron#Production). This amounts to more than 7×10^{13} moles of ^{11}B. A calculation, as before, shows that p–^{11}B fusion could provide all of our energy needs for a period of

$$(0.4) \times (6.02 \times 10^{23}\,\mathrm{mol^{-1}}) \times (7 \times 10^{13}\,\mathrm{mol}) \times (8.7\,\mathrm{MeV}) \\ \times (1.6 \times 10^{-13}\,\mathrm{J\,MeV^{-1}})/(1.7 \times 10^{21}\,\mathrm{J\,y^{-1}}) = 1.4 \times 10^{4}\,\mathrm{y} \tag{7.31}$$

In order to understand the overall utilization of boron resources it is important to consider the ways in which this element is currently used. Boron has a significant number of applications in a variety of areas. Some of the more important are:

- advanced composite materials;
- borosilicate glass;
- boron carbide ceramics;

- semiconductors;
- magnetic materials;
- metallic alloys;
- abrasives;
- detergents and bleaches;
- pharmaceuticals;
- insecticides;
- reactor control rods;
- reactor shielding.

These applications make use or several of boron's unique properties. In some cases (e.g. composites), the high strength of boron fibers is utilized. In other cases (e.g. semiconductors and pharmaceuticals), the electronic or chemical properties of boron are important.

The last two applications in the list above are of particular interest to the field of nuclear energy. ^{10}B has an enormous thermal neutron absorption cross section, 3837 b. Therefore, boron is one of a small number of materials that are suitable for absorbing neutrons from thermal fission reactors (the other commonly used material is cadmium). Applications include the use of boron in control rods to regulate the flow of neutrons between fuel elements. Boron is also added to structural steels used to surround the reactor core in order to minimize the radiation hazard from the reactor and to prevent the neutrons from activating other reactor materials, as described above in the section on neutron activation. Boron is particularly important for neutron control and shielding because it reacts with thermal neutrons by the reaction,

$$\mathrm{n} + {}^{10}\mathrm{B} \rightarrow {}^{7}\mathrm{Li} + \alpha \tag{7.32}$$

producing only stable (non-radioactive) by-products.

With the exception of the applications related to fission reactors, the use of boron as noted above does not depend on the specific isotope. As ^{11}B atoms constitute the majority of naturally occurring boron atoms, the use of ^{11}B as a fusion fuel must compete directly with these other applications. At present, commercial production of boron is about 4×10^9 kg y^{-1}. Compared to known terrestrial boron resources, this would suggest a lifetime of boron resources of a few hundred years. Thus, the non-fusion applications of boron would deplete the supply of this element long before its use as a fusion fuel would. However, as with lithium, there are substantial amounts of boron in the oceans. In fact, boron is the 10th most abundant element dissolved in sea water. It most commonly exists as a borate ion, $B(OH)_4^-$, or as undissociated boric acid $B(OH)_3$. The average concentration of boron in the oceans is 416 μmol kg^{-1}. This translates to a total boron content in the oceans of about 6.2×10^{15} kg. If fully utilized for non-fusion applications, this would mean a lifetime for boron resources in excess of 10^6 y. At present, practical technology for extracting boron from sea water does not exist. Most work to date on removing boron from aqueous solution has been for the purpose of purifying the water rather than recovering the boron.

Helium resources

A final consideration concerning the sustainability of fusion energy has to deal with the availability of helium. While helium (that is, primarily ^{4}He) is not a fuel for nuclear fusion reactions, it is an important fusion resource. At present, magnetic confinement fusion appears to be one of the most likely candidates for the construction of a commercial power reactor. Magnetic confinement fusion, as described in chapter 4, requires substantial magnetic fields which are provided by liquid helium cooled superconducting magnets.

World helium resources are limited, with a total estimated amount of about 8.6×10^9 kg. While helium used for magnet cooling is not consumed, a small fraction is inevitably lost. It is, therefore, important to look at the amount of helium that is produced on an annual basis. At present, annual helium production is about 3×10^7 kg worldwide giving an estimated lifetime of helium resources of about 300 years. The vast majority of helium that is produced is extracted from natural gas deposits. If fusion energy, or any other major approach to non-fossil fuel energy, is implemented, then the production of natural gas, and the subsequent extraction of helium would significantly decrease.

The sustainability of fusion energy would therefore begin with the efficient use of helium as a cryogen and the minimization of its loss. However, in the long term, the sustainability of nuclear fusion as an energy source will likely depend on the development of one (or both) of the following technologies:

- viable magnetic field technologies that do not make use of helium as a cryogen, such as high temperature superconducting magnets cooled by liquid nitrogen;
- other fusion confinement technologies that do not require large magnetic fields.

Certainly, on a time scale that is much less than the anticipated longevity of nuclear fusion as an energy source, helium availability concerns will need to be resolved.

7.5 Economic factors

The decision to commercialize a new energy source must include a consideration of a large number of relevant factors including environmental consequences, longevity of the necessary resources, total power available and versatility of the resulting energy. These factors have all been considered, at least to some extent, in the discussion of fusion energy presented in this book. However, it is also important to consider how fusion energy competes with other carbon-free energy sources from an economic standpoint. As fusion energy is not yet at a point where it is technologically viable, it is difficult to consider all relevant aspects of its economics. However, we can understand some of the basic principles of how to approach such a comparison with other energy sources and assess those aspects of fusion energy that are known from the current status of research.

The cost per unit of electricity, C (i.e. per MWh) generated by a facility may be expressed as

$$C = F + \frac{M + DR}{Pkh} \tag{7.33}$$

where F is the cost of fuel per MWh of electricity generated, M is the annual maintenance and operating costs, P is the total rated capacity (in MW), k is the average annual capacity factor, h is the number of hours per year (8760 h y^{-1}), D is the capital cost of the facility and R is the capital recovery factor. The cost per unit electricity as calculated in equation (7.33) is referred to as the 'levelized cost of electricity'. The capital recovery cost accounts for the capital cost of the facility amortized over the payback period of the facility. Typically, the capital recovery factor is expressed as

$$R = \frac{i(1+i)^n}{(1+i)^n - 1} \tag{7.34}$$

where n is the payback period in years and i is the annual interest rate expressed as a fraction. Thus, R is the annual fraction of the capital cost plus interest spread evenly over the payback period. The payback period is chosen to be a period less than the life expectancy of the facility. For example, for a nuclear fission power plant, the payback period might be chosen to be 15 years, compared to a typical life expectancy of 40 years. As an example, for a payback period of 15 years and an annual interest rate of 3% ($i = 0.03$), then equation (7.34) gives $R = 0.084$, meaning that 8.4% of the capital cost contributes to the payback each year for 15 years.

For each type of facility, the three components that contribute to the net cost of electricity, that is, fuel, operational costs and capital costs, are different. Some renewable energy facilities, such as wind energy, have no fuel costs and relatively small operational costs, meaning that capital recovery costs are the most significant component of the cost of electricity. Other energy generating facilities, such as coal-fired generating stations and nuclear fission plants, have obvious fuel costs and, typically, more significant operational costs than wind turbines. The levelized cost of electricity per MWh from different sources in the United States is shown in figure 7.7. The graph shows the maximum, minimum and average cost for the different technologies. With the exception of solar thermal and offshore wind, which are developing technologies, costs for all sources fall within certain limits, mostly between $50 and $150 per MWh of electricity.

In the early 2000s there was considerable debate concerning the economic viability of fusion energy. In more recent years, cost overruns at major fusion research facilities, such as ITER, have adversely affected the perception of the cost of fusion reactor construction and operation. At present, however, it is difficult to make any long-term predictions concerning the economy of future commercial fusion power plants. However, we can make some comparisons with other energy production technologies. Future fusion power stations will be most similar to fission power stations as both require fuel, operational expenditures and are technically advanced facilities. For nuclear fission generating stations, Parisi and Ball (2019) have broken down the components of the levelized cost of electricity in equation (7.33) as 10% from fuel costs, 15% from operating expenses and 75% from capital cost recovery. Thus, from figure 7.7 we expect fuel costs for nuclear fission generating stations to be in the order of $10 USD per MWh.

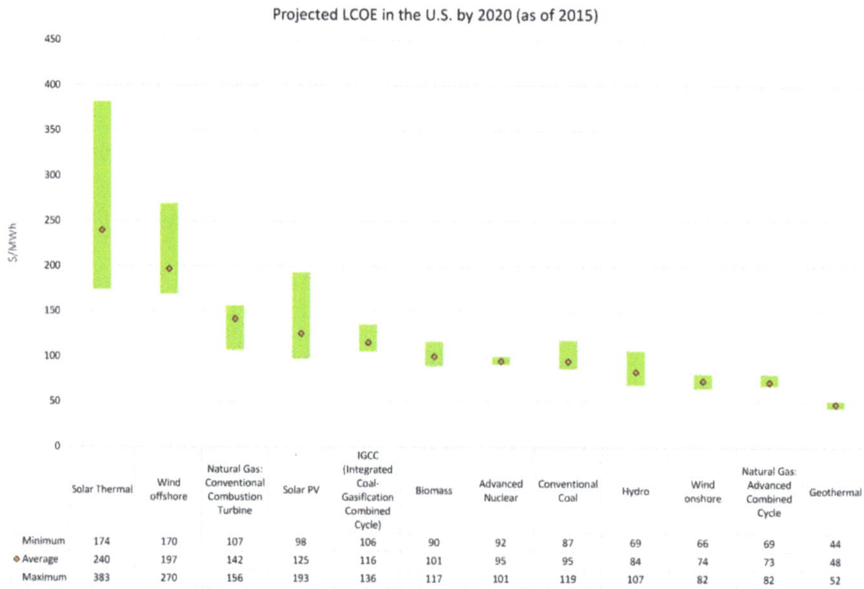

Figure 7.7. Levelized cost of electricity in the United States from different sources. This Levelized Cost of Electricity image has been obtained by the author from the Wikimedia website https://commons.wikimedia.org/wiki/File:Projected_LCOE_in_the_U.S._by_2020_(as_of_2015).png where it was made available by Andynct under a CC BY-SA 4.0 license. It is included within this book on that basis. It is attributed to Andynct.

We can compare fission reactor fuel costs to anticipated fuel costs for a d–t fusion reactor. The fuel input into the d–t reactor will be deuterium and ^{6}Li. The cost of these fuels is highly variable, depending on purity and quantity. At present, the bulk price for heavy water (D_2O) is around \$300 USD kg^{-1}, corresponding to \$1500 USD kg^{-1} for deuterium (\$3 USD mol^{-1}). The price of ^{6}Li is around \$2500 USD kg^{-1} (\$15 USD mol^{-1}) (Shimwell *et al* 2016). Thus, the fuel cost will be dominated by the price of ^{6}Li. The sustainable utilization of energy from d–t fusion would require the extraction of the relevant fuel nuclides from sea water. This technology is well established for deuterium, and the technology for extracting ^{6}Li from sea water is in the early stages of development.

Each d–t fusion gives off 17.6 MeV. 14.1 MeV is given up as kinetic energy to the neutron, while 3.5 MeV is given up as kinetic energy to the alpha particle. If we assume that the neutron energy is not recovered, then we use 3.5 MeV per fusion as the recoverable energy. Finally, we use 40% as the thermal-to-electric conversion efficiency. This estimate gives the fuel cost per MWh as

$$\begin{aligned}(18\ \$\ \text{mol}^{-1}) \times [(0.4) \times (6.02 \times 10^{23}\ \text{mol}^{-1}) \times (3.5\ \text{MeV}) \\ \times (4.45 \times 10^{-23}\ \text{MWh MeV}^{-1})]^{-1} = 0.48\ \$\ \text{MWh}^{-1}\end{aligned} \tag{7.35}$$

That is, less the one tenth of the cost of fuel per MWh for a thermal neutron fission reactor. Certainly, we might expect fuel costs to increase if lithium needs to be extracted from sea water. However, this calculation makes it obvious that the principal components to the cost of electricity produced by nuclear fusion will be capital expenditures and operating costs.

A detailed evaluation of capital and operating costs for fusion generated electricity is difficult at the present, since a technologically viable reactor has not yet been constructed. However, Entler *et al* (2018) recently provided a detailed assessment of the levelized cost of electricity from different energy sources, including nuclear fusion. Figure 7.8 shows their comparison of costs for several different energy sources. This shows that fusion energy is slightly more expensive than fission energy, but that it competes well with most other energy sources. The figure shows that offshore wind energy is more expensive, as is consistent with the analysis previously illustrated in figure 7.7.

In addition to the direct costs associated with electricity production by the various energy sources as represented by the levelized cost of electricity, we can define additional ‘external costs’. These are not costs that are incurred directly during energy production but are the result of the impact of production or use of the energy. There are three main classifications of external costs related to the following:

- adverse effects on human health that result from pollution or global climate change;
- adverse environmental effects such as radioactive or chemical contamination or loss of biodiversity;
- adverse effects resulting from the depletion of resources such as reduction or loss of water or mineral resources.

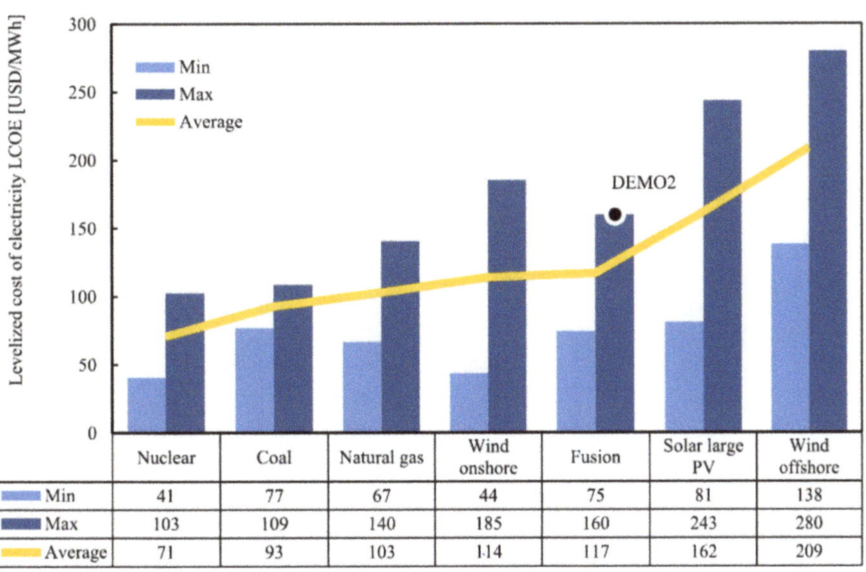

	Nuclear	Coal	Natural gas	Wind onshore	Fusion	Solar large PV	Wind offshore
Min	41	77	67	44	75	81	138
Max	103	109	140	185	160	243	280
Average	71	93	103	114	117	162	209

Figure 7.8. Levelized cost of electricity generated by different methods. Reprinted from Entler *et al* (2018). Copyright (2018) with permission from Elsevier.

The external costs associated with some energy production methods is fairly high, e.g. coal, due to chemical pollution and its contribution to climate change, or nuclear fission, due to safety risks and production of radioactive waste. On the other hand, external costs associated with (e.g.) onshore wind, are minimal. In this respect, fusion energy benefits from the fact that it emits little or no chemical pollution or greenhouse gases and produces minimal radioactive by-products. Figure 7.9 shows a comparison of levelized cost of electricity including external costs. The graph shows that there is substantial economic motivation for the development and implementation of nuclear fusion as a means of generating electricity.

There have been detailed and careful cost analyses for producing electricity using nuclear fusion and these studies have, for the most part, suggested that future commercial fusion power would be competitive from an economic standpoint with other energy options. However, there is still a long way to go before nuclear fusion becomes commercially viable and there is still uncertainty when, or perhaps even if, this will happen. There are many technical issues to be resolved, but a major concern for the development of fusion power is the availability of adequate funding. Nuclear fusion facilities are expensive to construct, often with significant cost overruns, and the timescale for construction and testing is very long. As well, results are not completely certain. In fact, a number of experimental facilities, e.g. Shiva, Nova, NIF, TFTR and JET, have not achieved performance that is consistent with their design expectations. This situation makes it difficult for private industry to invest in future fusion development. Government funding for fusion research is also uncertain. The costs associated with ITER have caused Canada to withdraw as a sponsor and the United States has scaled-back its financial commitment to the project.

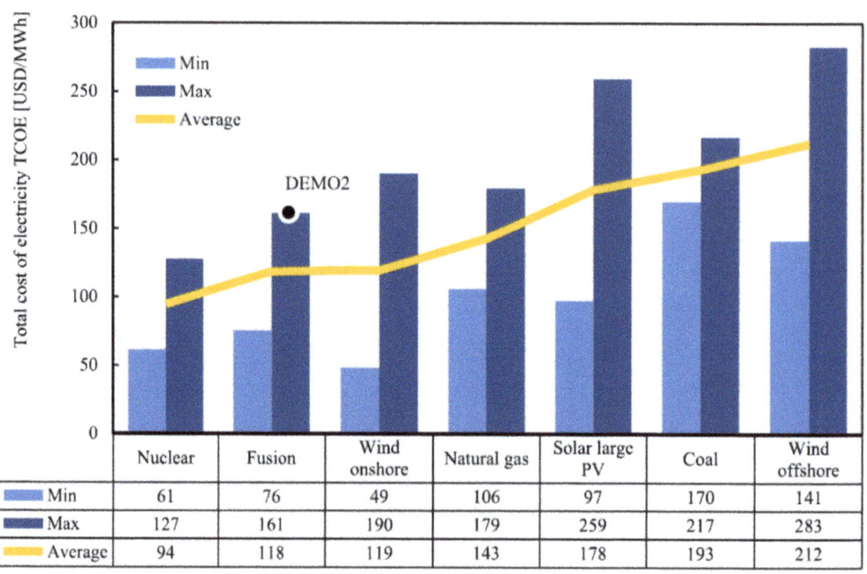

	Nuclear	Fusion	Wind onshore	Natural gas	Solar large PV	Coal	Wind offshore
Min	61	76	49	106	97	170	141
Max	127	161	190	179	259	217	283
Average	94	118	119	143	178	193	212

Figure 7.9. Levelized cost of electricity including external costs generated by different methods. Reprinted from Entler et al (2018). Copyright (2018) with permission from Elsevier.

Fusion energy research must compete with other energy approaches for research and development funds as well as investment capital. Other sustainable carbon-free energy sources, particularly solar and wind are already well-established technologies with clearly defined financial requirements for further expansion. As well, the investment into relatively simple technologies, such as pumped hydroelectric storage and batteries, will complement existing and future renewable energy facilities and help provide carbon-free baseload capability. Public opinion also plays an important role in political decisions. Recent studies (e.g. Turcanu *et al* 2020) have considered such factors with respect to nuclear fusion energy. Survey results indicate that there is relatively limited public awareness of fusion energy. While there is generally a positive attitude toward fusion from those who have some familiarity with it, there is also a tendency to view fusion with some apprehension because of a perceived connection with potentially unsafe fission energy. Small-scale projects, such as some of those discussed in chapter 6, may be more attractive from a funding standpoint than major initiatives such as ITER and DEMO, but have less certain outcomes, as discussed further in section 7.6.

Thus, the road to a viable commercial fusion reactor is a difficult one, not only from a technical standpoint but also, and perhaps predominantly, from an economic one.

7.6 Scientific and technical challenges of fusion energy

In the last three chapters we have presented an overview of approaches to nuclear fusion in three areas—magnetic confinement reactors, inertial confinement reactors and several smaller scale experiments that fall mostly outside these more conventional approaches. The current status of fusion energy development in these three areas as well as the way forward are described in the present section.

Magnetic confinement fusion

Certainly, the most highly developed approach to fusion energy is magnetic confinement. It also has the longest history, in one form or another, and has been an area of active research since the 1950s. At present the two mainstream approaches that have attracted the most interest are the tokamak and the stellarator and significant new, projects are underway in both these areas. The common challenge to the development of both of these devices is the attainment of a sufficient triple product that would provide a gain compatible with economical net energy production. Those areas in which technological development will be required in order to make a viable power reactor are discussed here.

Magnet technology

The utilization of superconducting magnets is essential for the construction of a large-scale magnetic confinement reactor. This technology will be utilized for the ITER and JT-60SA tokamaks and the Wendelstein 7-X stellarator. All three of these reactors utilize conventional superconductors cooled by liquid helium. For the large-scale utilization of fusion energy, high temperature superconductors cooled by liquid nitrogen will be required.

First wall and divertor technology

The first wall, that is, the inner wall of the plasma chamber, is subject to some of the most extreme condition in the reactor, although the divertor will likely experience even harsher conditions. A major factor in the design of the first wall and the divertor is the choice of appropriate materials. Although the extreme thermal conditions would suggest that materials with as high a melting point as possible would be the best choices. However, this is not necessarily the case. As first wall atoms will inevitably interact with plasma ions and some will become impurities in the plasma, a consideration of the detrimental effects of different plasma impurities is relevant. In general, heavy impurity ions have more adverse effects on plasma properties than light impurity ions, so making the first wall out of a low atomic number material is advantageous. Typically, beryllium or carbon is considered an appropriate choice. Divertors are exposed to the highest plasma temperatures, so tungsten, with the highest melting point of the elemental metals, is commonly used. Commercial reactors, compared to present experimental reactors, are expected to operate at greater power levels and for longer periods of time. Robustness of first wall and divertor technologies under such conditions are not known.

Breeder blankets and shielding

One of the essential design criteria for a viable commercial nuclear fusion reactor is tritium self-sufficiency. As tritium is not available in nature, the reactor must breed at least as much tritium as it consumes. This is accomplished by means of a lithium containing breeder blanked located outside the plasma vessel, as discussed in chapter 4. While it is important that fusion neutrons are incident on lithium in the breeder blanket, it is also important that the blanket and additional shielding material are effective at minimizing neutron flux outside the reactor. The experiments planned for ITER will be the first to seriously address these problems. This is partly due to the fact that most previous magnetic confinement reactor experiments have dealt with d–d fusion which neither requires tritium nor produces the high intensity high energy neutron flux that results from d–t fusion.

Resource sustainability

The most significant progress toward a viable power reactor has utilized the d–t fusion reaction. Since this reaction has the greatest fusion cross section at any given combination of plasma parameters, and, therefore, requires the most easily attained triple product, it is most likely that this reaction will be used for fusion power reactors, at least in the foreseeable future. The widespread utilization of fusion energy will, therefore, require a corresponding program in the development of resources, specifically lithium needed for breeding tritium, by non-traditional methods. As noted in section 7.4 above, the development of technology for extracting lithium from ocean resources would be the most likely approach to this problem.

Tokamak designs are certainly more advanced than those of stellarators. Many of the points raised in the present section will be considered in detail for tokamaks with the construction and operation of ITER and later with DEMO, if that project proceeds. Wendelstein 7-X will consider some of these points for stellarator design, as well. It is important to note, that many of the concerns related to tokamak and stellarator design have to deal with materials issues, particularly with respect to the adverse effects that result from extreme thermal and radiation conditions in and around the reactor. The planned International Fusion Materials Irradiation Facility (IFMIF) will provide information on materials properties, in particular on the effects of intense neutron irradiation, that will be crucial for the design of future magnetic confinement fusion reactors.

Inertial confinement fusion

Since the earliest experiments that used lasers as a driver to create nuclear fusion reactions, much effort has been directed to improving laser properties such as laser power and wavelength, uniformity of irradiation and pulse profiles. These investigations have sought to increase energy transfer to the fusion fuel, minimize instabilities and optimize compression before heating, and over the years significant progress has been made in all of these areas. While frequency tripled Nd-lasers have come close to providing sufficient power at a suitable wavelength to achieve scientific breakeven, they have not, in actuality, reached ignition conditions. Two fundamental deficiencies remain in existing inertial confinement fusion devices that preclude the construction of a viable fusion reactor—laser efficiency and laser repetition rate. As a result, many major laser fusion installations, such as the National Ignition Facility, are now largely devoted to fusion research in support of nuclear weapons development programs.

The development of new laser technology that provides greater efficiencies and greater repetition rates is essential to the construction of a laser fusion reactor with net energy production. At present there are two directions to the research on new laser development—diode pumped solid state lasers (DPSSL) and KrF lasers. DPSSLs for potential inertial confinement fusion research are based on Nd-laser technology and require frequency tripling. Details of both DPSSLs and KrF lasers have been presented in chapter 5. A number of experimental facilities have been constructed to investigate both these possible inertial confinement drivers. While experimental results have been encouraging in terms of repetition rates for both laser designs, power levels, at this time, are typically less than 1% of that required for ignition. Certainly, technical developments in this area are necessary before laser fusion could be considered a viable alternative to the current state of magnetic confinement fusion.

It is also likely that the implementation of suitable advanced ignition techniques, such as fast ignition or shock ignition, will be necessary to achieve sufficient gain to exceed breakeven conditions. Refining direct drive and/or indirect drive approaches to inertial confinement fusion will be an additional challenge to developing this technology.

Other fusion approaches

Several approaches to nuclear fusion that have been described in chapter 6 are scientifically sound. The polywell, magneto-inertial fusion, field reversed configuration, *z*-pinch machines and magnetized liner inertial confinement fusion, for example, would fall into this category. These devices cover a wide range of scales, from near table-top devices, such as the polywell, to devices that make use of fairly large-scale facilities, such as the MagLIF device. Research efforts in some of these areas, such as the polywell, have existed for many years, while others are relatively new. As a result, various experimental methods are at somewhat different stages of development and face different technical challenges. In general, these experiments have yielded results that fall far short of those produced by large-scale mainstream fusion devices, such as the tokamak, although many have clearly exhibited fusion reactions, as evidenced by appropriate neutron emission and with further development could show promise for commercial viability.

Ultimately, financial considerations are a crucial factor in any research activity. Some of the fusion approaches discussed in chapter 6 make use of existing facilities, such as the MagLIF experiments at Sandia National Laboratories, and others utilize very modest equipment, such as the polywell. In either case, these experiments involve a much smaller financial investment than major experimental facilities such as ITER. Although achieving a viable approach to nuclear fusion that could lead to a commercial reactor is probably unlikely, a breakthrough could occur that would lead to a new direction for fusion reactor development. If this is the case, development times and expenses will be much less than for traditional magnetic or inertial confinement fusion. Thus, in many cases continued research in these areas is warranted and even if it does not lead to commercialization of fusion energy there are some potential side benefits, such as:

- Results that increase our fundamental understanding of plasmas and fusion reactions. This may particularly be the case for aneutronic fusion, which is being investigated in some of these small-scale approaches.
- Application as neutron sources for the preparation of isotopes for medical and scientific applications.

7.7 The future of nuclear fusion energy

Although much has been accomplished in the quest for commercial nuclear fusion energy, much still remains to be done. While there are no obvious fundamental scientific reasons that would prevent the development of a viable fusion reactor, there are substantial technical and economic challenges to be resolved.

Magnetic confinement reactors, in particular tokamaks, have come the closest to producing net energy (that is commercial breakeven) from fusion reactions. Certainly, the ITER initiative and the subsequent DEMO initiative will presumably provide definitive evidence one way or the other for the ability of fusion energy to make a significant contribution to our energy needs, at least on a time scale of the remainder of this century. While these projects will, first of all, try to establish a gain from fusion reactions that is compatible with net commercial energy production, they will, as well, establish design criteria for power reactors in terms of components such as divertors, breeder blankets, fueling systems and superconducting magnets. A significant component of this work will involve the development of appropriate materials that will provide the best combination of durability, economy and environmental sustainability. The construction of the IFMIF facility will be an important factor in this aspect of fusion power development. The eventual implementation of efficient and economical technologies for the extraction of lithium from sea water is an additional requirement for the widespread application of nuclear fusion reactors.

While the major direction of nuclear fusion research is in the area of tokamaks, it is possible that future developments using other fusion approaches may ultimately show that the tokamak is not the best device for commercial power production. At present, the stellarator follows the tokamak in terms of the magnitude of research efforts for magnetic confinement reactors. Differences in the complexity (and expense) of reactor construction, along with limitations, such as the Greenwald limit and the Troyon limit, which may be more important factors for the tokamak than for the stellarator, may ultimately be factors influencing the choice between these two designs.

Inertial confinement fusion utilizing lasers is the principal alternative to magnetic confinement fusion. At present significant advances in laser technology will be needed in terms of attaining sufficient beam power with greatly improved efficiency and repetition rate.

Perhaps equal to or even greater than the scientific and technical challenges of successfully developing fusion power are the economical challenges. Fusion energy turned out to be much more than the twenty-year development project that was envisioned in the 1950s. Construction of major fusion research facilities such as ITER is both time consuming and expensive. The continuation of funding either from government (often international), corporate or private supporters is a concern that needs to be considered very seriously. The main difficulty from a commercial standpoint is the large expense combined with a long payback period and an uncertain possibility of success. These adverse factors from an investment standpoint are further exacerbated by the sometimes less than successful results obtained from previous experimental fusion devices. This is true of both magnetic confinement reactors such as TFTR and inertial confinement reactors such as Nova and the National Ignition Facility. Smaller scale fusion experiments are also not exempt from this trend. For example, in 2014 Lockheed Martin announced their plans to construct a high β compact fusion reactor based on a linear mirror geometry using magnetic confinement. They projected a functioning prototype reactor within five years. The most recent results have shown less than a 1% ionization fraction of the gas. Measured peak electron density and temperature were 10^{16} m^{-3} and 20 eV, respectively, well below that achieved in large-scale machines.

The economic perspective of fusion energy can be compared to that for other carbon-free energy alternatives, most notably solar energy and wind energy. Both of these energy sources are based on well-established technologies that have been shown to be viable and economic. Although certainly technological advances in both these fields are possible and are the subject of ongoing research, such advances are fairly incremental and build on a technology that already works. Compared to nuclear fusion, solar and wind energy development is inexpensive and has a well-established payback period. The success of ITER (or something equivalent) to live up to its design goals, or at least close to it, is an essential step to progressing further with the development of fusion power and is likely a prerequisite for maintaining continuing financial support and political approval. This progress must be in parallel with programs such as IFMIF and a serious consideration of lithium resources. The continuation of fusion development with the construction and successful operation of DEMO is essential for the future of fusion energy.

From the standpoint of dealing with the adverse environmental effects of our energy use and the diminishing fossil fuel resources, the development of substantial carbon-free energy resources on a time scale of a few decades is essential. The time scale for development of fusion power means that we must rely on other energy sources in the meantime. Developing carbon-free alternative energy sources such as solar and wind, as well as hydroelectric and geothermal, along with the necessary storage capabilities to maintain reliable baseload power, is essential to dealing with our energy needs and minimizing adverse environmental consequences on a time scale of a small number of decades. The most optimistic scenario for fusion power, would suggest a functioning DEMO reactor sometime around the middle of the present century, with the construction of operational power reactors in the second half of the century. Thus, it would seem that the successful implementation of an alternative energy strategy must proceed with the development of more traditional resources such as solar, wind, geothermal and hydroelectric, with the hope that, at some point, fusion energy would provide a reliable and economical means of satisfying baseload power needs.

References and suggestions for further reading

Bobin J L 2014 *Controlled Thermonuclear Fusion* (Singapore: World Scientific)

Cardozo N J L 2019 Economic aspects of the deployment of fusion energy: the valley of death and the innovation cycle *Phil. Trans. R. Soc.* **A377** 20170444

Dunlap R A 2004 *An Introduction to the Physics of Nuclei and Particles* (Belmont, CA: Brooks/Cole)

Dunlap R A 2019 *Sustainable Energy* 2nd edn (Boston, MA: Cengage)

Dunlap R A 2020 *Renewable Energy—Volume 1: Requirements and Sources* (San Rafael, CA: Morgan & Claypool)

Dunlap R A 2020 *Renewable Energy—Volume 2: Mechanical and Thermal Energy Storage Methods* (San Rafael, CA: Morgan & Claypool)

Dunlap R A 2020 *Renewable Energy—Volume 3: Electrical, Magnetic and Chemical Energy Storage Methods* (San Rafael, CA: Morgan & Claypool)

Entler S, Horacek J, Dlouhy T and Dostal V 2018 Approximation of the economy of fusion energy *Energy* **152** 489–97

Evans J H 2008 Fusion needs a realistic cost assessment *Nature* **454** 397

Evans J C, Lepel E L, Sanders R W, Wilkerson C L, Silker W, Thomas C W, Abel K H and Robertson D R 1984 *Long-Lived Activation Products in Reactor Materials* U.S. Nuclear Regulatory Commission report NUREG/CR-3474 available at https://osti.gov/servlets/purl/6776358

Ikeda K 2010 ITER on the road to fusion energy *Nucl. Fusion* **50** 014002

International Fusion Research Council (IFRC) 2005 Status report on fusion research *Nucl. Fusion* **45** A1–28

Komoda S I and Igarasi S I 1978 Neutron cross sections of Li-6 *J. Nucl. Sci. Eng.* **15** 79–81

Mlynar J 2007 Focus on: JET *European Centre of Fusion Research* Report EFD-R(07)01

Moir R W and Barr W L 1973 'Venetian-blind' direct energy converter for fusion reactors *Nucl. Fusion* **13** 35–45

Oliphant T A, Ribe F L and Coultas T A 1973 Direct conversion of thermonuclear plasma energy by high magnetic compression and expansion *Nucl. Fusion* **13** 529–32

Parisi J and Ball J 2019 *The Future of Fusion Energy* (London: World Scientific)

Parkins W E 2006 Fusion power: will it ever come? *Science* **311** 1380

Shimwell J, Kovari M, Lilley S, Zheng S, Morgan L W G and Packer L W 2016 A parameter study of time-varying tritium production in solid-type breeder blankets *Fusion Eng. Des.* **104** 34–9

Smith D L, Bretscher M M and Meadows J W 1981 Measurement of the cross section for the ^{7}Li (n,n't)^{4}He reaction in the 7- to 9-MeV energy range *J. Nucl. Sci. Eng.* **78** 359–69

Steiner D, Embrechts M J, Varsamis G, Vesey R, Gierszewski P and Wong C P C 1989 Helium-3 blankets for tritium breeding in near-term fusion devices *Fusion Eng. Des.* **8** 121–25

Takada E, Iguchi T and Nakazawa M 1988 New tritium breeding blanket using ^{3}He gas loop instead of Li-compounds for fusion reactor system *J. Nucl. Sci. Technol.* **25** 315–17

Turcanu C, Prades A, Sala R, Perko T and Oltra C 2020 Fusion energy: a deeper look into attitudes among the general public *Fusion Eng. Des.* **161** 111891

Windsor C 2019 Can the development of fusion energy be accelerated? An introduction to the proceedings *Phil. Trans. R. Soc.* **A377** 20170446

Wolska J and Bryjak M 2013 Methods for boron removal from aqueous solutions—a review *Desalination* **310** 18–24

Yoshikawa K, Noma T and Yamamoto Y 1991 Direct-energy conversion from high-energy ions through interaction with electromagnetic fields *Fusion Sci. Technol.* **19** 870–75

Zarnstorff M C and Goldston R J 2017 Magnetic fusion energy *Energy from the Nucleus* ed G M Crawley (Singapore: World Scientific) pp 165–88

www.ingramcontent.com/pod-product-compliance
Ingram Content Group UK Ltd.
Pitfield, Milton Keynes, MK11 3LW, UK
UKHW051900100926
8556IPUK00005B/702

9 780750 333054